METHODS IN MOLECULAR BIOLOGY

Series Editor
John M. Walker
School of Life and Medical Sciences
University of Hertfordshire
Hatfield, Hertfordshire, AL10 9AB, UK

For further volumes:
http://www.springer.com/series/7651

Statistical Analysis in Proteomics

Edited by

Klaus Jung

Department of Medical Statistics, University Medical Center Göttingen, Göttingen, Germany

Editor
Klaus Jung
Department of Medical Statistics
University Medical Center Göttingen
Göttingen, Germany

ISSN 1064-3745 ISSN 1940-6029 (electronic)
Methods in Molecular Biology
ISBN 978-1-4939-7987-5 ISBN 978-1-4939-3106-4 (eBook)
DOI 10.1007/978-1-4939-3106-4

Springer New York Heidelberg Dordrecht London

Softcover re-print of the Hardcover 1st edition 2016

Cover illustration: For the complete image, please see Figure 2 of Chapter 3.

Printed on acid-free paper

Humana Press is a brand of Springer
Springer Science+Business Media LLC New York is part of Springer Science+Business Media (www.springer.com)

Preface

Among the high-throughput technologies that are currently used in biomedical research, those used in proteomics are perhaps the oldest. While mass spectrometry and 2-D gel electrophoresis were already used in the 1980s for simultaneous measuring of the abundance of multiple proteins, statistical methods for the analysis of high-throughput data experienced their great evolution first with the development of DNA microarrays in the mid-1990s.

Although there is a large overlap between statistical methods for the different "omics" fields, methods for analyzing data from proteomics experiments need their own specific adaptations. Therefore, the aim of this book is to provide a collection of frequently used statistical methods in the field of proteomics. This book is designated to statisticians who are involved in the planning and analysis of proteomics experiments, beginners, as well as advanced researchers. It is also designated to biologists, biochemists, and medical researchers who want to learn more about the statistical opportunities in the analysis of proteomics data.

The different chapters of this book focus on the planning of proteomics experiments, the preprocessing and analysis of the data, the integration of proteomics data with other high-throughput data, as well as some special topics. For statisticians who are new in the area of proteomics, the first chapter provides a detailed overview of the laboratory techniques used in this exciting research area.

Göttingen, Germany *Klaus Jung*

Contents

Part IV Data Integration

Part V Special Topics

Contributors

SURUCHI AGGARWAL • *Immunology Group, International Centre for Genetic Engineering and Biotechnology, New Delhi, India*
TIM BEIẞBARTH • *Department of Medical Statistics, University Medical Center Göttingen, Göttingen, Germany*
CHRISTIAN BENDER • *TRON-Translational Oncology at the University Medical Center Mainz, Mainz, Germany*
LIONEL BLANCHET • *Analytical Chemistry-Chemometrics, Institute for Molecules and Materials, Radboud University Nijmegen, Nijmegen, The Netherlands; Department of Biochemistry, Nijmegen Centre for Molecular Life Sciences, Radboud University Medical Centre, Nijmegen, The Netherlands*
ANTONELLA CHIECHI • *Department of Medicine, Indiana University School of Medicine, Indianapolis, IN, USA*
HYUNGJUN CHO • *Department of Statistics, Korea University, Seoul, South Korea*
DAVID CONDE • *Departamento de Estadística e Investigación Operativa, Facultad de Ciencias, Universidad de Valladolid, Valladolid, Spain*
HASSAN DIHAZI • *Clinic of Nephrology and Rheumatology, University Medical Center Göttingen, Göttingen, Germany*
SOO-HEANG EO • *Department of Statistics, Korea University, Seoul, South Korea*
MIGUEL A. FERNÁNDEZ • *Departamento de Estadística e Investigación Operativa, Facultad de Ciencias, Universidad de Valladolid, Valladolid, Spain*
YAN FU • *National Center for Mathematics and Interdisciplinary Sciences, Key Laboratory of Random Complex Structures and Data Science, Academy of Mathematics and Systems Science, Chinese Academy of Sciences, Beijing, China*
XIN GAO • *Department of Mathematics and Statistics, York University, Toronto, ON, Canada*
SVEN H. GIESE • *Research Group Bioinformatics (NG4), Robert Koch-Institute, Berlin, Germany; Department of Bioanalytics, Institute of Biotechnology, Technische Universität Berlin, Berlin, Germany; Wellcome Trust Centre for Cell Biology, School of Biological Sciences, University of Edinburgh, Edinburgh, UK*
SILVIA VON DER HEYDE • *Department of Medical Statistics, University Medical Center Göttingen, Göttingen, Germany; IndivuTest GmbH, Hamburg, Germany*
YUAN JI • *Department of Health Studies, The University of Chicago, Chicago, IL, USA*
KLAUS JUNG • *Department of Medical Statistics, Georg-August-University Göttingen, Göttingen, Germany*
ULRIKE KORF • *Division of Molecular Genome Analysis, German Cancer Research Center (DKFZ), Heidelberg, Germany*
FRANK KRAMER • *Department of Medical Statistics, University Medical Center Göttingen, Göttingen, Germany*
JOCHEN KRUPPA • *Department of Medical Statistics, Georg-August-University Göttingen, Göttingen, Germany*
JULIA KULIGOWSKI • *Neonatal Research Centre, Health Research Institute La Fe, Valencia, Spain*

Juhee Lee • *Department of Applied Mathematics and Statistics, Santa Cruz, CA, USA*
Christof Lenz • *Bioanalytical Mass Spectrometry, Max Planck Institute for Biophysical Chemistry, Göttingen, Germany; Core Facility Proteomics, Institute of Clinical Chemistry, University Medical Center, Göttingen, Germany*
Peter Müller • *Department of Mathematics, Austin, TX, USA*
David Pérez-Guaita • *Centre for Biospectroscopy, School of Chemistry, Monash University, Clayton, Australia*
Guillermo Quintás • *Safety and sustainability Division, Leitat Technological Center, Valencia, Spain; Analytical Unit, Health Research Institute La Fe, Valencia, Spain*
Bernhard Y. Renard • *Research Group Bioinformatics (NG4), Robert Koch-Institute, Berlin, Germany*
Habtom W. Ressom • *Department of Oncology, Lombardi Comprehensive Cancer Center, Georgetown University Medical Center, Washington, DC, USA*
Nadège Richard • *CCMAR, Centre of Marine Sciences of Algarve, University of Algarve, Faro, Portugal*
Cristina Rueda • *Departamento de Estadística e Investigación Operativa, Facultad de Ciencias, Universidad de Valladolid, Valladolid, Spain*
Bonifacio Salvador • *Departamento de Estadística e Investigación Operativa, Facultad de Ciencias, Universidad de Valladolid, Valladolid, Spain*
Frank-Michael Schleif • *School of Computer Science, University of Birmingham, Edgbaston, Birmingham, UK*
Tomé S. Silva • *SPAROS Lda., Olhão, Portugal*
Agnieszka Smolinska • *Department of Toxicology, Nutrition and Toxicology Research Institute Maastricht (NUTRIM), Maastricht University, Maastricht, The Netherlands*
Johanna Sonntag • *Division of Molecular Genome Analysis, German Cancer Research Center (DKFZ), Heidelberg, Germany*
Tsung-Heng Tsai • *Department of Oncology, Lombardi Comprehensive Cancer Center, Georgetown University Medical Center, Washington, DC, USA; Bradley Department of Electrical and Computer Engineering, Virginia Tech, Arlington, VA, USA*
Minkun Wang • *Department of Oncology, Lombardi Comprehensive Cancer Center, Georgetown University Medical Center, Washington, DC, USA; Bradley Department of Electrical and Computer Engineering, Virginia Tech, Arlington, VA, USA*
Amit Kumar Yadav • *Drug Discovery Research Center (DDRC), Translational Health Science and Technology Institute, Faridabad, Haryana, India*
Irene Sui Lan Zeng • *The Department of Statistics, The University of Auckland, Auckland, New Zealand*
Yitan Zhu • *Program for Computational Genomics and Medicine Research Institute, NorthShore University HealthSystem, Evanston, IL, USA*
Franziska Zickmann • *Research Group Bioinformatics (NG4), Robert Koch-Institute, Berlin, Germany*

Part I

Proteomics, Study Design, and Data Processing

Chapter 1

Introduction to Proteomics Technologies

Christof Lenz and Hassan Dihazi

Abstract

Compared to genomics or transcriptomics, proteomics is often regarded as an "emerging technology," i.e., as not having reached the same level of maturity. While the successful implementation of proteomics workflows and technology still requires significant levels of expertise and specialization, great strides have been made to make the technology more powerful, streamlined and accessible. In 2014, two landmark studies published the first draft versions of the human proteome.

We aim to provide an introduction specifically into the background of mass spectrometry (MS)-based proteomics. Within the field, mass spectrometry has emerged as a core technology. Coupled to increasingly powerful separations and data processing and bioinformatics solution, it allows the quantitative analysis of whole proteomes within a matter of days, a timescale that has made global comparative proteome studies feasible at last. We present and discuss the basic concepts behind proteomics mass spectrometry and the accompanying topic of protein and peptide separations, with a focus on the properties of datasets emerging from such studies.

Key words Proteomics, 2-DE, Electrophoresis, Mass spectrometry, Separations

1 Introduction

The term "proteomics" in its original meaning denotes the study of the entire observable protein complement (or proteome) of a biological system, be it a relatively homogeneous microbial cell culture or a tissue sample obtained from a hospital patient. When Marc Wilkins first coined the term "proteome" in 1994, however, proteomics was a distant goal rather than a tangible technological reality. Even the identification of a few tens of proteins would take researchers weeks to months of work, let alone the assessment of their quantities or modification status. Over the past 20 years however proteomics has grown from a promise into a mature set of technologies that has allowed for example the publication of first full draft versions of the human proteome in 2014 [1, 2]. Virtually all aspects of proteome analysis have seen huge improvements, from sample preparation, protein and peptide separations, detection and quantitative analysis especially by mass spectrometry which has

Klaus Jung (ed.), *Statistical Analysis in Proteomics*, Methods in Molecular Biology, vol. 1362,
DOI 10.1007/978-1-4939-3106-4_1,

emerged as a core proteomics technology, to the statistical and bioinformatic analysis of the large and multilayered datasets that a global "omics" approach produces.

Following technological progress, Tyers and Mann in 2003 redefined proteomics as "almost everything post-genomic: the study of the proteome in any given cell and the set of all protein isoforms and modifications, the interactions between them, the structural description of proteins and their higher-order complexes" [3]. While the genome of an organism is considered to be mostly static, the proteome shows dynamic properties with protein profiles changing in dependence of time and a variety of extracellular and intracellular stimuli (i.e., cell cycle, temperature, differentiation, stress, apoptotic signals). The realization that the proteome is highly dynamic in turn led to an increased demand for quantitative information, as information about the detectability of a protein was superseded by information about relative changes in its abundance, modification status, localization, and interaction partners [4].

Finally, an increased appreciation of the complexity of the proteome led to a refinement of our understanding what defines a protein. The seemingly simple concept of "DNA makes RNA makes proteins" does not describe the observed complexity of proteins in its entirety. While the huge success of genome sequence projects over the past decades has certainly been a prerequisite for the progress observed in proteomics [4], there is a plethora of parameters defining the biological role of a protein that are not determined by the gene(s) encoding its sequence, e.g., splicing events, enzymatic processing, or posttranslational modifications [5]. Consequently the term "protein species" is finding increased use as it more accurately describes protein diversity [6, 7].

In addition, there is currently no amplification technology for proteins comparable to PCR. The huge dynamic range observed for protein quantities in biological samples immediately translates into dynamic range requirements for any analytical approach to proteomics samples, necessitating elaborate separation and enrichment strategies to simplify biological specimens [8].

In this introduction we discuss some of the major technical and experimental approaches that are taken in proteomics research today, and discuss how the structure of the resulting data influences bioinformatics approaches to generate knowledge from these data. Special focus is given to protein and peptide separations, and to mass spectrometry which has emerged as a key protein detection and quantitation technology in proteomics research.

2 Separation Technologies in Proteomics

2.1 Bottom-Up Versus Top-Down Proteomics

Separations are a central feature of all analytical strategies in proteomics. The proteins contained in any biological specimen may be separated and analyzed either on the intact protein level or on the peptide level following endoproteinase digestion. Digestion to peptides has many analytical benefits that have improved the performance of proteomics workflows, especially if mass spectrometry is used for detection. On the level of sample handling and separations, peptides generated by for example trypsin digestion of proteins are a far more homogeneous group of analytes than the underlying proteins with regard to molecular weight, hydrophobicity and solvent solubility, since they do mostly not exhibit any significant higher order structure. In addition they show a much more controlled charging behavior under controlled pH conditions, and will in the majority not be modified by for example glycosylation sites. Consequently many peptide separations show much higher resolution than protein separations, especially where chromatography-based separations are concerned.

In addition, mass spectrometry as the most frequently used detection principle in proteomics heavily favors peptides over proteins. Peptides show a more uniform and efficient ionization and charging behavior than proteins, produce better response on several types of mass spectrometer detectors and, most importantly, can be routinely fragmented by ion activation techniques to provide sequence and structure information. Taken together the detection of for example tryptic peptides in complex mixtures by modern mass spectrometry equipment is orders of magnitude more sensitive than the detection of proteins. Therefore the most common approach in proteomics is to prepare and separate proteins, digest them with endoproteinases, separate the resulting peptides yet again, and analyze them for identity, modification state, and quantity by mass spectrometry. In addition, enrichment strategies used to target low abundance subpopulations maybe employed. This so-called "bottom-up" approach comes with its own challenges: digestion multiplies the number of analytes in the sample (e.g., 2000 protein will produce an average of >100,000 peptides on tryptic digestion), and it is not always straightforward to back-assign a digest peptide to the protein it originated from, a problem referred to as "protein inference" [5]. Still, the benefits outweigh these challenges by far, making "bottom-up" analysis the prevalent approach in proteomics as compared to the "top-down" approach where proteins are treated and analyzed in their intact state throughout [9, 10]. After discussing options for protein and peptide separations, we will then focus on the "bottom-up" approach and the principles applied to peptide analysis by mass spectrometry.

2.2 Protein Level Separations

2.2.1 Sodium Dodecyl Sulfate Polyacrylamide Gel Electrophoresis (SDS-PAGE)

The approach most frequently taken for protein separation is still SDS-PAGE. Proteins are dissolved in a buffer containing sodium dodecyl sulfate (SDS), and the resulting negatively charged adducts are pulled through a gel of a defined polymerization degree (or pore size range) by electrophoretic migration. The separation is achieved according to the apparent molecular weight, or rather the hydrodynamic radius of the resulting protein-SDS adducts [11].

SDS-PAGE is compatible with a very wide range of protein solubilization and sample handling requirements, making it a very good choice for the separation of for example very hydrophobic integral membrane proteins. After staining with Coomassie or silver staining, entire lanes covering a broad range of apparent molecular weight can be investigated. Depending on the scientific task, only proteins from defined MW regions can be investigated, and results can easily be correlated with for example Western blot analysis. One of the shortcomings of SDS-PAGE as a one-dimensional separation is its limited resolution, which does not allow to detect and separate more than a few tens of bands at best. Consequently it has to be combined with other separations strategies either on the protein or—after endoproteinase digestion—on the peptide level to successfully analyze complex proteome samples.

2.2.2 Two-Dimensional Gel Electrophoresis (2-DE)

High-resolution two-dimensional polyacrylamide gel electrophoresis (2D PAGE) is a commonly applied separation technique in proteomics, and has been one of its driving forces for decades [12, 13]. 2D PAGE allows the separation of proteins according to two largely orthogonal parameters (Fig. 1): their isoelectric point (pI) and their apparent molecular weight (Mr), enabling the separation of complex protein mixtures and their visualization on a single high-resolution gel [14–17]. Depending on the gel size and pH gradient used, 2D PAGE can resolve up to 5000 different proteins simultaneously and detect and quantify <1 ng of protein/spot [17]. 2D PAGE can thus be used to generate protein expression profiles from different samples, e.g., healthy versus diseased,

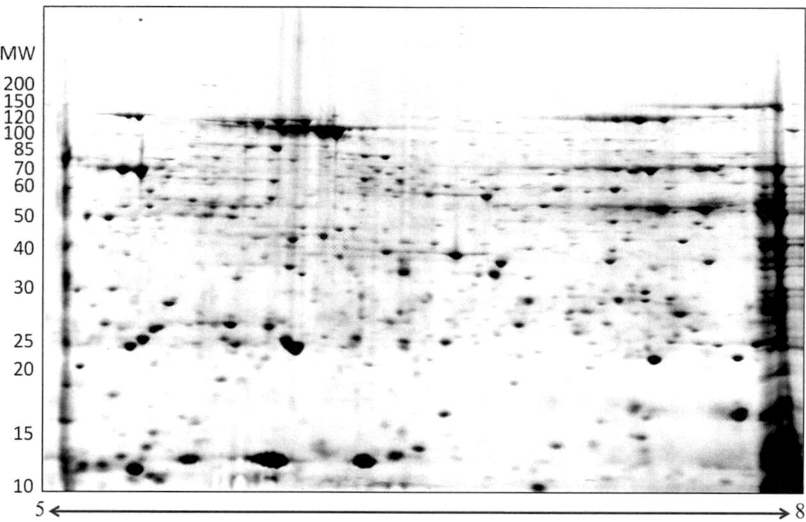

Fig. 1 2-DE reference maps of tissue extract proteins. 150 μg protein was loaded on an 11-cm IPG strip with a linear pH gradient pI 5–8 for isoelectric focusing; a 12 % SDS-polyacrylamide gel was used for the SDS-PAGE. Proteins were stained with fluorescent stain

knockout versus wild type. Sample solubilization is a critical step for the reproducibility of the 2D PAGE to get as many proteins solubilized as possible in reproducible manner, to disrupt (in most cases) their non-covalent bonds, and to obtain them in a defined charge state without modification of the polypeptide [17]. Following solubilization, isoelectric focusing (IEF) is used as the first level of protein separation. IEF is very sensitive to charge alterations and therefore to sample contaminations, e.g., by salts that may alter the protein charge. The second dimension of separation is an SDS-PAGE where the proteins are separated according to their apparent molecular weight. Finally, the separated proteins are then visualized using a staining technique. Several staining methods are commonly used: (1) Coomassie Blue is used to visualize proteins separated by 2D PAGE but suffers from low sensitivity. (2) Silver staining is more sensitive than Coomassie Blue but involves a complex multistep staining protocol, which limits gel-to-gel reproducibility [18]. (3) Fluorescence-based staining is highly sensitive and has a wide linear range of detection between staining intensity and protein volume, enabling accurate quantitation of high and low abundance proteins [19, 20]. The fluorescence technologies offer also the possibility of multicolor labeling and detection. If samples have to be compared, they can be labeled with different dyes corresponding to different excitation and emission wavelengths, mixed and separated on a single gel allowing a differential analysis of protein expression [21–24]. After the staining step, gel images are captured and the resulting 2D maps are analyzed using dedicated image processing software. Spots of interest are then excised and processed for identification by mass spectrometry.

There are several limitations to 2D PAGE as a separation method for proteomic studies. For example, hydrophobic proteins hardly enter the gel and are often lost during 2D PAGE, limiting its used for the analysis of for example integral membrane proteins. Very high or very low molecular weight proteins, highly acidic or highly basic proteins may also be lost during gel separation. Due to the often limited staining sensitivity, 2D PAGE also requires relatively large amounts of protein.

In addition 2D PAGE involves many manual processing steps and is therefore not easily automated. Moreover, the heterogeneity of cell types in tissue samples makes their analysis with 2D-gels to a challenge. Due to these limitations of two-dimensional electrophoresis, separations techniques, such as multidimensional liquid chromatography and capillary electrophoresis upstream from the mass spectrometer offer solid alternatives that can overcome the 2D PAGE limitations.

The first separation step in 2D PAGE, i.e., isoelectric focusing (IEF), is sometimes also used as a standalone method to fractionate complex protein samples, and has been commercialized for example in the OFFGEL system. When used for protein separations it suffers from the same limitations as mentioned above, and is therefore much more widely used as another dimension of peptide separation (pIEF) downstream [25].

2.2.3 Chromatography-Based Protein Separations

As an alternative to electrophoretic separations such as SDS-PAGE or 2D PAGE, a wide range of chromatographic separation approaches have been established to separate and purify intact proteins, and have found their way into proteomics research [26, 27]. The separation is most frequently based on one of the three major physicochemical properties that describe proteins: hydrophobicity (reversed phase chromatography), charge (ion exchange chromatography), and molecular weight (size exclusion chromatography). In addition, affinity for example by noncovalent protein–protein interactions can be used as a separation principle [28–34].

Chromatography-based separations are relatively straightforward to scale up and automate, and are therefore especially suited for multi-stage separation workflows. Most separation principles suffer from a limited range of available buffer conditions which only allow to focus on subgroups of proteins for example in a certain molecular weight or charging region, but are less suitable for generic proteomics approaches. Consequently these separation principles are mostly employed where the enrichment or purification of a single protein or class of proteins is desired. Size exclusion chromatography (SEC) of proteins has emerged as a suitable prefractionation method for generic proteomics approaches as it is compatible with SDS-containing buffers that allow the solubilization of hydrophobic as well as hydrophilic proteins. It is inherently of low chromatographic resolution, but can be used to great success in multidimensional separation approaches as it has a high loading capacity.

2.2.4 Chips and Arrays

In addition to the classical separation methods, the array technology provides an ideal tool to study enriched subsets of proteins or protein domains. Various protein array technologies have emerged over the last decades that promise rapid examination of different samples on a protein scale offering better perspectives for proteomics. Antibodies based arrays are highly promising in this case the antibodies are immobilized on a specially treated array surfaces. The samples of interest are then applied to the arrays and only the proteins that bind to the relevant antibodies remain bound to the chip and be analyzed and quantified. The immobilized molecules can also be peptides or other small molecules [35–37]. Readouts for protein-based arrays can derive from protein interactions, protein modifications or enzymatic activities. The quality of the immobilized molecule, e.g., antibody, is critical for the readout of the system. Once developed it could provide for a convenient proteome analysis.

2.3 Peptide Level Separations

2.3.1 Chromatography-Based Peptide Separations

There is a multitude of available peptide separation approaches that can be used to simplify the hundreds of thousands of peptides produced by enzymatic digestion of a complex protein sample. Similar to proteins, peptides may be separated according to a range of physicochemical properties, such as hydrophobicity, charging at defined pH or polarity. One prerequisite has forced the development of peptide separations for proteomics over the past years, i.e., that the last chromatographic separation step should be readily coupled to mass spectrometry to allow for highly automated LC-MS/MS analyses. This makes several demands on an ideal separation strategy: it should work with volatile buffer systems at low flow rates that do not interfere with the mass spectrometer's ionization process. In addition, it should be readily miniaturizable as the sensitivity of the electrospray ionization process used in most of today's proteomics mass spectrometers is concentration-dependent, i.e., flow rates in the low nanoliter/min regime are highly desirable [38, 39]. Finally, this separation should be highly resolving for, for example, tryptic peptides, which typically have a length of 6–25 amino acid residues. All these requirements are best met by capillary diameter (50–75 μm) reversed phase-C18 chromatography under acidic conditions, e.g., with volatile formic or acetic acid buffer systems, at corresponding flow rates of 150–400 nl/min [40]. Indeed this chromatography regime seems to present a "sweet spot" when coupled to mass spectrometry. Together with improvements in chromatography materials, high-pressure liquid chromatography hardware and the use of long columns, many proteomics workflows today use this as the only separation step at all, an approach referred to as "single-shot proteomics" [41, 42].

In many cases however it can still be beneficial to add another dimension of chromatographic separation to the overall workflow to achieve greater simplification of the sample prior to for example mass spectrometric analysis. Any second chromatographic dimension preceding the final reversed phase separation hyphenated to the mass spectrometer only or offline should ideally be highly orthogonal to the latter, i.e., separate peptide analytes by a different physicochemical principle to ensure efficient separation, and be readily integrated into the overall workflow with regard to buffer systems without causing need for, for example, additional desalting or concentration steps. Examples of first dimension separations that are frequently used in proteomics research include strong cation exchange chromatography or reversed phase chromatography at neutral pH [43].

2.3.2 Electrophoresis-Based Peptide Separations

Several gel free separation techniques found their way to proteomics, among them capillary electrophoresis (CE) is a rapid and efficient technique used to separate a variety of compounds including proteins. In CE proteins are driven by electric field through electrolyte solution and are separated according to their ion mobility. The advantage of this separation method is that it requires a low sample load [44, 45]. Coupled to the mass spectrometry as detection/identification method, the CE becomes an attractive separation method in proteome analysis. The on-line coupling of CE and MS offered an interesting alternative to 2D PAGE and to common chromatographic separation techniques, the protein mixtures can be analyzed within short time and with high resolution [46].

3 Mass Spectrometry-Based Proteomics

Mass spectrometry (MS) has emerged as a key technology in proteomics as it presents the most versatile high sensitivity detection system for peptide and protein analysis today. Contrary to for example antibody-based detection, MS is unbiased in principle, although mass spectrometry response is greatly influenced by the physicochemical properties of peptides and proteins. In addition, there are a number of different mass spectrometry techniques or "flavors" that will be described in the next paragraphs.

3.1 Ionization Techniques

Mass spectrometry involves the manipulation of ionized peptides and proteins under high vacuum conditions. Consequently, ways have to be found to get rid of the solvent and adduct shells that usually surround these analytes in any solution, and transfer charge(s) onto them in a controlled and reproducible fashion without inducing analyte decomposition. Two so-called "soft" ionization techniques have emerged over the past 25 years that allow for this largely non-destructive transfer from solution to the gas phase: Matrix Assisted Laser Desorption/Ionization (MALDI) and Electrospray Ionization (ESI). To recognize the almost revolutionary contribution that these techniques have had on the Life Sciences, the Nobel Prize in Chemistry 2002 was awarded to key inventors John Fenn (ESI) and Koichi Tanaka (MALDI) jointly with Kurt Wüthrich (for NMR).

3.1.1 Matrix Assisted Laser Desorption/Ionization (MALDI)

In MALDI (Matrix Assisted Laser Desorption/Ionization), peptides or proteins are mixed in solution with a large excess (roughly 10E4) of a small, UV-absorbing organic molecule, the so-called matrix. Microliter volumes of this solution are then deposited on a flat target made of conductive material, and the droplet dried by slow evaporation. Under suitable conditions, co-crystallization occurs where the analyte molecules are embedded in matrix crystals. The sample plate with the dried spots is then introduced into the vacuum chamber of the mass spectrometer, where it is irradiated with short nanosecond pulses of UV laser light. In the resulting process of rapid desorption and ionization, positively charged analyte ions are formed which can then be extracted from the source region using electrostatic fields and further manipulated for mass analysis [47–50].

During the MALDI process, peptides and proteins are ionized mainly as singly charged $[M+H]^+$ species. This introduces both benefits and limitations: on the one hand, MALDI mass spectra are usually straightforward to interpret as in most cases each observed signal corresponds to a single peak. On the other hand, this necessitates the use of mass analyzers with a large "mass range" (more precisely: m/z range) as high molecular weight ions translate into high m/z signals. In addition, singly charged biomolecules are in many cases more problematic with regard to sequence analysis since the repulsion between multiple charges on the same molecule is one of the driving forces for efficient fragmentation. Finally, MALDI is a discontinuous technique where usually several hundred to thousand individual laser shot experiments have to be accumulated to obtain high quality spectra. Even at the kHz laser frequencies available in modern instrumentation, this makes the process of for example peptide sequencing slow compared to ESI-based--> instrumentation. Finally MALDI cannot be directly hyphenated to chromatographic separations. While the latter limitation can be moderated by offline coupling ("LC-MALDI"), the combination of limiting factors has led to a decrease in the use of MALDI-based mass spectrometers in proteomics research. It still finds significant use in defined applications that require rapid fingerprinting from a non-separated sample, e.g., for microbial identification [51].

3.1.2 Electrospray Ionization (ESI)

ESI (Electrospray Ionization) today is the standard ionization technique in proteomics research. For ESI, a volume or stream of an aqueous analyte solution usually containing organic modifiers is sprayed from a sharp (μm diameter) needle tip towards the orifice (i.e., the entry to the vacuum section) of a mass spectrometer. The process is driven by application of a kV electrostatic potential differential between the needle and the orifice and happens at atmospheric pressure, making ESI an instance of the larger group of ionization techniques referred to as Atmospheric Pressure Ionization, or API. The thin liquid filament produced from the needle is quickly broken up into small droplets containing a small number of analyte ions preformed in solution. Through a combination of electrostatic repulsion (leading to "Coulomb explosions" that break droplets apart) and evaporation of solvent molecules, droplet of diminishing size that contain less and less analyte ions are produced until finally single analyte ions are produced either through droplet shrinking ("charge residue model") or by emission from highly charged droplets containing other analyte molecules ("ion evaporation model"). The produced analyte ions usually contain two to five charges for peptide analytes (e.g., $[M+2H]^{2+}$), or tens of charges in the case of intact protein analytes [38, 52–54].

The higher charging observed in ESI compared to MALDI has both advantages and disadvantages. Multiple charges compress the m/z range required from the mass analyzer, since, for example, for peptide produced by trypsination the majority of m/z values observed fall into the range of 350–1250. In addition, multiple charges on an analyte help drive fragmentation through charge repulsion or are actually (in the case of Electron Transfer Dissociation, or ETD) a prerequisite for some fragmentation techniques. In addition, multiple charge states (or m/z values) of the same analyte provide multiple readouts of the analyte's mass and thus potentially more accurate mass determinations. On the downside the presence of multiple charge states for each analyte in a complex mixture requires algorithms to properly assign ("deconvolute") these charge states, and often complicates spectra. The main benefit of ESI as a continuous ionization technique is that it is readily hyphenated to chromatographic or electrophoretic separations, providing a readout of the separation eluent in real time. Provided that the mass analyzer is fast enough to perform sequencing events at sufficient speed this leads to a very high sequencing capacity of the resulting hyphenated LC-ESI-MS setups.

3.2 Mass Analyzers

Following ionization, peptides and proteins of different mass and charge are separated in the vacuum region of the mass spectrometer by their mass-to-charge (m/z) ratio and detected. The m/z separation by different mass analyzers follows very different physical principles. Their performance can be characterized by the following parameters: (1) m/z range (or "mass range"), i.e., the range of m/z values for which ions can be transmitted at all; (2) transmission, i.e., the percentage of ions successfully transmitted through the mass analyzer in a given mode of operation. Transmission is invariably dependent on m/z value; (3) resolution, i.e., the ability to separate ions of similar m/z. Today, the most common definition used for resolution is the m/z value of a peak divided by its width at half height (FWHM, Full Width Half Height); (4) mass accuracy, i.e., the deviation of observed m/z values from their theoretically expected values, which is usually specified in parts per million (ppm). In this section we focus on the most common analyzer types used in proteomics mass spectrometry, and discuss their features and benefits rather than principles of operation.

Quadrupole and quadrupole ion trap mass analyzers are inherently low resolution, low mass accuracy analyzers which are often operated at "unit" resolution, i.e., a constant peak width of ~0.7 FWHM that translates into resolution values of 500–1500 for typical peptide peaks in the range of m/z 400–1000. In addition, they are relatively slow when operated in scanning mode, i.e., when covering a wide m/z range. To make up for this low resolution they possess excellent transmission characteristics with transmission values in excess of 90 % for wide m/z ranges. Consequently they are often used to filter for specific ions, e.g., when selecting for MS/MS precursors, or for manipulating ion packages, e.g., when used as collision cells for inducing MS/MS fragmentation (*see* below) [55].

Time-of-Flight (ToF) mass analyzers are of moderate resolution (10,000–40,000 FWHM) and exhibit mass accuracies in the range of 5–25 ppm with frequent calibration. To achieve good resolution, ions are usually accelerated in a direction orthogonal to their initial motion, and reflected on a so-called reflectron, or mirror stage, before hitting the detector. As a consequence of orthogonal acceleration and reflecting the ion beam, transmission is usually low, on the order of a few percent. The low transmission is partially recovered by the high speed of acquisition [56]. Modern Time-of-Flight analyzers operate at frequencies of up to 5 kHz, i.e., 5000 individual experiments per second. Even when these are accumulated before writing the data to disk, acquisition speed of up to 100 Spectra-to-Disk can be obtained. Through data accumulation the signal-to-noise ratio can be improved even at weak absolute signal strength. Its discontinuous mode of operation makes Time-of-Flight the perfect match for the equally discontinuous MALDI. Indeed, MALDI-ToF mass spectrometers were one of the first high resolution instrument class introduced into proteomics research [57, 58]. Today, however, ESI-ToF mass spectrometers are as common.

Orbitrap mass analyzers are high resolution (15,000–140,000 FWHM), high accuracy (0.5–5 ppm) analyzers that have almost become a standard in proteomics mass spectrometry. Ions are introduced into a small spindle-shaped electrostatic cell, and the imaging current recorded from their axial motion recorded in a non-destructive fashion. From the observed frequency transient, the m/z spectrum is then calculated by Fourier Transformation. Same as for the similarly operated Fourier Transform-Ion Cyclotron Resonance (FT-ICR) mass analyzers, mass resolution is a function of transient duration and decay to higher m/z values, so practical resolution values obtained are similar to those obtained for ToF instruments [59, 60]. The Orbitrap mass analyzer does not require frequent recalibration, making it a very good choice for instrument operated in high throughput environments.

3.3 Tandem Mass Spectrometry (MS/MS)

Proteomics samples are highly complex mixtures of very similar analytes. Following the most commonly employed bottom-up approach that involves tryptic digestion, a sample containing for example 2000 protein species will produce an estimate 100,000 peptides on digestion [publication Matthias Mann]. Consequently it is not enough to determine the accurate mass of a digest peptide to unambiguously determine its identity. Even when combined with chromatographic retention time information, an accurate mass tag (AMT) will only serve to identify a tryptic peptide in proteomes of limited size, and only when information about for example posttranslational modifications is excluded [Lit]. In most cases, information about the peptide's sequence has to be obtained within the mass spectrometer to allow for unambiguous identification. This usually requires tandem mass spectrometry, i.e., the use of two mass analyzers in combination with an event causing sequence-specific degradation of the peptide.

3.3.1 Product Ion Scanning

The most common tandem mass spectrometry implementation is the product ion scan. A peptide ion of defined m/z value is filtered from the whole population of ions using a first MS stage, often achieved using a quadrupole mass filter. This isolated precursor ion is then fragmented in the mass spectrometer to produce sequence-specific ions, which are then separated by their m/z and detected in a second stage mass analyzer, e.g., a ToF or an Orbitrap. Each peptide is thus characterized by its time of introduction to the MS (i.e., its retention time when the MS coupled to a chromatographic separation), its precursor m/z value and a set of fragment m/z values. In the positive ion mode and when using suitable fragmentation techniques (*see* below), peptides fortunately produce a defined set of largely sequence-specific fragments which can be denominated using a system devised by Roepstorff and Fohlman as early as 1984 [61, 62].

3.3.2 Precursor Ion and Constant Neutral Loss Scanning

In product ion scanning, all fragments derived from a single precursor are recorded. In some instances it can also be useful to alternatively record all precursors producing a single fragment, or marker ion, e.g., when this is predictive for a structural feature, e.g., a posttranslational modification. For these so-called precursor ion scans, the first stage mass analyzer is scanned across the precursor m/z range while the second stage mass analyzer is set to a fixed m/z to filter for the marker ion. Precursor ion scans have been successfully employed to screen for, for example, phosphorylated or glycosylated peptide precursor ions in complex mixtures using either Triple Quadrupole (QqQ) or Quadrupole-Time-of Flight (QqToF) mass spectrometers [63, 64].

A related experiment is the Constant Neutral Loss Scan, where both mass analyzers are scanned simultaneously but at an m/z offset to detect precursors specifically losing neutral molecules indicating for example phosphorylation. Neither precursor ion scanning nor constant neutral loss scanning are much used in proteomics studies today, since specific detection of for example phosphopeptides may be achieved much more efficiently by for example affinity enrichment.

3.3.3 Data-Dependent Versus Data-Independent Acquisition

In a typical proteomics mass spectrometry experiment, in excess of 100,000 peptide precursors need to be sequenced in a few hours of mass spectrometer acquisition time. Consequently the selection and sequencing of peptide precursors has to be a fully automated process, with the required sequencing speeds being on the order of 25 peptides/s [65]. Modern mass spectrometers achieve this through Data Dependent Acquisition (DDA) routines implemented in their acquisition software. In DDA mass spectrometer first performs an MS scan to detect all peptide precursors coming from the ion source at a specific time. Up to 25 suitable precursors are the identified using criteria as intensity, charge state and *m/z*, and sequentially submitted to a corresponding number of product ion scans for obtaining sequence-specific fragmentation. Once finished, another cycle is started with the next MS scan [66, 67]. Current instrumentation is capable of sequencing speed of ten product ion spectra per second, producing a capacity of up to 36,000 sequencing events/h. As not all sequencing events are successful and nonredundant, around five to six peptide identifications/s of acquisition time represent the current state of the art [65, 68, 69].

The discrepancy between the required and the achieved sequencing speeds and the resulting undersampling of complex samples has prompted researchers and instrument manufacturers to look for fundamentally different data acquisition strategies especially for reproducible quantitative comparison of large numbers of samples.

If undersampling in DDA renders the detection and quantitative analysis of analytes of interest irreproducible, one alternative is to forego a dynamic selection of peptide precursors and rather target sets of peptides that carry the desired information, e.g., about the quantity of a set of proteins. Selected Reaction Monitoring (SRM, also frequently called Multiple Reaction Monitoring, or MRM) on Triple Quadrupole Mass Spectrometers is the most popular targeted acquisition strategy. In SRM, the two quadrupole mass analyzers of the spectrometer are set to preprogrammed fixed *m/z* values that filter for a combination of a peptide precursor and, after fragmentation in a collision cell, a sequence-specific fragment. While this so-called transition does not carry full spectral information, it can be seen as a highly specific detection channel for these peptides. Several hundred of these channels can be monitored sequentially in a single LC-ESI-SRM experiment to provide quantitative information on dozens of peptides of interest [70].

Targeted mass spectrometry methods require upfront knowledge and specification of the analytes of interest, and are limited by the number of transitions that can be monitored in a single experiment. Newer developments in Data-Independent Acquisition (DIA), e.g., SWATH acquisition (Sequential Window Analysis of All Theoretical Fragment ion Spectra) [71], allow the simultaneous detection and quantitation of a principally unlimited number of analytes in a single LC-ESI-MS experiment. All peptide precursors undergo fragmentation at less stringent filtering, and traces for sequence-specific fragments contained in a previously obtained spectral library are extracted from the data to provide quantitative information. From a single experiment, 10,000 s of fragment ion traces can be extracted that allow consistent quantitation of for example 2500 proteins derived from 15,000 peptides from *S. cerevisiae* [72]. For brevity we refer the reader to the literature for details of the implementation.

3.3.4 Ion Activation

All MS/MS experiments and approaches require techniques for a controlled, reproducible and reproducible activation of precursor ions to obtain structure-specific decomposition in the mass spectrometer's vacuum [73]. While there are a multitude of techniques for ion activation available, only a handful of them are suitable for the large scale analysis of peptides for proteomics.

Collision-Induced Dissociation (or CID, sometimes referred to as Collisionally Activated Dissociation, or CAD) is a so-called ergodic, even-electron ion activation technique where excess vibrational energy is deposited in peptide precursors through multiple collisions with small neutral gas molecules, e.g., nitrogen, in a collision cell of defined gas pressure in the mass spectrometer, leading to eventual breaking of covalent bonds. CID is by far the most commonly used ion activation technique in proteomics mass spectrometry, and is highly reproducible even across different instrumental platforms and laboratories. It provides excellent sequence information especially on non-modified peptides generated from the trypsination of proteins for bottom-up proteome analysis [74, 75]. For large peptide precursors or peptides carrying labile modifications, e.g., glycosylation, it often produces only limited sequence information. Electron Transfer Dissociation (or ETD, related to the less-often applied Electron Capture Dissociation, or ECD) can be used as an alternative or even complementary ion activation technique in these cases. In ETD, a single, odd electron is transferred from a reactant gas onto the peptide precursor in the mass spectrometer. The resulting odd-electron fragmentation mechanisms are quite different from those produced in CID [76], e.g., labile modifications are often retained on peptides. ETD requires the peptide precursor to be higher charged ($n \geq 3$) for efficient fragmentation though, making it a better match for larger peptides produced using enzymes other than trypsin, or even for small intact proteins [77].

3.4 Analysis of MS and MS/MS Data

3.4.1 Peptide Identification

How to exploit now the information contained in LC-ESI-MS data sets obtained from complex peptide mixtures? Each fragmented peptide precursor is characterized by (1) it retention time, (2) its intact mass-to charge ratio, (3) its charge state which can in most cases be deduced from the isotopic pattern and (4) a set of more or less structure-specific fragment ions. A typical LC-ESI-MS data set will today encompass in excess of 100,000 such precursor "feature sets."

At the beginning of peptide mass spectrometry, sequence was often derived from MS/MS fragment ion patterns by de novo sequencing [78]. By reading out amino acid-specific mass differences between ions of either C- or N-terminal fragment ion series, partial stretches of a peptide's sequence can in many cases be derived from the spectrum. By combining several such stretches and information about for example the presence of individual amino acids or the C-terminal amino acid which may be derived from individual marker ions, the complete sequence of a peptide can be obtained in select cases. The process is highly error-prone though and hampered by incomplete fragmentation, overlay of different ion series, or additional non-sequence-specific fragmentation events. It is therefore usually used as a last resort in cases where other approaches fail, e.g., in the case of proteins from organisms which are poorly covered in genome and proteome sequence databases. Related to full-blown de novo sequencing is the peptide sequence tag approach [79] where a short sequence tag of as little as three to four consecutive amino acids together with information on the remaining masses (or tag) required to combine to the peptide's full mass are often sufficient for unambiguous identification of the peptide sequence in a full proteome sequence database. Same as de novo sequencing, the approach is still relatively error-prone and computationally expensive.

Today, protein identification in the majority of cases is achieved by Peptide Fragment Fingerprint (PFF) matching. Here the set of fragments characterizing a peptide precursor is not interpreted at all, but is pattern-matched against fragment patterns predicted in silico for peptides generated from a theoretical digest of all proteins in a protein sequence database. Each match is then scored based on the agreement between the observed and the predicted pattern. In the most commonly used probabilistic approach, the score reflects the chances of a random assignment against the background of the whole database. PFF matching is implemented in a significant number of both academic and commercial algorithms, or database search engines, such as SEQUEST [80], Mascot [81], OMSSA [82], Paragon [83], or Andromeda [84].

In case of peptide modifications, the exact position of the modification on the primary sequence of the peptide may be as important as its presence in itself, e.g., in the case of phosphorylation where peptides may contain more than one serine/threonine/tyrosine residue that can be phosphorylated. This so-called site localization problem can also be addressed, often by comparing the search engine scores obtained for different theoretically present positional modification isomers of the same primary peptide sequence and deriving a metascore. The most popular implementations of this concept are the AScore [85], the MASCOT Delta Score [86], and phosphoRS [87].

All scoring-based approaches for peptide identification or site localization suffer from the presence of false positive/negative identifications, a fact that is easily recognized when different search algorithms are compared against one another. Individual scores cannot be validated per se, except by comparison with results obtained on synthetic standards, a concept that is prohibitively expensive for global analyses. Ways must therefore be found to estimate the validity of results on the basis of the whole ensemble. This can be achieved in two ways. The most widely taken approach is based on the estimation of False Discovery Rates [88]. The sequence database used for Peptide Fragment Fingerprint matching is extended by (or concatenated with) sequences generated through for example scrambling or reversing the individual protein sequences. Sequence reversal is usually preferred as it will not change the amino acid composition, the number of available trypsin cleavage sites or the overall length distribution of the resulting tryptic peptides. When the ensemble of fragment ion spectra is searched against the resulting forward/reverse database, all hits recorded against the reverse part are considered random, with the same number of random matches expected from the forward part of the database, and a False Discovery Rate be estimated. The resulting lists of forward and reverse matches can be used to truncate the results list to a specified FDR level, both on the peptide and on the protein level.

An alternative approach relies on a semi-supervised machine learning approach that uses both high-scoring PSMs ("positive PSMs") and negative PSMs obtained against shuffled protein sequence databases to derive a model that improves differentiation between correct and false positives. The approach is implemented in the Percolator algorithm which has been widely implemented in a number of database search pipelines [89].

3.4.2 Protein Inference, In-Depth Proteomics and Quantitation

Another challenge in bottom-up proteomics is that even the correct identification of a peptide sequence does not necessarily lead to correct identification of a protein, or even its functional state. Peptide sequences may be conserved across whole families of proteins or different splice isoforms; function might be mediated by single or multiple posttranslational modifications, e.g., phosphorylation cascades in case of cell signaling; and finally most proteins do not function in isolation, but rather in the context of for example protein–protein complexes. What is more, single or even multiple experimentally validated peptide sequences cannot necessarily be linked to a single set of genes coding for a protein, making the correlation of genomics, transcriptomics and proteomics data challenging [5–7]. It is therefore of utmost importance not only to identify and quantitate all functionally relevant structural features of a single proteoform in each experiment, but to do so and follow changes across different cell compartments, functional states or isoforms, and with a number of biological and technical replicates that allow their visualization on a base of statistical significance.

These requirements have several consequences. First, the implementation of algorithms that derive the most plausible set of protein properties (e.g., identity, modification state, and quantity) from an observed set of peptide properties. The approach followed by most algorithms—and implemented in all relevant commercial and academic software packages—follows the principle of Occam's Razor: to find and use the most concise explanation to explain all relevant observations. While this approach is widely accepted in the community, researchers should still be aware that a list of protein identification or quantitation results may actually represent more proteoforms than apparent, and any mechanism or software implementation used to communicate and discuss proteomics data should allow mining multilayered data of this type.

Second, there is still a need to improve proteomics workflows further so that they provide the highest possible amount of information with moderate effort regarding sample preparation and instrument time, and at high technical reproducibility. The ideal workflow should provide full information (sequence coverage, modification state, quantity) about all proteoforms [6] in the sample, not require more than a few hours of instrument time to allow acquisition and analysis of relevant numbers of biological and technical replicates for improved statistical significance, and involve as few sample preparation and fractionation steps as these are potential sources of non-reproducibility. This trend to what is often referred to as in-depth proteomics has been a significant driver of both mass spectrometer technology and proteomics workflow development over the past years [8].

Finally, it has been realized that all successful proteomics experiments need to involve suitable strategies for quantitation and quantitative standardization. If a protein's concentration is just above detection level in state A, and just below detection level in state B of a biological system, this might reflect small changes in the efficiency of sample preparation or instrument performance on a given day as much as its actual concentration. The often used Venn diagrams that represent the sets of peptides or proteins either detected or not detected in the different states are thus rather a reflection of analytical reproducibility than of biological meaning. Quantitative experimentation should include direct information about either relative concentration changes, or absolute information about protein concentration in relation to the attainable limits of detection and quantitation.

3.4.3 Quantitation from MS and MS/MS Data

Which properties of proteins acquired in a proteomics experiment can be used for quantitation? And what are practical strategies for the introduction of either relative or absolute quantitation standards? If gel staining techniques are used for detecting and resolving the different proteins in a sample then the quantitation can be decoupled from the identification or characterization of the protein in question, which has significant implications for the workflow. For example, if 2DE is used to visualize and quantitate proteins then it is frequent practice to only process those spots by excision, in-gel digestion and staining that exhibit differential staining behavior. Identity and differential quantity will thus be only established for a subset of the available protein complement. One of the shortcomings here is the fact that even from a seemingly pure 2DE spot visualized by for example silver or fluorescent staining, often 10–20 unique proteins may be identified by nanoLC-ESI-MS analysis. The correlation between the "identity" of a spot and the protein actually causing the differential staining is therefore not always straightforward. Combined with the high manual effort required to produce 2DE gels in sufficient numbers of replicates, the approach is used less frequently, and more and more studies rather use global or targeted mass spectrometry techniques for peptide and protein quantitation [90].

More or less all mass spectrometry-based workflows use an upfront chromatographic separation. The peptides (or proteins) thus analyzed also undergo a two-dimensional separation of chromatographic retention time versus mass-to-charge ratio. In addition, MS/MS sequencing events are triggered at irregular frequency if data-dependent acquisition is applied. The number of MS/MS sequencing events per peptide can be used as a proxy to represent its quantity and, if the values for multiple peptides from the same protein are combined, for protein quantity. This general approach is referred to as spectral counting [91], and was one of the first routinely used for quantitating proteins from nanoLC-ESI-MS data since the number of sequencing events can be directly derived from regular protein identification experiments. Spectral counting thus comes "at no extra cost," and was shown early on to provide a stable if rough approach to estimate protein quantity. Several strategies have been proposed to in the meantime to improve the accuracy and dynamic range of spectral counting: weighting observed spectral counts by protein length and normalizing to the sum of all values (Normalized Spectral Abundance Factors, NSAF) [92], by the fraction of theoretically observable peptides without or with exponential weighting (Protein Abundance Index, PAI, exponentially modified spectral abundance index, emPAI) [93] or by a relational scoring of observed versus theoretically observable peptides (Absolute Protein Expression, APEX) [94]. Despite its limited accuracy, spectral counting is frequently employed especially for the analysis of Affinity Purification-Mass Spectrometry (AP-MS) experiments. Here, proteins isolated by affinity capture against a bait protein linked to a solid phase support are tryptically digested and analyzed by nanoLC-ESI-MS. To differentiate between binding which is specific to the bait protein and nonspecific binding to for example the solid phase support, control experiments for example without bait protein are required. As the observed enrichment factors for specifically bound proteins are usually high but sample amounts are limited, spectral counting is a logical choice as the relative quantitation is directly available from the protein identification data.

More accurate approaches to peptide and protein quantitation use the peak intensity or peak area observed for a peptide's isotopic pattern during chromatographic separation. A number of approaches have been developed for label-free quantitation [95, 96] that then correlate the observed peak area with peptide and protein concentration. For a relative determination of peptide and protein concentration it is then sufficient to calculate the ratios of the observed peak areas versus a reference sample. Multiple peptides will translate into multiple peak area ratios for a given protein, making the quantitation more robust. More developed approaches have also tried to correlate observed peak areas with absolute concentration. Due to the inherent differences between individual peptides with regard to ionization efficiency, the peak area of a single peptide will usually not be sufficient to accurately reflect a protein's concentration. Silva et al. established in 2006 that the sum of the three most abundant peptides per protein provide a relatively stable readout of a protein's absolute concentration in LCMSE, a precursor of current Data-Independent Acquisition Strategies, and provided that quantitated reference proteins are spiked into the sample will allow for estimation of the absolute protein concentration across several orders of magnitude ("Top3" approach) [97]. In the meantime multiple algorithms have been developed for label-free quantitation to enable both relative and absolute quantitation of proteins. Of note is the combination of iBAQ, a quantitation algorithm implemented in the popular MaxQuant software package [98], with the use of a quantitated protein standard encompassing 48 proteins across 5 orders of magnitude linear range. From the iBAQ quantitation values of the standard proteins a linear calibration curve can be calculated, which in turn allows the estimation of all proteins in the sample for, for example, determination of copy numbers across several orders of magnitude linear range. The approach has found great utility in the biological research community, from the first studies that correlated protein expression with transcriptome data down to the determination of protein stoichiometry in protein–protein complexes [99].

Label-free quantitation relies on the reproducible parallel analysis of the samples to be compared. The technical variability of the total analysis workflow has to be lower than the sample-to-sample or state-to-state variability to allow for quantitative results to be obtained at statistical significance. While great strides have been made to increase the reproducibility of both chromatography and mass spectrometric acquisition, this still precludes label-free quantitation from being used in workflows that require multistep sample preparation, e.g., in the analysis of protein phosphorylation. In these instances more traditional workflows involving internal stable heavy isotope-labeled standards are still prevalent. An excellent review of different standardization strategies can be found at [90]. Heavy isotope-labeled standards can be introduced in several ways: as discrete synthetic peptides carrying one or more heavy isotope-labeled amino acid building blocks (Absolute Quantitation of Peptides, or AQUA [100]); as chemical labels that are reactive to either all peptides in a digested samples (iTRAQ [101], TMT [102], dimethyl labeling [103]) or only peptides carrying select amino acid residues, e.g., cysteine (iCAT [104]); as C-terminal labels introduced during the process of enzymatic protein digestion (O18 labeling [105]); or as metabolic labels introduced by growing a system, e.g., in media containing heavy isotope-labeled amino acids (SILAC [106, 107]) or even globally labeled media (15N labeling [108]). The choice of labeling strategy depends on a set of questions to be answered for experimental design: (1) Are labels for select peptides and proteins sufficient, or is a global labeling required? (2) Is relative quantitation sufficient, or rather absolute quantitation required? (3) Can the system be metabolically labeled at all (e.g., in cell culture), or is this not possible as in most biomedical systems? (4) What degree of multiplexing is required, i.e., how many different stable isotope labels (or "channels") are required? And finally (5) at which point can the label be introduced into the overall workflow? The earlier internal standards are introduced, the higher the number of experimental steps in the workflow that may be normalized.

A topic underlying protein quantitation by mass spectrometry is again the challenge of protein inference. Quantitation values obtained from multiple peptides per protein can be averaged to achieve more reliable quantitation, but only if all peptides actually belong to the same proteoform. Accurate quantitation thus relies on a detailed qualitative understanding of the proteomic sample under investigation.

4 Proteomics and Other Omics; A Short Summary

A detailed qualitative understanding requires a comprehensive analysis of the peptides and proteins contained in a biological sample. As today's mass spectrometry-driven workflows mostly rely on identification by sequence database matching, accurate protein sequence databases are usually regarded as a prerequisite for successful proteome analysis. While this holds true in most cases, it can also be beneficial to use highly customized sequence databases generated for example from transcriptome analysis to analyze genetic variability, or work with organisms where little or no high quality information is contained in the available protein sequence databases. Peptide and protein sequences derived from mass spectrometry data can in turn be used to annotate genome or transcriptome databases, refine gene models or validate gene expression on the protein level. This rapidly blossoming field of research is usually called proteogenomics, and is attracting increased attention as it can help improve our understanding of the interplay between genetic coding, transcription and phenotype [109].

Another research field that is still in its infancy is the combination of proteomics and metabolomics, i.e., the global analysis of small organic molecule metabolite concentration changes. While metabolomics in and of itself is receiving considerable attention for example as a potential diagnostic tool, the integration of proteomics and metabolomics data is still challenging, although there is great promise to improve for example our knowledge of metabolic flux or cell signaling.

All in all, mass spectrometry-based proteomics has reached the scientific mainstream. While the technology, workflows and data processing still require significant specialist expertise, the analytical results already support a major body of research in the life sciences, from basic research to biomedical applications. Using state-of-the-art equipment and knowhow it is today possible to perform an in-depth quantitative proteome analysis in a number of days.

References

1. Wilhelm M, Schlegl J, Hahne H et al (2014) Mass-spectrometry-based draft oft he human proteome. Nature 509:582–587
2. Kim MS, Pinto SM, Getnet D et al (2014) A draft map of the human proteome. Nature 509:575–581
3. Tyers M, Mann M (2003) From genomics to proteomics. Nature 422:193–197
4. Pandey A, Mann M (2000) Proteomics to study genes and genomes. Nature 405:837–846
5. Rappsilber J, Mann M (2002) What does it mean to identify a protein in proteomics? Trends Biochem Sci 27:74–78
6. Smith LM, Kelleher NL (2013) Proteoform: a single term describing protein complexity. Nat Methods 10:186–187
7. Schlüter H, Apweiler R, Holzhütter HG et al (2009) Finding one's way in proteomics: a protein species nomenclature. Chem Cent J. doi:10.1186/1752-153X-3-11
8. Lenz C, Urlaub H (2014) Separation methodology to improve proteome coverage depth. Expert Rev Proteomics 11:409–414
9. Catherman AD, Skinner OS, Kelleher NL (2014) Top Down proteomics: facts and perspectives. Biochem Biophys Res Commun 445:683–693
10. Ahlf DR, Thomas PM, Kelleher NL (2013) Developing top down proteomics to maximize proteome and sequence coverage from cells and tissues. Curr Opin Chem Biol 17:787–794
11. Weber K, Osborn M (1969) The reliability of molecular weight determinations by dodecyl sulfate-polyacrylamide gel electrophoresis. J Biol Chem 244:4406–4412
12. Hanash SM (2001) 2D or not 2D is there a future for 2D gels in proteomics? Proteomics 1:635–637
13. Dihazi H, Müller GA (2007) The urinary proteome: a tool to discover biomarker of kidney diseases. Expert Rev Proteomics 4:39–50
14. O'Farrell PH (1975) High resolution two-dimensional electrophoresis of proteins. J Biol Chem 250:4007–4021
15. O'Farrell PZ, Goodman HM, O'Farrell PH (1977) High resolution two-dimensional electrophoresis of basic as well as acidic proteins. Cell 12:1133–1141
16. Klose J (1975) Protein mapping by combined isoelectric focusing and electrophoresis of mouse tissues. A novel approach to testing for induced point mutations in mammals. Humangenetik 26:231–243
17. Lilley KS, Razzaq A, Dupree P (2002) Two-dimensional gel electrophoresis: recent advances in sample preparation, detection and quantitation. Curr Opin Chem Biol 6:46–50
18. Switzer RC III, Merril CR, Shifrin S (1979) A highly sensitive silver stain for detecting proteins and peptides in polyacrylamide gels. Anal Biochem 98:231–237
19. Steinberg TH, Jones LJ, Haugland RP, Singer VL (1996) SYPRO Orange and SYPRO Red

protein gel stains: one-step fluorescent staining of denaturing gels for detection of nanogram levels of protein. Anal Biochem 239:223–237
20. Patton WF (2002) Detection technologies in proteome analysis. J Chromatogr B Analyt Technol Biomed Life Sci 771:3–31
21. Unlu M, Morgan ME, Minden JS (1997) Difference gel electrophoresis: a single gel method for detecting changes in protein extracts. Electrophoresis 18:2071–2077
22. Gharbi S, Gaffney P, Yang A et al (2002) Evaluation of two-dimensional differential gel electrophoresis for proteomic expression analysis of a model breast cancer cell system. Mol Cell Proteomics 1:91–98
23. Dihazi H, Dihazi GH, Jahn O et al (2011) Multipotent adult germline stem cells and embryonic stem cells functional proteomics revealed an important role of eukaryotic initiation factor 5A (Eif5a) in stem cell differentiation. J Proteome Res 10:1962–1973
24. Dihazi H, Dihazi GH, Nolte J et al (2009) Differential proteomic analysis of multipotent adult germline stem cells and embryonic stem cells reveals high proteome similarity. J Proteome Res 8:5497–5510
25. Zuo X, Speicher DW (2002) Comprehensive analysis of complex proteomes using microscale solution isoelectrofocusing prior to narrow pH range two-dimensional electrophoresis. Proteomics 2:58–68
26. Lin D, Tabb DL, Yates JR III (2003) Large-scale protein identification using MS. Biochim Biophys Acta 1646:1–10
27. Link AJ, Eng J, Schieltz DM et al (1999) Direct analysis of protein complexes using MS. Nat Biotechnol 17:676–682
28. Issaq HJ, Chan KC, Janini GM et al (2005) Multidimensional separation of peptides for effective proteomic analysis. J Chromatogr B 817:35–47
29. Majors RE (1980) Multidimensional high performance liquid chromatography. J Chromatogr Sci 18:571–580
30. Giddings JC (1984) Twodimensional separations: concept and promise. Anal Chem 56:1258A–1264A
31. Cortes HJ (ed) (1990) Multidimensional chromatography. Techniques and applications. Marcel Dekker, New York
32. Anderegg RJ, Wagner DS, Blackburn RK, Opiteck GJ, Jorgenson JW (1997) A multidimensional approach to protein characterization. J Protein Chem 16:523–526
33. Neverova I, Van Eyk JE (2005) Role of chromatographic techniques in proteomic analysis. J Chromatogr B 815:51–63
34. Neverova I, Van Eyk JE (2002) Application of reversed phase high performance liquid chromatography for subproteomic analysis of cardiac muscle. Proteomics 2:22–31
35. Zhu H, Klemic JF, Chang S et al (2000) Analysis of yeast protein kinases using protein chips. Nat Genet 26:283–289
36. Lueking A, Horn M, Eickhoff H et al (1999) Protein microarrays for gene expression and antibody screening. Anal Biochem 270:103–111
37. MacBeath G (2002) Protein microarrays and proteomics. Nat Genet 32:S526–S532
38. Schmidt A, Karas M, Dülcks T (2003) Effect of different solution flow rates on analyte ion signals in nano-ESI MS, or: when does ESI turn into nano-ESI? J Am Soc Mass Spectrom 14:492–500
39. Luo Q, Gu Y, Wu S-L et al (2008) Two-dimensional strong cation exchange/porous layer open tubular/mass spectrometry for ultratrace proteomic analysis using a 10 μm id poly(styrene-divinylbenzen porous layer open tubular column with an on-line triphasic trapping column. Electrophoresis 29:1804–1811
40. Sandra K, Moshir M, D'hondt F et al (2008) Highly efficient peptide separations in proteomics. Part 1. Unidimensional high performance liquid chromatography. J Chromatogr B 866:48–63
41. Köcher T, Pichler P, Swart R et al (2012) Analysis of protein mixtures from whole-cell extracts by single-run nanoLC-MS/MS using ultralong gradients. Nat Protoc 7:882–890
42. Hsieh EJ, Bereman MS, Durand S et al (2013) Effects of column and gradient lengths on peak capacity and peptide identification in nanoflow LC-MS/MS of complex proteomics samples. J Am Soc Mass Spectrom 24:148–153
43. Sandra K, Moshir M, D'hondt F et al (2009) Highly efficient peptide separations in proteomics. Part 2. Bi- and multidimensional liquid-based separation techniques. J Chromatogr B 877:1019–1039
44. Haubitz M, Wittke S, Weissinger EM et al (2005) Urine protein patterns can serve as diagnostic tools in patients with IgA nephropathy. Kidney Int 67:2313–2320
45. Weissinger EM, Wittke S, Kaiser T et al (2004) Proteomic patterns established with capillary electrophoresis and mass spectrometry for diagnostic purposes. Kidney Int 65:2426–2434
46. Wittke S, Fliser D, Haubitz M et al (2003) Determination of peptides and proteins in human urine with capillary electrophoresis-mass spectrometry, a suitable tool for the establishment of new diagnostic markers. J Chromatogr A 1013:173–181
47. Karas M, Hillenkamp F (1988) Laser desorption ionization of proteins with molecular

masses exceeding 10,000 daltons. Anal Chem 60:2299–2301
48. Karas M, Glückmann M, Schäfer J (2000) Ionization in matrix-assisted laser desorption/ionization: singly charged molecular ions are the lucky survivors. J Mass Spectrom 35:1–12
49. Stevenson E, Breuker K, Zenobi R (2000) Internal energies of analyte ions generated from different matrix-assisted laser desorption/ionization matrices. J Mass Spectrom 35:1035–1041
50. Krüger R, Pfenninger A, Fournier I et al (2000) Analyte incorporation and ionization in matrix-assisted laser desorption/ionization visualized by pH indicator molecular probes. Anal Chem 73:5812–5821
51. Patel R (2015) MALDI-TOF MS for the diagnosis of infectious diseases. Clin Chem 61:100–111
52. Whitehouse CM, Dreyer RN, Yamashita M et al (1985) Electrospray interface for liquid chromatographs and mass spectrometers. Anal Chem 57:675–679
53. Fenn JB, Mann M, Meng CK et al (1989) Electrospray ionization for mass spectrometry of large biomolecules. Science 246:64–71
54. Emmett MR, Caprioli R (1994) Microelectrospray mass spectrometry: ultra-high-sensitivity analysis of peptides and proteins. J Am Soc Mass Spectrom 5:605–613
55. Schwartz JC, Jardine I (1996) Quadrupole ion trap mass spectrometry. Methods Enzymol 270:552–586
56. Chernushevich IV, Loboda AV, Thomson BA (2001) An introduction to quadrupole-time-of-flight mass spectrometry. J Mass Spectrom 36:849–865
57. Hines WM, Parker K, Peltier J et al (1998) Protein identification and protein characterization by high-performance time-of-flight mass spectrometry. J Protein Chem 17:525–526
58. Beinvenut WV, Daon C, Pasquarello C et al (2002) Matrix-assisted laser desorption/ionization-tandem mass spectrometry with high resolution and sensitivity for identification and characterization of proteins. Proteomics 2:868–876
59. Hardman M, Makarov AA (2003) Interfacing the orbitrap mass analyser to an electrospray ion source. Anal Chem 75:1699–1705
60. Zubarev R, Makarov AA (2013) Orbitrap mass spectrometry. Anal Chem 85:5288–5296
61. Roepstorff P, Fohlman J (1984) Proposal for a common nomenclature for sequence ions in mass spectra of peptides. Biomed Mass Spectrom 11:601
62. Steen H, Mann M (2004) The ABC's (and XYZ's) of peptide sequencing. Nat Rev Mol Cell Biol 5:699–711
63. Huddleston MJ, Bean MF, Carr SA (1993) Collisional fragmentation of glycopeptides by electrospray ionization LC/MS and LC/MS/MS: methods for selective detection of glycopeptides in protein digests. Anal Chem 65:877–884
64. Carr SA, Huddleston MJ, Annan RS (1996) Selective detection and sequencing of phosphopeptides at the femtomole level by mass spectrometry. Anal Biochem 239:180–192
65. Michalski A, Cox J, Mann M (2011) More than 100,000 detectable peptide species elute in single shotgun proteomics runs but the majority is inaccessible to data-dependent Lc-MS/MS. J Proteome Res 10: 1785–1793
66. Huang EC, Henion JD (1990) LC/MS and LC/MS/MS determination of protein tryptic digests. J Am Soc Mass Spectrom 1:158–165
67. Covey TR, Huang EC, Henion JD (1991) Structural characterization of protein tryptic peptides via liquid chromatography/mass spectrometry and collision-induced dissociation of their doubly charged molecular ions. Anal Chem 63:1193–1200
68. Zubarev A (2013) The challenge of the proteome dynamic range and its implications for in-depth proteomics. Proteomics 13:723–726
69. Hebert AS, Richards AL, Bailey DJ et al (2014) The one hour yeast proteome. Mol Cell Proteomics 13:339–347
70. Picotti P, Aebersold R (2012) Selected reaction monitoring-based proteomics: workflows, potentials, pitfalls and future directions. Nat Methods 9:555–566
71. Gillet LC, Navarro P, Tate S et al (2012) Targeted data extraction of the MS/MS spectra generated by data-independent acquisition: a new concept for consistent and accurate proteome analysis. Mol Cell Proteomics 11, O111.016717. doi:10.1074/mcp.O111.016717
72. Selevsek N, Chang CY, Gillet LC et al (2015) Reproducible and consistent quantification of the Saccharomyces cerevisiae proteome by SWATH-mass spectrometry. Mol Cell Proteomics 14:739–749
73. Sleno L, Volmer DA (2004) Ion activation methods for tandem mass spectrometry. J Mass Spectrom 39:1091–1112
74. Wells JM, McLuckey SA (2005) Collision-induced dissociation (CID) of peptides and proteins. Methods Enzymol 402:148–185
75. Olsen JV, Macek B, Lange O et al (2007) Higher-energy C-trap dissociation for peptide modification analysis. Nat Methods 4:709–712
76. Syka JE, Coon JJ, Schroeder MJ et al (2004) Peptide and protein sequence analysis by electron transfer dissociation mass spectrometry. Proc Natl Acad Sci U S A 101:9528–9533

77. Mikesh LM, Ueberheide B, Chi A et al (2006) The utility of ETD mass spectrometry in proteomic analysis. Biochim Biophys Acta 1764:1811–1822
78. Medzihradsky KF, Chalkley RJ (2015) Lessons in de novo peptide sequencing by tandem mass spectrometry. Mass Spectrom Rev 34:43–63
79. Mann M, Wilm M (1994) Error-tolerant identification of peptides in sequence databases by peptide sequence tags. Anal Chem 66:4390–4399
80. MacCoss MJ, Wu CC, Yates JR 3rd (2002) Probability-based validation of protein identifications using a modified SEQUEST algorithm. Anal Chem 74:5593–5599
81. Perkins DN, Pappin DJ, Creasy DM et al (1999) Probability-based protein identification by searching sequence databases using mass spectrometry data. Electrophoresis 20: 3551–3567
82. Geer LY, Markey SP, Kowalak JA (2004) Open mass spectrometry search algorithm. J Proteome Res 3:958–964
83. Shilov IV, Seymour SL, Patel AA et al (2007) The Paragon Algorithm, a next generation search engine that uses sequence temperature values and feature probabilities to identify peptides from tandem mass spectra. Mol Cell Proteomics 6:1638–1655
84. Cox J, Neuhauser N, Michalski A et al (2011) Andromeda: a peptide search engine integrated into the MaxQuant environment. J Proteome Res 10:1794–1805
85. Beausoleil SA, Villén J, Gerber SA et al (2006) A probability-based approach for high-throughput protein phosphorylation analysis and site localization. Nat Biotechnol 24: 1285–1292
86. Savitski MM, Lemeer S, Boesche M et al (2011) Confident phosphorylation site localization using the Mascot Delta Score. Mol Cell Proteomics 10, M110.003830. doi:10.1074/mcp.M110.003830
87. Taus T, Köcher T, Pichler P et al (2011) Universal and confident phosphorylation site localization using phosphoRS. J Proteome Res 10:5354–5362
88. Jeong K, Kim S, Bandeira N (2012) False discovery rates in spectral identification. BMC Bioinformatics 13 Suppl 16:S2. doi: 10.1186/1471-2105-13-S16-S2
89. Käll L, Canterbury JD, Weston J (2007) Semi-supervised learning for peptide identification from shotgun proteomics datasets. Nat Methods 4:923–925
90. Bantscheff M, Schirle M, Sweetman G et al (2007) Quantitative mass spectrometry in proteomics: a critical review. Anal Bioanal Chem 389:1017–1031
91. Lundgren DH, Hwang SI, Wu L et al (2010) Role of spectral counting in quantitative proteomics. Expert Rev Proteomics 7:39–53
92. Florens L, Carozza MJ, Swanson SK (2006) Analyzing chromatin remodeling complexes using shotgun proteomics and normalized spectral abundance factors. Methods 40:303–311
93. Ishihama Y, Oda Y, Tabata T (2005) Exponentially modified protein abundance index (emPAI) for estimation of absolute protein amount in proteomics by the number of sequenced peptides per protein. Mol Cell Proteomics 4:1265–1272
94. Vogel C, Marcotte EM (2012) Label-free protein quantitation using weighted spectral counting. Methods Mol Biol 893:321–341
95. Neilson KA, Ali NA, Muralidharan S et al (2011) Less label, more free: approaches in label-free quantitative mass spectrometry. Proteomics 11:535–553
96. Wong JW, Cagney G (2010) An overview of label-free quantitation methods in proteomics by mass spectrometry. Methods Mol Biol 604:273–283
97. Silva JC, Gorenstein MV, Li GZ et al (2006) Absolute quantification of proteins by LCMSE: a virtue of parallel MS acquisition. Mol Cell Proteomics 5:144–156
98. Schwanhäusser B, Busse D, Li N et al (2011) Global quantification of mammalian gene expression control. Nature 473:337–342
99. Smits AH, Jansen PW, Poser I et al (2013) Stoichiometry of chromatin-associated protein complexes revealed by label-free quantitative mass spectrometry-based proteomics. Nucleic Acids Res 41, e28
100. Gerber SA, Rush J, Stemman O et al (2003) Absolute quantification of proteins and phosphoproteins from cell lysates by tandem MS. Proc Natl Acad Sci U S A 100:6940–6945
101. Ross PL, Huang YN, Marchese JN et al (2004) Multiplexed protein quantitation in Saccharomyces cerevisiae using amine-reactive isobaric tagging reagents. Mol Cell Proteomics 3:1154–1169
102. Liang HC, Lahert E, Pike I et al (2015) Quantitation of protein post-translational modifications using isobaric tandem mass tags. Bioanalysis 7:383–400
103. Hsu JL, Huang SY, Chow NH et al (2003) Stable-isotope dimethyl labeling for quantitative proteomics. Anal Chem 75:6843–6852
104. Smolka MB, Zhou H, Purkayastha S et al (2001) Optimization of the isotope-coded affinity tag-labeling procedure for quantitative proteome analysis. Anal Biochem 297:25–31

105. Fenselau C, Yao X (2009) 18O2-labeling in quantitative proteomic strategies: a status report. J Proteome Res 8:2140–2143
106. Ong SE, Blagoev B, Kratchmarova I et al (2002) Stable isotope labeling by amino acids in cell culture, SILAC, as a simple and accurate approach to expression proteomics. Mol Cell Proteomics 1:376–386
107. Geiger T, Cox J, Ostasiewicz P et al (2010) Super-SILAC mix for quantitative proteomics of human tumor tissue. Nat Methods 7:383–385
108. Krijgsveld J, Ketting RF, Mahmoudi T et al (2003) Metabolic labeling of C. elegans and D. melanogaster for quantitative proteomics. Nat Biotechnol 21:927–931
109. Nesvizhskii A (2014) Proteogenomics: concepts, applications and computational strategies. Nat Methods 11:1114–1125

Chapter 2

Topics in Study Design and Analysis for Multistage Clinical Proteomics Studies

Irene Sui Lan Zeng

Abstract

This chapter discusses the design issues in clinical proteomics study and provides specific suggestions for addressing these questions when using the standard guidelines for the planning. It provides two methods for the sample size estimation in study design. The first method is used for the planning of a clinical proteomic study at the discovery or verification stage; the second method is proposed for the systematic planning of a multistage study. The second part of the chapter introduces three approaches to analyzing the clinical proteomic study and provides analyses for two case studies of clinical proteomic discoveries.

Key words Checklist for planning, Study design for discovery, Verification and validation, Sample size estimation, Statistical analysis for a clinical proteomic study

1 Introduction

Biomarker research investigates molecule, cell and genetic markers which can be used for disease diagnosis, prognosis, and monitor. System biology, a branch of biology that utilizes systematic approaches to studying hundreds and thousands of molecules, cells, and genes simultaneously, provides platforms for discovering unknown biomarkers. After the scrutiny of verification and validation process, some of these discovered biomarkers are useful for pharmacology industry to produce new tools or therapies that will be used ultimately in clinical practice. Clinical proteomics study investigates proteins species of humans in particular. One of the most common systematic approaches in proteomics is mass spectrometry (MS) coupled with liquid chromatographic (LC). LC separates the partitioned proteins and peptides into smaller molecules. MS records quantitation of peptides from the studied tissues, such as plasma, serum or lymphocytes. Other systematic approaches include MALDI-TOF (Matrix Assist Laser Desorption/Ionization-Time of Flight) mass spectrometry, SELDI-TOF (Surface Enhance

Klaus Jung (ed.), *Statistical Analysis in Proteomics*, Methods in Molecular Biology, vol. 1362, DOI 10.1007/978-1-4939-3106-4_2, © Springer Science+Business Media New York 2016

Laser Desorption/Ionization) mass spectrometry [1, 2], and two-dimensional gel electrophoresis technology [3].

1.1 Subjects of a Clinical Proteomic Study

A clinical proteomic study investigates proteins of human tissues. It involves samples from diseased or healthy subjects. The studied samples are collected from body fluid such as plasma, serum, cell lysates, saliva, or urine. The fluid samples are used for comparisons in proteome or candidate proteins between different physiological groups in a case–control, cohort, or a cross-sectional study. In a cohort study, fluid samples are being investigated over time to validate proteins that can be used for monitoring a studied disease.

1.2 Scope of a Clinical Proteomic Study

Clinical proteomic study with objectives of identifying disease-associated protein markers can be classified according to stages. In a typical proteomic lab, different types of proteomic platforms are designed for different stages of a clinical proteomic study. Amongst these platforms, 2D gel and mass spectrometer are the primary devices for the quantitation of protein abundance. In studies using mass spectrometers (MS), they can be further divided into label-free or labeled experiment. Labeled experiments utilize one of the biochemical techniques [4, 5] that enable multiple samples being analyzed in a same MS run (experiment). Both labeled and label-free experiments require suitable statistical experimental designs to eliminate potential confounding effects.

1.3 Stages of Clinical Proteomic Studies

1.3.1 Discovery

A discovery clinical proteomic study aims for a global identification in a specified human tissue proteome. It starts with a reasonable size of samples from diseased and healthy human subjects. The samples are generally either being collected at the same time or are collected over a short time course; examples are CPTAC-time course ovarian cancer and breast cancer studies [6].

1.3.2 Verification

A verification clinical proteomic study verifies the discoveries by using different platforms and different human subjects. The proteomic analysis at this stage is more focusing and only for the selected candidate proteins or protein groups. The study samples in verification are independent of the discovery samples.

1.3.3 Validation

A validation clinical proteomic study validates finding from verification study or discovery study by using assay that is clinically available and feasible to be used on a larger scale of human subjects. It involves with multiple clinical centers and a considerable number of human subjects. The study may use various laboratories. The findings of validation study will be ultimately used to guide the clinical utility.

1.4 Solution of the Study

The solutions of these clinical proteomic studies will discover unknown proteins and further advance our understandings of molecular functions of all protein species. The discovered proteins are useful to assist the diagnostic and treatment guidelines for diseases. The solutions are also applied to the prognosis and monitor of health outcomes among hospital or community patients. Examples are using protein markers as hospital discharging criteria and using them as criteria for further post-discharge referrals.

2 Planning and Design of a Clinical Proteomic Study

2.1 Issues and Guidelines for Planning

2.1.1 Potential Design Issues

The particular design issues related to biomarker-based studies that researchers need to cope with are: validity and reliability of biomarker measurements, special sources of bias, reverse causality and false positives (STROBE-ME [7]). These issues are all applicable in the planning of a clinical proteomic study. The validity issues of a clinical proteomic study can be addressed in the validity phase of a multistage study, through using the clinical outcomes (i.e., an established diagnostic gold standard for disease or mortality). The reliability issues are important to be addressed at every stage; it can be resolved by including reproducibility assessment whenever necessary. A proteomic study, as any epidemiological and clinical studies, has potential biases including selection bias, information bias, and confounding bias [8]. The particular sources of biases in relation to the proteomic studies can arise from a wrong selection of controls in a case–control study and an improper choice of fluid tissue that lacks biological plausibility. The bias can also be owing to using non-reproducible measures and non-validated clinical outcomes for the validation. The confounding bias may be contributed to the failure of using a rigorous experimental design to control for experimental confounding effects (i.e., label) and clinical confounding effects (i.e., physiological conditions). It can also cause by a lack of confounding clinical information for the adjustment in the analysis.

The causality and reversed causality problem arises from the question of whether the irregulated protein being a cause or an effect. A protein is a source of the disease if changes in the abundance are followed by alterations in frequency of occurrence of the disease when controlling for other factors. Reversed causality concerns the observed change in protein abundance is not a cause but a consequence of the disease [9]. In clinical proteomic study, the observed irregulated protein abundance can be a cause and/or an effect. Applying biological pathway analysis in the validation stage can illuminate the dark side of this issue. Mendelian randomization, which uses genotype variants as the instrument variable in an observational study, is another approach to further illuminating the myth of reverse causality in conventional observational studies [10, 11]. This method can be useful in the validation stage and will be illustrated in Subheading 2.3.

The false positive discovery issues are addressed by correcting for multiple comparisons in conventional clinical or biomedical studies that involve a large number of responses (i.e., whole genome sequencing). One preferred method is to control the false discovery rate (FDR) [12–14] which is estimated by proportion of false discoveries out of all significant results. Apart from these conventional correction methods, using an optimization algorithm that maximizes number of true positive discoveries in a multi-stage design will give better balance of false positives and false negatives [15].

2.1.2 Guidelines for Planning and Design

Clinical proteomic study is one emerging area in biomarker-based research. It commonly uses observational study in the discovery, verification, and validation of new protein markers for diagnosis, monitor and prognosis. In 2011, due to increasing needs for improving reporting of scientific research using biomarkers, guideline that concerns data collection, handling and storage of biological samples, study design and analysis were published. This guideline of STrengthening Reporting of OBservational studies in Epidemiology (STROBE)-Molecular Epidemiology statement (STROBE-ME) is an extension from the STROBE [16]. Both STROBE and STROBE-ME can be used as generic guidelines for the planning of a clinical proteomic study.

Tables 1, 2, 3, and 4 provide specific considerations in different stages of clinical proteomic studies that are parallel to the design items of checklist 4–8 and 12 of STROBE and STROBE-ME. These considerations include:

Study design: the study designs methods that are used in conventional clinical research and epidemiology studies can be adopted and adapted to the clinical proteomic studies after careful considerations in their properties. Not all of the designs are suitable for the discovery phase, such as case–cohort study. Case–cohort study has the potential confounding effect of time-to-failure in controls, and is not considered as an optimal design for the discovery.

Participants: Heterogeneity of participants shall be taken into account either in the design or the analysis of the study. Using efficient matching method to control for confounding effects, or recording the potential confounding effects to be adjusted in the analysis are essential for achieving unbiased discovery, verification, and validation.

Data source/measurements: Using rigorous laboratory experimental design (i.e., using blinded label of group allocation and a robust statistical design), accurate and precise data recording and data management are critical to the achievement of an information-bias-free proteomic study.

Table 1
The checklists for clinical proteomic studies that is parallel to STROBE and STROBE-ME (design item 4: study design)

STROBE	STROBE-ME (molecular epidemiology)	STROBE-CP (clinical proteomics)
Item 4. Study design		
Present key elements of study design early in the paper	ME-4 Describe the special study designs for molecular epidemiology (In particular nested case–control and case–cohort) and how they were implemented	Discovery: Case–control design is considered to be a primary method for discovery clinical proteomic study, of which healthy subjects form the control group. Nested case–control and case–cohort study design should be used with cautions in discovery study
		Verification/validation: Case–control and comparative cohort design are considered to be appropriate at this stage
	Biological sample collection: ME-4.1 Report on the setting of the biological sample collection; amount of sample; nature of collecting procedures; participant conditions; time between sample collection and relevant clinical or physiological endpoints	
	Biological sample storage: ME-4.2 Describe sample processing (centrifugation, timing, additives, etc.)	
	Biological sample processing: ME-4.3 Describe sample storage until biomarker analysis (storage, thawing, manipulation, etc.)	
	Biomarker biochemical characteristics: ME-4.4 Report the half-life of the biomarker, and chemical and physical characteristics (e.g., solubility)	

Variables, validity/reliability of measurements: In discovery research, Mass Spectrometry is the most common technique in experiment, but it is more platform-dependent [17]. Defining instrumental features of the platform, such as label, mass-to-charge-ratio, quality measures, as well as including reproducibility assessments will reduce the bias and increase the true discovery rate.

Table 2
The checklists for clinical proteomic studies that is parallel to STROBE and STROBE-ME (design item 6: participants)

STORBE	STROBE-ME (molecular epidemiology)	STROBE-CP (clinical proteomics)
Item 6. Participants		
For case–control study: Give the eligibility criteria and the sources and methods of case ascertainment and control selection Give the rationale for the choice of case and control; for matching case–control, give the matching variables and number of controls per case For cross-sectional study: Give the eligibility criteria and the sources and methods of selection of participants	ME-6 Report any habit, clinical conditions, physiological factor or working or living condition that might affect the characteristics or concentrations of the biomarker	Discovery/verification/validation: report any clinical conditions/interventions, physiological and working /living conditions that might affect the abundances and post translations of the proteins

Table 3
The checklists for clinical proteomic studies that is parallel to STROBE and STROBE-ME (design items 7–8: variable and data source)

STORBE	STROBE-ME (molecular epidemiology)	STROBE-CP (clinical proteomics)
Item 7. Variable		
Clearly define all outcomes, exposures, predictors, potential confounders and effect modifiers. Give diagnostic criteria, if applicable		Define any instrumental and experimental features of the platform, such as labels, mass-to-charge-ratio, retention time, quality measures, if applicable, as potential experimental confounders and effect modifiers
Item 8. Data source/measurement		
For each variable of interest, give sources of data and details of methods of assessment (measurement) Describe comparability of assessment methods if there is more than one group	ME-8 Laboratory methods: report type of assay used, detection limit, quantity of biological sample used, outliers, timing in the assay procedures (when applicable) and calibration procedures or any standard used	Use blinded labels for grouping (i.e., case vs. control) in the laboratory analysis if applicable. Use robust statistical method for experimental design

Table 4
The checklists for clinical proteomic studies that is parallel to STROBE and STROBE-ME (Validity/reliability and sample size)

STORBE	STROBE-ME (molecular epidemiology)	STROBE-CP (clinical proteomics)
Validity/reliability of measurement and internal/external validation		
	ME-12.1 Report on the validity and reliability of measurement of the biomarker(s) coming from the literature and any internal or external validation used in the study	Consider including reproducibility assessment in the design as a compulsory component for discovery, verification, and validation In discovery and verification study, use internal reference for comparison In validation study, multiple clinical outcomes can be considered to use as external validation standards
Sample size estimation		
		Any multiple comparisons adjustment method used needs to be specified in the sample size estimation

For clinical proteomic study involving tumor proteins, the REMARK [18] can also be used as a generic guideline for the validation of prognostic protein markers.

2.2 Study Design of a Discovery Clinical Proteomic Study

In a discovery clinical proteomic study, there are five specific items that worth researchers paying attentions to in the statistical planning:

Heterogeneities of the studied patients: for cancer patients, the heterogeneity can be classified by histological diagnosis of disease stages and metastasis [19]. For other patients, the heterogeneity is classified by using different acknowledged diagnostic techniques. The subtypes of patients may have different patterns and effect sizes for the protein markers. Given the latent heterogeneity of studied subjects, grouping them together without specification may distort the real picture.

The control subjects to be used for the comparison: healthy subjects are the best option as controls. In some studies, patients are used when they had known distinctive clinical features compared to the studied patients. One example is to have benign patients as control to compare with malignant patients. In these studies, the discovered markers will be useful for clinical diagnosis, instead of population screening. In general, using patients as controls in the discovery is discouraged as there will be potential unknown confounding factors introduced by the disease. Careful selection of the controls makes considerable impacts on the success of a study [20].

The matching variables and matching ratios: matching variables are those factors known to confound with protein abundance but not in the center of investigations. Matching ratio of a case–control or matched cohort study is determined by incidence of the diseases and feasibility of subjects' participations. In a study involving patients with a rare disease, $1{:}n$ ratio is applied to patients and controls in order to achieve the statistical power within a foreseeable study period.

Number and frequency of samples collection: when the study has stimulators (interventions), the number of samples will be influenced by the frequency of sample collections. The sample collections are affected by the known effects of stimulator, known or predictable half-life of the protein and the budget for the study.

Reproducibility of the experiment: if the method of tissue sample collection is new, to include the reproducibility assessment as a standard procedure in the discovery stage will be important. A reproducibility assessment includes 8–30 patients and controls samples [21]. Biological and technical duplicates of these subgroup samples will respectively provide evidences of clinical and technological repeatability for proteomic research. The reproducibility assessment will also provide qualitative information about feasibility of launching a discovery experiment.

Several typical study designs of clinical and epidemiological research that can be adopted in the discovery phase are introduced as followed:

(a) *Comparative cross-sectional study*: comparative cross-sectional study simultaneously collects tissue samples from different groups of subjects at a same time-point. These groups are known to be physiologically different.

(b) *Matched case–control study*: matched case–control design select confirmed disease cases to compare with non-cases (control). Case and non-cases are chosen from the same population. The control group is commonly formed by healthy subjects and used matching variables age, gender and races. The matching scheme uses a ratio of $1{:}n$, where n is an integer ≥ 1. Frequency matching uses combinations of ordinal (i.e., age group) and nominal (gender and race) variables to form strata for the matching.

(c) *Comparative prospective cohort study*: comparative cohort study prospectively collect tissue samples from different groups of subjects at different time-points. It sometimes includes healthy subjects as a control group.

2.3 Study Design of a Verification Clinical Proteomic Study

A verification clinical proteomic study has a known number of candidate markers for the evaluation. It requires different platforms and a larger number of subjects [3]. These differences between verification and discovery studies reflect the statistical features of a verification study. These features are related to the false negative discoveries and confounding bias. At verification stage, type II errors are more concerning than type I errors, and the confounding effects from different platforms will need to be identified. These confounding effects shall be specified either prior to the study or through the study.

Several typical study designs in clinical and epidemiological research can be adopted in verification clinical proteomic study:

(a) Comparative cross-sectional study can be used in verification study when it suits the purpose of the research. An example is when the study is to replicate the cross-sectional discovery study and evaluate a known number of candidate proteins using a different platform.

(b) Matched case–control study is a good option for the verification study.

(c) Nested case–control study: nested case–control study is a variation of case–cohort study; controls are selected from the same source population and matched with cases using time-to-disease and other variables such as age, gender and race [8]. The tissue samples of case and matched control/s need to be collected at the time when case is developing the disease. Due to this method's lacking representativeness of controls, it shall be used with cautions. Case–cohort study has potential confounding effect of time-to-failure in controls. It is not considered as an optimal design in discovery and verification study.

(d) Comparative prospective cohort study is another good option for the verification study.

2.4 Study Design of a Validation Clinical Proteomic Study

A clinical validation proteomic study is similar to a conventional epidemiology study with involvements of biomarkers. The STARD [22] and other guidelines in molecular epidemiology study and genetic epidemiology are useful references for the design of a validation study. There are differences in the statistical properties of a proteomic validation study when we compare it with conventional epidemiology study, discovery and verification proteomic study. Firstly, the number of measurements as responses in a model will be larger than those of the conventional epidemiology study. Secondly, the type II error and confounding effects introduced by the environmental and physiological changes in a longitudinal study will be more concerning than that in a discovery or a verification study. Thirdly, the integration of bioinformatics information, which may not be required in a

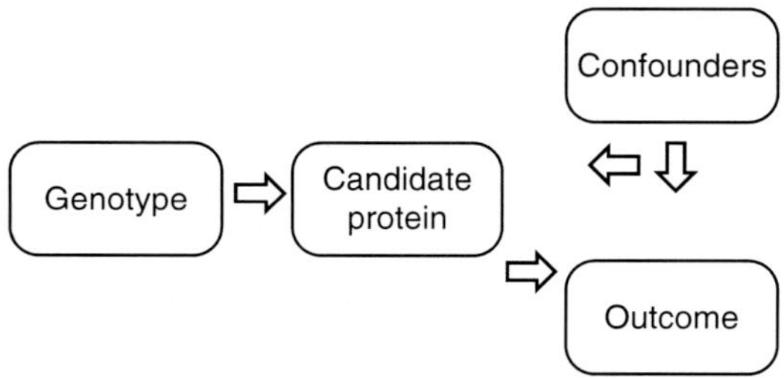

Fig. 1 Mendelian randomization in a validation clinical proteomic study

conventional epidemiology study, is a critical procedure for the proteomic biomarker validation.

Matched case–control and prospective cohort design are mostly suitable for a validation clinical proteomic study. In molecular epidemiology study, prospective cohorts design is considered to be the gold standard for validating biomarkers [23]. Matched case–control studies including healthy controls may not be feasible as the validation study requires a large number of participants from multiple centers. To include a large number of healthy cohorts may be problematic. Nested case–control study can be considered if having healthy control is not feasible.

In both the aforementioned designs, the reverse causality issue will need to be addressed. Mendelian randomization (MR), which is attributed to Katan's introduction to epidemiology [10], can be considered. The Mendelian randomization introduces one or more gene variants that link with the discovered protein as the instrumental variables (IV) [24] in a validation clinical proteomic study (Fig. 1). The selected germ-line genotypes will be independent of confounders (i.e., environmental exposures and clinical characteristics); they need to associate with the candidate proteins and only associate with the disease (outcome) through the candidate proteins. These above conditions of using Mendelian randomization make the reverse causation problem-free. Nevertheless, the conditions of the selection of genotype [25] in biomarkers validation may not be easily met. The limitations and assumptions of MR/IV [25] need to be checked before applying it in the study design and data analysis.

The loop of Mendelian randomization in a validation clinical proteomic study demonstrates the conditions of genotype, candidate proteins, confounders and outcome (Fig. 1).

2.5 Sample Size Estimation

2.5.1 Sample Size Estimation for a Discovery Study

When we need to estimate sample size for a case–control discovery study but having no information of the design parameters (i.e., mean and standard deviation), we can base on a biologically or clinically meaningful change in protein abundance for calculation. This change is the difference in the abundance or any transformed abundance (i.e., natural-log of abundance) between patients and controls. We can then base on a sequence of effect sizes to estimate the corresponding standard deviation using formula:

$$\lambda = \frac{\mu}{\delta},$$

where λ, μ and δ denote the effect size, change and standard deviation of the change respectively. A typical range of effect size is between 0.5 and 2.0.

The sample size estimation will use the formula for a two-sample t test or paired sample t test:

$$n_{\text{case}} = \frac{\left(\left|z_{\alpha/2}\right| + \left|z_{\beta}\right|\right)^2 \delta^2}{\mu^2} \times \left(1 + \frac{1}{r}\right),$$

where $z\alpha_{/2}$ and $z\beta$ are the critical values for type I and type II error respectively and δ^2 denotes the variance of the change. In the scenario of a two-sample test, δ^2 is the pooled variance estimated from the variance of patients and controls. Whereas in a paired sample test, δ^2 can be estimated from the variance of the derived random variable—the difference between the paired samples, or from the variance of patients and controls and their correlation coefficient. r is the ratio of control to case and n_{case} denotes the sample size for cases.

If we know the range of number of proteins to be identified in the discovery, we can adjust the sample size for multiple tests. The sample size adjustment uses a new type I error $\frac{\alpha}{m}$ to control for Family wise error (FWE) rate (the probability of having at least one type I error), where α and m represents the conventional type I error and number of tests respectively. Another less conservative and preferable approach is to use $\frac{\alpha}{m} \times k$ to adjust for the false discovery rate (FDR), where k represents the number of true positives (real significant proteins) may be detected by the discovery experiment. Controlling for FDR, which is the probability of having k type I errors of all significant tests and also called false positive rate, achieves more compromises between type I and type II errors than controlling for FWE [13, 14]. Table 5 provides sample sizes using the aforementioned formula for n_{case}, adjusted for FWE and FDR respectively. The final sample size also needs to adjust for missing data by multiplying (1 + proportion of missing) where appropriate. In the similar setting, this approach can be applied to a verification study.

Table 5
Sample size table for an effect size of 0.5–1.5 and folder change of 1.25 in the log (abundance)

Std	Effect size	n-case[a] (r = 1)	n-case[a] (r = 2)	n-case (r = 1) correcting for FWE[b]	n-case (r = 1) correcting for FDR[c]
0.446	0.5	84	63	228	181
0.319	0.7	43	32	116	92
0.248	0.9	26	19	70	56
0.203	1.1	17	13	47	37
0.172	1.3	12	9	34	27
0.149	1.5	9	7	25	20

[a]The number of cases is required to detect a fold change of 1.25 (which was transformed to the log scale in the calculation). Type I error is set to be 0.05 and type II error is 0.10 with critical values of −1.96 and −1.28 in the calculation respectively

[b]The multiple comparison correction controls Family-Wise-Error (FWE) rate using Bonferroni method, assumed the maximal number of proteins being discovered is 1000. The critical value for type I error z_{2}/α is −4.06 in the calculation

[c]The multiple comparison controls False-Discovery-Rate (FDR) using Benjamini and Hochberg method [14], assumed we know the maximal number of proteins being discovered is 1000, and the discovery will identify at least 10 % true positives. The critical value for type I error z_{2}/α is −3.48 in the calculation

2.5.2 Sample Size Estimation for Optimizing the Number of True Discoveries from a Multistage Study

In a multistage study, we consider the number of true positives after the discovery, verification, and validation. The sample size solution will include the sample sizes of these three phases while controlling for the overall false discovery rate. An algorithm that can systematically join the statistical tests of these three phases and estimate the sample size for each phase simultaneously will be useful. In the real applications, the budget and a statistical false discoveries control also need to be considered. The following section demonstrates a case study using a multistage design algorithm to derive the design solutions under the constraints of overall cost and number of false positives [15]. The algorithm provides the design solution to maximizing the number of true discovery through the multistage study. The algorithm is available via the R package *proteomicdesign*, which is designed to estimate the sample sizes for verification and validation given the result of a discovery experiment. The prior information required for the program is: change of each protein, a common standard deviation of protein abundance, sample size of the discovery stage, and the lab costs at verification and validation. The total laboratory assay cost for verification is expressed as a function of sample size (n) and number of protein (p) to be tested. The algorithm also includes a vector to correct for known artifact in the discovery. It is set to be one if there is no artifact for a protein.

This case study has 17 discovered proteins that are of interest to the researchers and needs to verify and validate into the next two stages. Researchers need to know the sample size for the verification and validation, given a budget of 500,000. R codes for the sample size estimation are shown in the following paragraphs based on *proteomicdesign* 2.0. **Note 1** gives the assumed study designs of the current version.

```
#step 1: assign stage II total lab assay cost.
Stage II total lab assay cost requires number of proteins p and sample size n.
>c1=280; c2=1015; cost2=function(n,p){c1*p+c2*n}
#step 2: assign stage III assay cost per subject.
stage III lab assay cost function per subject only requires the number of protein p.
>c3=200; cost3=function(p){c3*p}
#step 3: assign data and use the optim.two.stage.single() function.
Item.number assigns the number of local search in nested simulating annealing [15].
> p r o t e i n i d=c(100,101,102,103,104,105,106,107,108,109,110,111,112,113,114,115,116)
> b e t a=c(-11.43,0.83,5.30,3.86,3.43,0.13,-1.3,-1.23,-1.15,-1.11,-1.06,-1.02,7.82,4.17,4.39,0.10,0.05)
> s i g m a=c(1.91,0.38,2.10,0.73,2.16,0.05,0.16,0.17,0.17,0.17,0.17,0.17,2.23,0.98,0.92,0.09,0.17)
>protein<-data.frame(proteinid,beta,sigma)
>optim.two.stage.single(budget=500000,artifact=rep(1,17),protein=protein,n1=50,iter.number=20,assaycost2.function=cost2,assaycost3.function=cost3,n2.min=30,n2.max=100,n2.step=10)
```

The output solution matrix is shown in the GUI window; the first value −16.68 is the minimal negative number-of-proteins detected at the final validation stage; the following two values of 0.235 and 0.035 are the corresponding significant decision thresholds at discovery and verification stage respectively. The fourth value 100 is the sample size for verification.

```
[,1][,2] [,3]  [,4]...
[1,] -16.525 0.1420172 0.02011206 92
[2,] -16.684 0.2350000 0.03500000 100
......
[19,] -16.684 0.2350000 0.03500000 100
```

```
#step 4: use power.single.cost() to estimate the costs at each stage and sample size of stage III
>initial=c(0.235,0.035,100)
```

```
>power.single.cost(initial,protein=protein,artifact=rep(1,17),n1
=50,budget=500000,s=1000,assaycost2.function=cost2,assaycost3.
function=cost3,recruit=100,optimize=FALSE)
        [1] cost at stage II: 116211
        [1] cost at stage III: 382789
      [1]111.46 116211.00 382789.00
```

power.single.cost() provides the sample size of stage III to be 111.5 and costs at stage II and III as 116,211 and 382,789 respectively.

When utilizing the biological functioning group information of proteins to estimate the sample size, there are two functions *optim.two.stage.group* and *optim.two.stage.appr* available in the package. In **Note 1**, it specifies that the study design for the current version of proteomic design is limited to paired sample and matched case–control study only.

3 Data Analysis Strategy of a Discovery Clinical Proteomic Study

3.1 Statistical Methods

3.1.1 Derive Protein Ratio from a Single Run Case–Control Experiment

In a balanced designed single run multiplex or multi-label proteomic experiment, half of the labeled samples are cases and half samples are controls. The data analysis can be performed by deriving the ratio of the intensity at the peptide level between cases and controls. The peptide ratios can then be summarized into protein ratios by choosing one of the summary methods. Two of these summary methods are, (1) normalization by the median of all peptide ratios in this single run experiment [26]; (2) assigning a specified distribution of the peptide ratios, and using the distribution parameters to derive the protein ratio parametrically or non-parametrically [27].

3.1.2 Analysis of Variance for Each Protein or Feature Separately (Single Protein Model)

Analysis of variance has been suggested to use in multi-runs proteomic experiments by Oberg and Mahoney [28]. The variance of a protein's intensity (normally is the log transformation of the intensity) thus is proposed to be contributed to by the experimental factors and the physiological factors in the model. If the abundance data are hierarchical with peptides or precursor ions nested within proteins, this single-protein model can be defined using a multi-level structure.

We start to use the most common structure in a labeled experiment of which peptides nest within protein. Peptides are the level one unit. Since the protein is analyzed one by one, the protein unit does not exist; we can equivalently treat peptides as if they nest within biological samples (subjects), which are the level two units. The level one of the model describes relationship between the relative abundance of peptide and the experimental factors. The level two of the model represents relationship between the random intercepts of subjects and those effects at the subject level such as demographics and condition (disease or healthy).

This model can be formulated as below:

1. Define level one of the 2-level model. In the following equations, fixed effect coefficients use the Greek letters, and random coefficients use the Roman letters.

$$y_{i,l} = b_{0,l} + \beta_1 mz_{i,l} + \sum_{h=1}^{w} \beta_{2,h} \text{label}_{h,i,l} + \sum_{r=1}^{v} \beta_{3,r} \text{run}_{r,i,l} + \sum_{h=1}^{w} \beta_{4,h} mz_{i,l} \times \text{label}_{h,i,l} + \sum_{r=1}^{v} \beta_{5,r} mz_{i,l} \times \text{run}_{r,i,l} + \sum_{k=1}^{wv} \beta_{6,k} \text{label}_{h,i,l} \times \text{run}_{r,i,l} + e_{i,l}, \quad (1)$$

where $y_{i,l}$ denotes the log transformed intensity for peptide i and subject l, i has a range of 1 and n, l has a range of 1 and m; b_0, l denotes the random intercept for subject l; $mz_{i,l}$ denotes the m/z ratio for peptide i and subject l; β_1 denotes the regression coefficient for m/z ratio; $\text{label}_{h,i.l}$ is a dummy variable (0,1) for the multiplex label h in a labeled proteomic experiment of the response $y_{i,l}$ for peptide i subject l, h is an integer number ranged between 1 and w; $\beta_2 h$ denotes the regression coefficient for label h; $\text{run}_{r,i,l}$ is a dummy variable (0,1) for the identity of run or batch r that peptide i is identified in subject l; r is an integer number with range between 1 to v; $\beta_4 h, \beta_5 r, \beta_6 k$ denote the regression coefficients for the interactions terms of m/z ratio and label, m/z ratio and run, and label and run respectively; $e_{i,l}$ denotes the unexplained residual error for peptide i and subject l.

Of note, the response $y_{i,l}$ represents the intensity value of a unique reporter ion for a protein in the model. The aforementioned peptide i is referred to the observed reporter ion i.

At level one of the model, the response is the log transformed intensity value $y_{i,l}$. The explanatory variables include the experimental factors as fixed effects, namely the multiplex $\text{label}_{i.l}$, run$i.l$, $mz_{i,l}$, and their two-way interaction terms $mz_{i,l} \times \text{label}_{i,l}$, $mz_{i,l} \times \text{run}_{i,l}$, $\text{label}_{i,l} \times \text{run}_{i,l}$. It also includes a random effect, a random intercept $b_0 l$ for different subject l. Equation 1 defines an intercept term that is different across subjects, other terms such as run, label, m/z ratio and their two-way interactions that are the same for all *subjects*.

2. Define the level two of the 2-level model: The level two of the model describes the effects of variations in the subject level through the random intercept $b_0 l$ at level one. The response is the random intercept $b_0 l$. The explanatory variables of this level include the condition (i.e., diseased vs. normal) and other

subject level variables such as demographics and clinical characteristics.

$$b_{0,l} = \gamma_{0,0} + \sum_{b=1}^{q} \gamma_{1,b} \times z + \sum_{c=1}^{q} \gamma_{2,b} \times \text{condition}_{l,c} + u_{0,l}, \tag{2}$$

where $\gamma_{0,0}$ represents the fixed intercept term; $z_{l,b}$ denotes the covariate b of the subject l; $\gamma_{1,b}$ represents the regression coefficient for the subject level covariates $z_{l,b}$; $\mathbf{condition}_{l,c}$ is a dummy variable denoting the biological or psychological conditions (i.e., different interventions, disease vs. normal state) for the condition c and subject l; $\gamma_{2,c}$ represents the regression coefficient for the condition c; $u_{0,l}$ represents the random residual term of subject l.

Substituting Eq. 2 into Eq. 1 yields

$$y_{i,l} = \begin{bmatrix} \gamma_{0,0} + \sum_{b=1}^{q} \gamma_{1,b} \times z_{l,b} + \sum_{c=1}^{g} \gamma_{2,c} \times \text{condition}_{l,c} + \\ \beta_1 mz_{i,l} + \sum_{h=1}^{w} \beta_{2,h} \text{label}_{h,i,l} + \sum_{r=1}^{v} \beta_{3,r} \text{run}_{r,i,l} + \\ \sum_{h=1}^{w} \beta_{4,h} mz_{i,l} \times \text{label}_{h,i,l} + \sum_{r=1}^{v} \beta_{5,r} mz_{i,l} \times \text{run}_{r,i,l} + \sum_{k=1}^{wv} \beta_{6,k} \text{label}_{h,i,l} \times \text{run}_{r,i,l} \end{bmatrix} + \left[u_{0,l} + e_{i,l} \right] \tag{3}$$

where random variable $u_{0,l} \sim N\left(0, \tau_{0,0}^2\right)$ represents the between-subject variation, which is normal distributed with mean 0 and variance $\tau_{0,0}{}^2$, and random variable $e_{i,l} \sim N\left(0, \sigma^2\right)$ represents the conventional residual error term of reporter ion intensity and is also normal distributed with mean 0 and σ^2.

The fixed and random effects are grouped by separate square $[\cdot]$ brackets. The model defined in Eq. 3 assumes the regression coefficients $\beta_1, \ldots, \beta 6_{,w \times v}$ of experimental factors at the peptide level are constant across subjects. The run effects can be treated as a random variable when it is hypothesized that there are variations introduced by different runs.

The variance-covariance matrix for the random effect at level two can be represented as **G**, and the variance-covariance matrix for the level one random residual is defined as **R**. Since the model defined in Eq. 3 only has one random residual variable at level two and one random residual at level one. The level two variance-covariance matrix **G** only has one variance term $\tau_{0,0}{}^2$. The level one variance-covariance matrix **R** for level one residual also only has one variance term σ^2.

The variance-covariance matrix of the response $y_{i,l}$ is different from the variance-covariance matrix for the random effects.

When we assume that the level one error terms are independent, the variance and covariance of the response are derived as follows:

$$\begin{aligned}&\operatorname{var}\left(y_{il}\,\middle|\,\beta_0,,,,\ldots,,,,\beta_{w\times p},,,,u_{0,l},,,,X_{il}\right)=\operatorname{var}\left(u_{0,l}+e_{i,l}\right)=\tau_{0,0}^2+\sigma^2\\&\operatorname{cov}\left(u_{0,l}+e_{i=j,l},u_{0,l}+e_{i=k,l}\right)=\operatorname{cov}\left(u_{0,l}+u_{0,l}\right)=\tau_{0,0}^2\end{aligned}\tag{4}$$

where $e_{i=j,l}$ and $e_{i=k,l}$ are the error term for peptide j and k respectively. The covariance in Eq. 4 only has one term.

When the covariance term for the paired residual term within a subject is not zero, in another word, when the level one error terms are not independent, the covariance term for the response as defined in Eq. 4 will become

$$\operatorname{cov}\left(u_{0,l}+e_{i=j,l},u_{0,l}+e_{i=k,l}\right)=\operatorname{cov}\left(u_{0,l},u_{0,l}\right)+\operatorname{cov}\left(e_{i=j,l},e_{i=k,l}\right)=\tau_{0,0}^2+\delta_{j,k},$$

where $\delta j{,}k$ is the covariance between the paired residuals of the reporter ion intensities.

The block diagonal matrix for the variances-covariance matrix $\mathbf{V}_{n\times n}$ of the response $y_{i,l}$ is

$$\begin{bmatrix}A_{l=1}&0&0\\0&\ddots&0\\0&0&A_{l=m}\end{bmatrix}\tag{5}$$

where $A_l=\sigma^2\otimes I_s+\mathbf{B}\otimes J_s$, s is the number of peptide observations of subject l, Is represents the $s\times s$ identity matrix, and Js represents the $s\times s$ matrix of ones. $\mathbf{B}$ is the $s\times s$ matrix of $\left\{\tau_{0,0}^2+\delta_{j,k}\right\}$, $\delta j{,}k$ equals to σ^2 when $j=k$. $\mathbf{V}$ expands to a full matrix as

$$\begin{bmatrix}\tau_{0,0}^2+\sigma^2&\cdots&\tau_{0,0}^2+\sigma_{j=1,k=s}&0&\cdots&0&\\\vdots&\ddots&\vdots&\vdots&&\vdots&\\\tau_{0,0}^2+\sigma_{j=1,k=s}&\cdots&\tau_{0,0}^2+\sigma^2&0&\cdots&0&\\0&\cdots&0&\ddots&&&\\\vdots&\vdots&\cdots&0&\tau_{0,0}^2+\sigma^2&\cdots&\tau_{0,0}^2+\sigma_{j=1,k=s}\\\vdots&\vdots&\cdots&0&\vdots&\ddots&\vdots\\0&0&\cdots&0&\tau_{0,0}^2+\sigma_{j=1,k=s}&\vdots&\tau_{0,0}^2+\sigma^2\end{bmatrix}\tag{6}$$

where all diagonal terms are $\tau_{0,0}^2+\sigma^2$ and off-diagonal terms within the block sub-matrices are $\tau_{0,0}^2+\sigma_{j,k}$, all the other off-diagonal terms outside the block sub-matrices are zeros. For estimating the covariance term, $\mathbf{V}$ can be further decomposed as $\mathbf{V}=\mathbf{Z}_{n\times q}\mathbf{G}_{q\times q}\mathbf{Z}'_{q\times n}+\mathbf{R}_{n\times n}$, where Z denotes the design matrix for the random effects excluding level one error term and q denotes the number of random effects at the subject level.

A sub-matrix of the block diagonal variance-covariance matrix of response, for example, for a subject with three peptide observations can be defined as

$$\begin{bmatrix} \tau_{0,0}^2+\sigma^2 & \tau_{0,0}^2+\sigma_{1,2} & \tau_{0,0}^2+\sigma_{1,3} \\ \tau_{0,0}^2+\sigma_{1,2} & \tau_{0,0}^2+\sigma^2 & \tau_{0,0}^2+\sigma_{2,3} \\ \tau_{0,0}^2+\sigma_{1,3} & \tau_{0,0}^2+\sigma_{2,3} & \tau_{0,0}^2+\sigma^2 \end{bmatrix} \quad (7)$$

which has three extra unknown covariance terms. There are various covariance structures for the estimation of the unknown covariance terms $\sigma_{1,2}$, $\sigma_{1,3}$ and $\sigma_{2,3}$. Each structure is determined by the assumptions of the within-subject errors. More details of the formulation are referred to Zeng [29].

3.1.3 Multilevel Multivariate Analysis (A Multiple-Protein Model)

Similar to the single protein model, the multiple-protein models can also be explicitly formulated in a multilevel framework [29]. Proteins can be analyzed in a group using a multivariate approach. The advantages of a multiple-protein model over a single protein model can be explained statistically and biologically. Statistically, multivariate model with correlated responses achieve higher efficiency [30], analyzing multiple proteins that are identified by the same proteomic experiment also increases the precision for the platform/instrument parameter estimators. Biologically, proteins belong to the same biological group may demonstrate similar movement triggered by the intervention or over a time course. Analyzing multiple proteins will be able to identify any systematic change that is not easily detected in a single protein model.

3.2 Packages for Reading Mass Spectrometer Raw Data, Peak Detection, and Producing Matched Peptide and Protein Reports

The available commercial software to read raw spectrum data and match them to generate the peptide and protein reports includes protein pilot™, PEAK and others. Protein pilot program uses Paragon algorithm for search, Pro Group™ algorithm and Mascot algorithm for data processing [26]; and it is designed to process labeled experiment.

The available free packages for detecting peak from spectrum data include R bio-conductor package PROess [31], QUANT [32], and others.

When performing statistical analysis, R, SAS, and other statistical software that can deal with large datasets will be the suitable tools for the data management and analysis of clinical proteomic studies. The following Subheading 3.3 includes two case studies being analyzed by using R version 2.15.3 and SAS 9.3.

3.3 Two Case Studies

3.3.1 Case Study 1: Discovery Study

The first case study is a case–control study including 50 ovarian cancer patients and 50 healthy controls [33]. The proteomic profiling data were generated via surface-enhanced laser/desorption ionization (SELDI)-time of flight (TOF) mass spectrometer.

The proteomic experiment was conducted without having any biochemical labeling.
The analysis procedures include:

step1: Identifying peaks for each spectrum file using the R Bioconductor package *PROcess* [31].

The peak identification process includes baseline subtraction and peak identification for each spectrum file. Visualization of the spectrum and peaks are also recommended at this step.

step2: Pre-processing and quality assessment for a batch of spectra file;

After the visualization of each spectrum, we can use the procedure in *PROcess* to locate the peaks from the batch of all spectra data. Before the peaks being located, a pre-processing procedure which includes normalization and quality assessment is implemented for the batch of spectra data. After the pre-processing, the mean intensities from all post-processed spectra are used to locate the peaks via Signal-To-Noise (S-T-N) filtering and censored interval method.

step 3: Use ANOVA to analyze the intensity (or transformed intensity) of each identified peak.

The final *step 4* is a multiple comparison adjustment process.

A heat map R program was also provided for this case study to visualize the pattern of all ion intensities within the selected range of the spectrum with which to compare between patients and controls.

Figure 2 visualizes the baseline subtracted intensity for patient A01. The visualization of 25 patients at *step 1* is realized by using R codes:

```
#all patients and controls files were saved in one folder
>filedir<-"PATH OF FILE"
>fs<- list.files(filedir, pattern="\\.*csv\\.*", full.
names=TRUE)
>par(mfrow=c(5,5),mar=c(2,3,2,3))
#Visualize the first 25 files of cancer patients
>for (i in 1:25)
>{samplefile=read.files(fs[i]) >baselinesub<->bslnoff(sam
plefile,method="loess",plot=TRUE,bw=0.1,xlab
="m/z",cex=0.>5)
>title(list.files(filedir)[i])
}
```

To show the identified peaks of 25 patients and controls can also be achieved by using the following R codes:

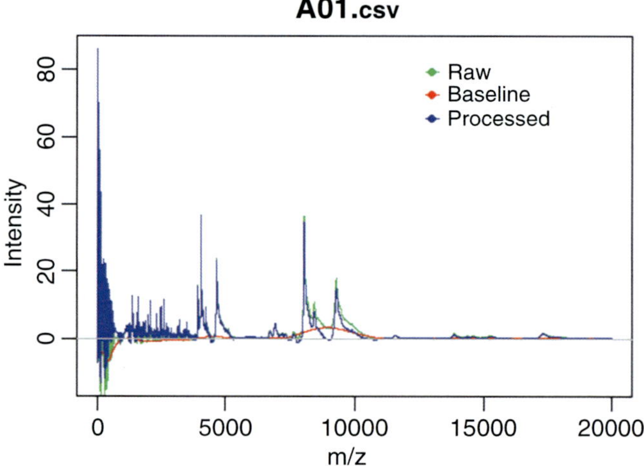

Fig. 2 The sample spectra after subtracting baseline of patient A01

```
#Patients
>for (i in 1:25)
>{samplefile=read.files(fs[i])
>baselinesub<-bslnoff(samplefile,method="loess",plot=FALSE,bw=0.1)
pkgobj <- isPeak(baselinesub,span=41,sm.span=11,plot=TRUE)
}
#Controls
>for (i in 51:75)
>{samplefile=read.files(fs[i])
>baselinesub<-bslnoff(samplefile,method="loess",plot=FALSE,bw=0.1)
pkgobj <- isPeak(baselinesub,span=41,sm.span=11,plot=TRUE)
}
```

Figure 3 visualizes the identified peaks using the mean intensity of patient group and control group and the S-T-N ratio filtering method; Fig. 4 shows the identified peaks using the mean intensity and the censor interval method to locate peaks at *step 2.*

```
#Baseline subtraction-batch processing
>removebase<-rmBaseline(filedir)
#Normalization
>rtM <- renorm(removebase, cutoff=1000)
#Identified peak
>peakfile <- paste(tempdir(), "testpeakinfo.csv", sep = "/")
>getPeaks(rtM, peakfile)
#Quality assessment: use the three criteria-quality < 0.4;
```

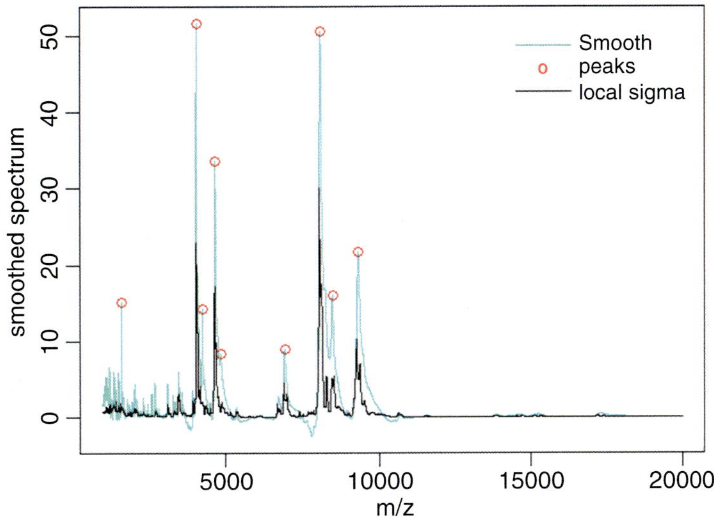

Fig. 3 Use mean spectrum and the Signal-to-Noise ratio filtering method (R function isPEAK()) to locate peaks

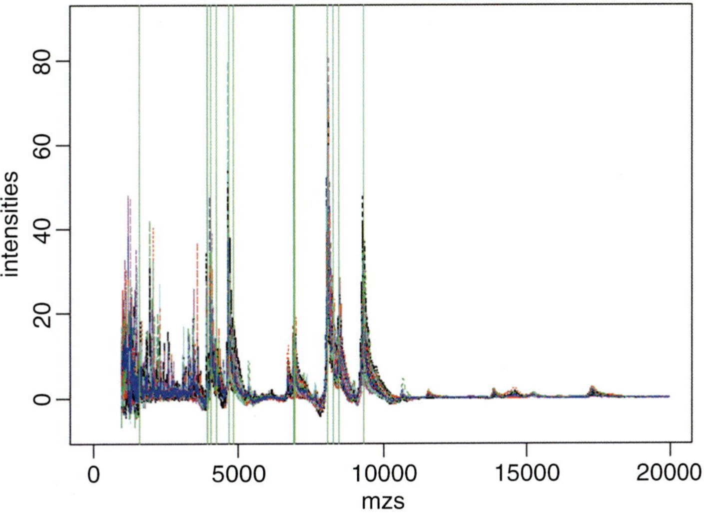

Fig. 4 Use mean spectrum and the censored interval method (R function pk2bmkr()) to locate peaks

```
#retain <0.1;peak < 1/2 of the mean peak number in the chip;
>qualRes <-quality(removebase, peakfile, cutoff=1000)
>print(qualRes) #all spectrum meet the criteria
#Use the mean spectrum to locate the peak
>grandAve <- aveSpec(fs)
>mzs <- grandAve[,1]
```

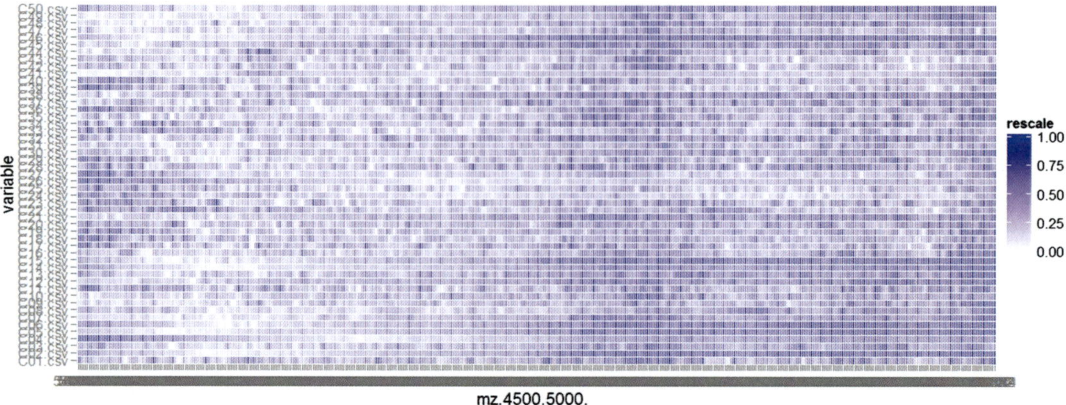

Fig. 5 Heat map of intensities for controls (m/z ranged between 4500 Th and 5000 Th)

```
>windows(4,5)
>grandOff <- bslnoff(grandAve[mzs>0,], method="loess",plot=T, bw=0.1)
>grandPkg <- isPeak(grandOff[grandOff[,1]>1000,], zerothrsh=1,plot=T, ratio=0.1)
>grandpvec <- round(grandPkg[grandPkg$peak, "mz"])
#Print location of the peaks
>print(as.vector(grandpvec))
[1] 1607 4052 4248 4666 4842 6907 8077 8467 9303
#Use the censored interval method to locate the peak
>bmkfile <- paste(tempdir(), "testbiomarker.csv", sep = "/")
testBio <- pk2bmkr(peakfile, rtM, bmkfile)
>mzs <- as.numeric(rownames(rtM))
>matplot(mzs, rtM, type = "l", xlim = c(0, 20000),ylab="intensities")
>bks <- getMzs(testBio)
#Print location of the peaks
>print(round(bks))
[1] 1607 3927 4053 4248 4667 4844 6904 6919 6922 6924 6925 6927 8074 8235 8461 9301
>abline(v = bks, col = "green")
#Use analysis of variance for peak i
>fit=glm(log(testBio[,i])~factor(group))
>qqnorm(fit$residuals)
```

Using the censored interval method achieved more peaks in this case study. The identified 16 peaks and their quantities were used in the analysis of variance. Some of the protein intensities were log transformed. The above analysis assumed that the abundance or logarithmic abundance is normal distributed, in

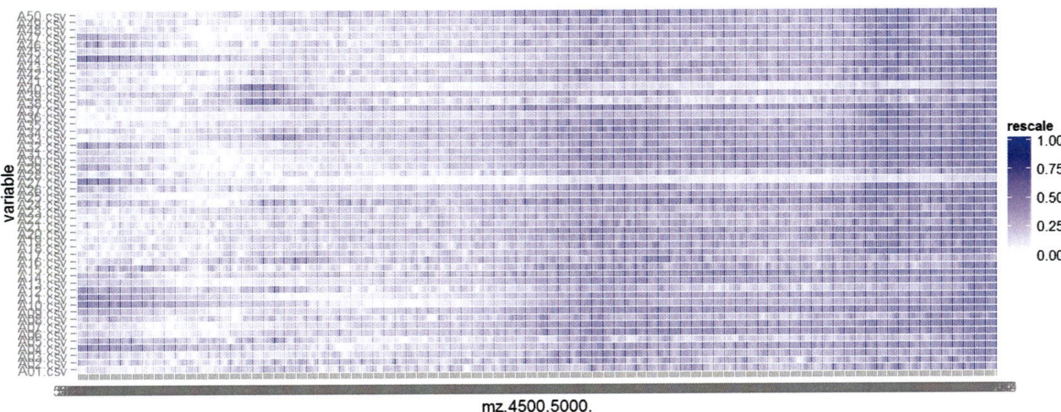

Fig. 6 Heat map of intensities for patients (*m/z* ranged between 4500 Th and 5000 Th)

Notes 2, an alternative distribution that treats the protein abundance as count data is discussed.

Figures 5 and 6 are heat maps of the spectrum ranged between 6500 Th and 7000 Th of the 50 patients and 50 controls respectively. Figures 7 and 8 are heat maps for spectrum ranged between 4500 Th and 5000 Th. The program uses R packages ggplot2, plyr, and reshape2 to draw the heat map.

```
>library(plyr)
>library(graphics)
>library(digest)
>library(ggplot2)
>library(reshape2)
>mz=factor(mzs)
>datam=data.frame(mz[6500:7000], rtM[6500:7000,1:100])
>datam=melt(datam)
>protein_m <-ddply(datam,.(variable),transform,rescale =rescale(value,c(0,1)))
>protein_m_cancer<-protein_m[1:25050,]
>protein_m_control<-protein_m[25051:50100,]
>windows(12,3)
>(p <- ggplot(protein_m_control, aes(mz.6500.7000.,variable))+ geom_tile(aes(fill=rescale),colour = "white") + scale_fill_gradient(low = "white",high = "blue"))
```

Table 6 summarizes the difference in intensity or log (intensity) of the 16 identified peaks. Eleven peaks had significant differences after the multiple comparison adjustment to control for FDR, using the method of Benjamini and Hochberg [14]. The FDR control procedure ranked all *p* values from smallest to largest (rank = *i*). The largest *i* less than the threshold controlling FDR is 11, therefore all *p* values of ranking less than or equal to 11 are

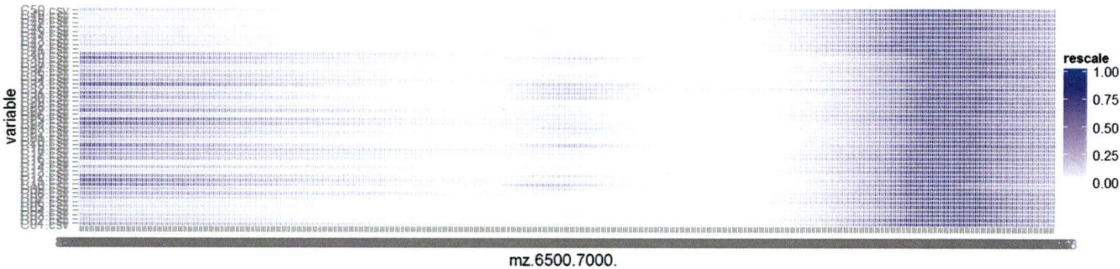

Fig. 7 Heat map of intensities for controls (*m/z* ranged between 6500 Th and 7000 Th)

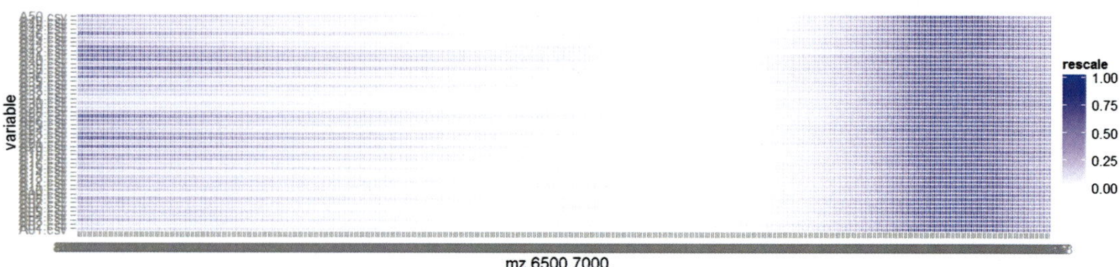

Fig. 8 Heat map of intensities for patients (*m/z* ranged between 6500 Th and 7000 Th)

considered to be significant after the adjustment. If all tests results have p values < 0.05 (all null hypotheses are rejected), an adaptive FDR control procedure Hochberg [34] need to be considered, where $p_{(i)}$ is compared to $\frac{\alpha}{m-i+1}$ instead.

Figure 9 demonstrates the residual plot from the ANOVA for each peak; some peaks used the transformed intensity as the results of their skewed residuals identified in the early analysis.

3.3.2 Case Study 2

The second case study of 95 colorectal samples from 90 cancer patients is used to demonstrate how to analyze the proteomic data through a multiple-protein model. This case study included 95 tumor samples of which 64 were from Colon Adenocarcinoma (COAD) samples, and 31 were from Rectum Adenocarcinoma (READ) samples [35, 36]. Patients had different cancer stages during the time of tissue collection. Amongst them 16 were *stage-I*, 42 were *stage-II* or *stage-II/II a*, 26 were *stage-III*, *-III a*, *-III b* or *-III c*, and 10 were *stage-IV*. The other clinical characteristics included age, gender, metastasis, number of lymph nodes, and tumor stage. The proteomic profiling was compared between COAD samples and READ samples, adjusted for clinical characteristics.

Data management: The peptide-spectrum-matches (PSMs) files were downloaded from The Clinical Proteomic Tumor Analysis

Table 6
Results of single feature model using ANOVA (case study one)

Peak location (*m*/*z* value)	Differences in intensity or log(intensity)# (s.e.)	*t* values	*p* values	Threshold controlling FDR	Rank of *p* value	a
1607	−11.43 (1.91)	−5.98	3.69E-08		5	*
3927	0.83 (0.38)#	2.2	0.0313	0.0094	14	
4053	5.30 (2.10)	2.53	0.013	0.0125	13	
4248	3.86 (0.73)	5.26	8.36E-07		8	*
4667	3.43 (2.16)	1.59	0.12	0.0063	15	
4844	0.13 (0.05)#	2.41	0.018	0.0156	12	
6904	−1.30 (0.16)#	−8.08	1.75E-12		1	*
6919	−1.23 (0.17)#	−7.19	1.29E-10		2	*
6922	−1.15 (0.17)#	−6.73	1.13E-09		3	*
6924	−1.11 (0.17)#	−6.49	3.59E-09		4	*
6925	−1.06 (0.17)#	−6.29	9.04E-09		6	*
6927	−1.02 (0.17)#	−6.03	2.89E-08		7	*
8074	7.82 (2.23)	3.5	*7.00E-04*	*0.0188*	*11*	*
8235	4.17 (0.98)	4.25	4.89E-05		10	*
8461	4.39 (0.92)	4.77	6.40E-06		9	*
9301	0.10 (0.09)	1.14	0.26	0.0031	16	

[a]*The significant peaks after the adjustment controlling for FDR Benjamini and Hachberg [14]
#The intensity values have been log transformed

Consortium (CPTA) data portal. SAS program was used to filter the peptide precursor ion observations and match with the clinical data. Only those peptide spectrum with unique matching (Ambiguous = 0) and those proteins had number of observed peptide spectrum > 10 in each subject were included in the following analysis.

Principal component analysis: According to the CPTA data pipeline description, the precursor area can be used for relative quantitation. After natural logarithmic transformation, the distribution of the area becomes symmetrical. The mean log (precursor area) of each patient × protein combination was derived to form a new dataset, of which each patient only has one record of each protein. Because not all proteins were observed in every patient, the missing values were imputed by the mean to enable the conduction of principal component analysis (PCA). PCA was performed to create a group for each protein according to its

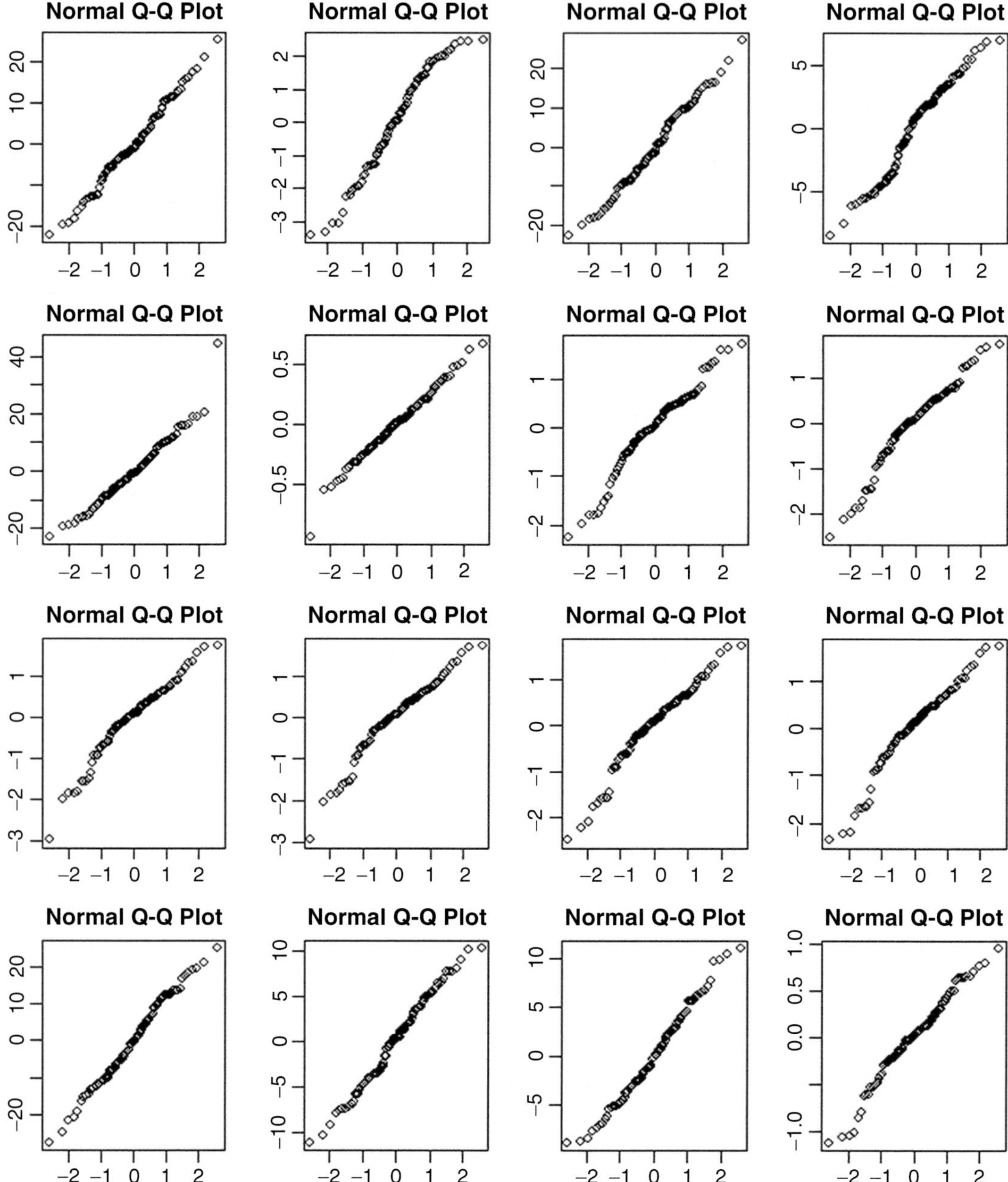

Fig. 9 The normal QQ plot for residuals from single feature analysis of variance models

maximum loading to which the principal component was attached. 463 proteins were identified as group one with a minimal protein pair-wise correlation of 0.32. The abundance of peptides from these 463 proteins and clinical data were fitted into a multiple-protein model which was used to identify any differentiating features between COAD and READ patients.

The analytical multiple-protein model is constructed as follows:

Level 1:

$$\gamma_{i,l,p} = \phi_{0,l} + \beta_{0,p}\text{protein}_{i,l,p} + \beta_{1,p}\text{cancer\_type}_{i,l,p} + e_{i,l,p}, \quad (8)$$

Level 2-protein level:

$$\begin{aligned} \beta_{0,p} &= \alpha_0 + b_{0,p} \\ \beta_{1,p} &= \alpha_1 + b_{1,p} \end{aligned} \quad (9)$$

Level 2-subject level:

$$\phi_{0,l} = \gamma_{0,0} + \gamma_2 \times \text{tumor\_stage}_{i,l,p} + \gamma_3 \times \text{age}_{i,l,p} + \gamma_4 \times \text{gender}_{i,l,p} + c_{0,l} \quad (10)$$

where $\beta_0{,}p$ and $\beta_1{,}p$ are the random effect coefficients for proteins; $\gamma_{0,0}, \gamma_2, \ldots, \gamma_4$ are the fixed effects coefficients for the grand intercept, tumor stage, age, and gender respectively. The random residuals terms include the random error residual term $e_{i,l,p}$, the subject level residual $c_{0,}l$, and the two protein level residual terms $b_0{,}p$ and $b_1{,}p$ which represents the protein intercept and cancer type respectively.

The model described in Eqs. 8–10 was computed using the R *lmer* function. The codes that were used to implement the model and derive the random effects and variance are:

```
#Disease_Code is an indicator for cancer type (CORD or READ)
>fit<-lmer(logArea~factor(tumor_stagecode)+factor(gender)+centralized_mz+(1 + centralized_mz + factor(Disease_Code)|protein) + (1|ID),data=proteincluster1)
#Derive the random effect coefficients of the cancer type
>raneff_cancertype=ranef(fit)$protein[,3];
#derive the random intercept of the protein
>raneff.protein=ranef(fit)$protein[,1]
#derive the variance of the random effect for the cancer type
>cancertype_var_mat=attr(ranef(fit,postVar = TRUE)[[1]], "postVar")
>cancertype_var=cancertype_var_mat[3,3,1:463]
```

The final model (model 1) included random effects of cancer type and centralized *m/z* at the protein level, and fixed effects of tumor stage, gender, and centralized *m/z*. The final model was selected based on a smallest REML criterion at the converge state (Tables 7 and 8). Including centralized *m/z* as a random effect had improved the fitness of the model. Although the tumor stage, gender is not significant in the final model, including them reaches a smaller REML criterion than excluding them (Table 7 model 1: 747992.6 vs. Table 8 model 2: 748581.9). The variance in the

Table 7
The results of multiple proteins model 1 from R function *lmer*

Model 1					
Random effect					
Groups	Name	Variance	Std. dev.	Corr.	
Protein	Intercept	7.88	2.81		
	Centralized m/z	3.92E-06	0.00198	0.22	
	Cancer type READ vs. COAD	0.0847	0.291	0.92	0.31
Patient	Intercept	7.88	2.81		
Residual		3.54	1.88		
Number of observation: 181,460, groups: protein, 463; patient, 90					
Fixed effect					
		Estimate	Std. error	*t* value	
	Intercept	15.63	0.928	16.85	
	Tumor stage				
	t2 vs. t1	0.125	1.003	0.12	
	t3 vs. t1	0.059	0.932	0.063	
	t4 vs. t1	−0.416	1.09	−0.382	
	Gender male vs. female	−0.185	0.336	−0.55	
	Centralized m/z	−0.00033	0.00012	−2.62	

relative abundance of proteins was also reduced when including these covariates. The random effect of cancer type is positively associated with protein random intercept which represents the average log (abundance) of the protein.

The difference in log (abundance) between READ and COAD samples were visualized by a caterpillar plot (Fig. 10, the top left picture) and a normal *qq* plot (Fig. 11). Both plots indicate that, there are more proteins in READ having upward shifting than downward shifting intensities when compared to COAD.

The above analysis for case study 2 assumed that the protein abundance or the log-transformed abundances are multivariate normal distributed. In **Notes 3** and **4**, more discussions are made with the model assumption and method of handling missing data.

Table 8
The results of multiple proteins model 2 from R function *lmer*

Model 2					
Random effect					
Groups	Name	Variance	Std. dev.	Corr.	
Protein	Intercept	1.047	1.023		
	Centralized m/z	7.32E-06	0.00271	0.33	
	Cancer type READ vs. COAD	5.325	2.31	0.55	0.72
Patient	Intercept	1.53	1.24		
Residual		3.54	1.88		
Number of observation: 181,460, groups: protein, 463; patient, 90					
Fixed effect					
		Estimate	Std. error	*t* value	
	Intercept	15.6	0.149	104.9	
	Centralized m/z	-0.000224	0.00015	-1.52	

4 Summary and Discussion

This paper starts with an introduction to clinical proteomic research using 4S (Subject, Scope, Stage, and Solution). It then discusses the design issues of the clinical proteomic study which is a new class of the conventional clinical studies involving biomarkers. Under the same spirit of STROBE and STROBE-ME, it proposed checklists for designing a proteomic study, which are parallel to the design items listed in STROBE-ME.

The second section (Subheading 2) of the paper introduces several design methods to clinical proteomic studies according to their different stages. It also demonstrates two methods for the sample size estimation when researchers consider each stage separately or systemically.

The third section (Subheading 3) covers topics in the analysis. This section introduced three analytical methods and demonstrated them by using two case studies. Case study 1 used ANOVA and treated the peak intensity as a continuous variable in the analysis. Case study 2 used the multiple-protein model to analyze a group of proteins identified by PCA. It can also apply to proteins that are grouped together according to their biological functions.

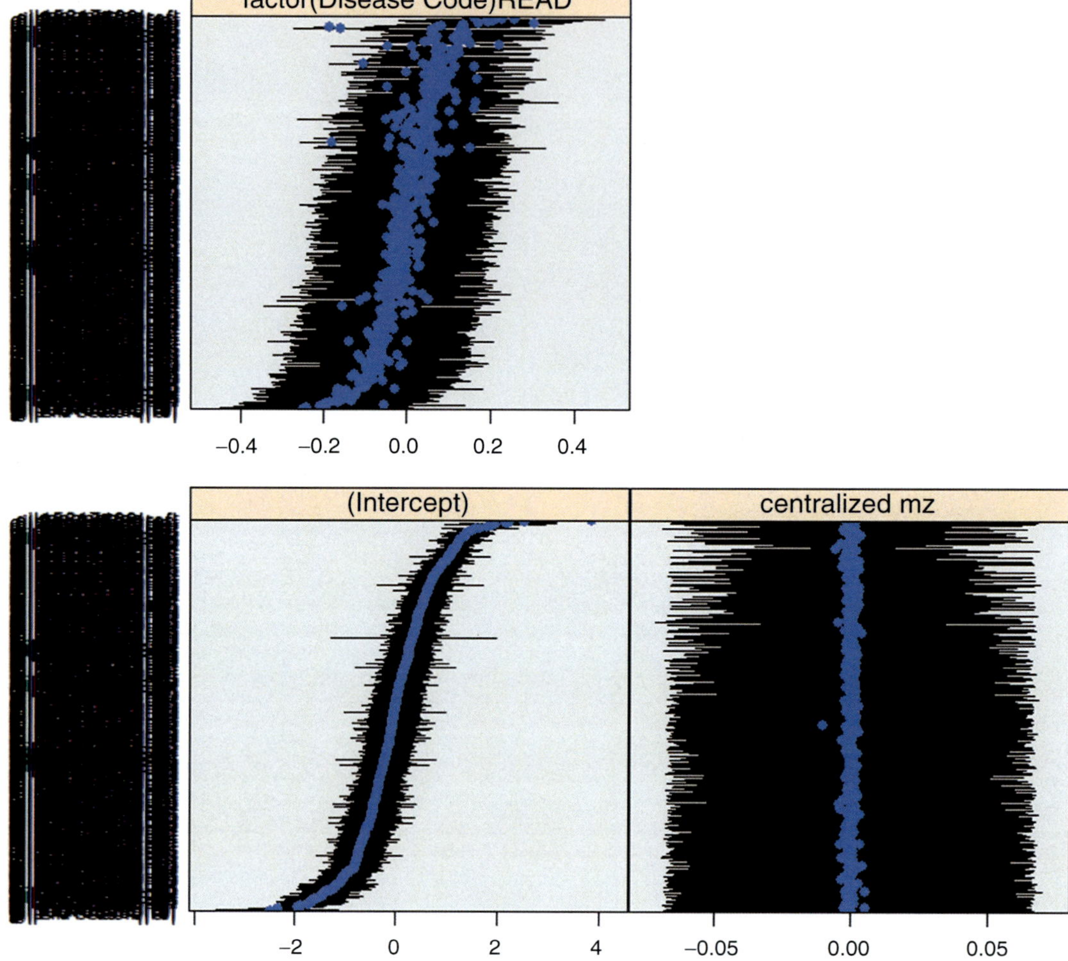

Fig. 10 Caterpillar plot of the random effect of differences in log (abundances) between COAD and READ cancer patients

5 Notes

1. The optimal design procedure. The R program provided in Subheading 2.4 for sample size estimation uses the proteomicdesign 2.0 package for a three-stage study design. The current version 2.0 only provides solutions for matched case–control or paired sample design. The solution provided in the algorithm is to use the entire assigned budget. If users need to seek optimal solutions within a smaller amount of the assigned budget, a serial of budget constraints can be defined by adding a small subtractive term accordingly.

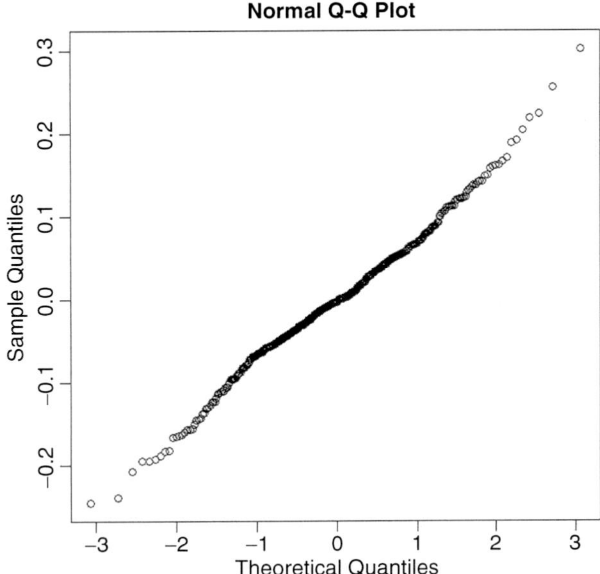

Fig. 11 Normal QQ plot for differences in log(abundance) between COAD and READ

2. Case study 1: analysis of variance. In case study 1, the other distribution for consideration to model the abundance data is the beta-binomial distribution that treats peptide abundance as count data. Under this assumption, the generalized linear model can be used for the analysis of the beta-binomial distributed responses.
3. Case study 2: the multiple- protein model. In cast study 2, the linear multilevel multiple-protein model works well for multivariate normal distributed protein abundance or their transformed abundances. For other distributions, the nonlinear multilevel model should be considered.
4. Missing values. The missing values problem needs to be considered in the analysis section, but it is not covered in this chapter. The missing data mechanism (i.e., missing at random or not at random) can be handled by using an empirical Bayesian approach.

Acknowledgements

The colorectal proteomic data used in this publication were generated by the Clinical Proteomic Tumor Analysis Consortium (NCI/NIH). The mass spectral intensity data of ovarian study are provided by the proteomic databank of Centre for Cancer Research (http://home.ccr.canccr.gov/ncifdaprotcomics/ppatterns.asp). Some codes of producing heat map are modified from the blog: http://learnr.wordpress.com/2010/01/26/ggplot2-quick-heatmap-plotting for NBA game. The author's first learning of Mendelian randomization method in 2010 is attributed to Professor Thomas Lumley.

References

1. Chapman JR (1996) Protein and peptide analysis by mass spectrometry. Humana Press Inc., Totowa, NJ
2. Palmblad M, Tiss A, Cramer R (2009) Mass spectrometry in clinical proteomics—from the present to the future. Proteomics Clin Appl 3:6–17
3. Anderson L (2005) Candidate-based proteomics in the search for biomarkers of cardiovascular disease. J Physiol 563:23–60
4. Shadforth IP, Dunkley TP, Lilley KS et al (2005) i-Tracker: for quantitative proteomics using iTRAQ. BMC Genomics 6:145
5. Corthals GL, Rose K (2007) Quantitation in proteomics, in proteome research: concepts, technology and application. Springer, Berlin
6. Mertins P, Yang F, Liu T et al (2014) Ischemia in tumors induces early and sustained phosphorylation changes in stress kinase pathways but does not affect global protein levels. Mol Cell Proteomics 13:1690–1704
7. Galloa V, Eggerc M, McCormack V et al (2011) STrengthening the Reporting of OBservational studies in Epidemiology-Molecular Epidemiology STROBE-ME: an extension of the STROBE statement. J Clin Epidemiol 64:1350–1363
8. Kleinbaum DG (2003) ActivEpi. Springer, New York
9. Lagiou P, Adami H-O, Trichopoulos D (2005) Causality in cancer epidemiology. Eur J Epidemiol 20:565–574
10. Katan M (1986) Apolipoprotein E isoforms, serum cholesterol, and cancer. Lancet 1:507–508
11. Qi L (2009) Mendelian randomization in nutritional epidemiology. Nutr Rev 67:439–450
12. Verhoeven KJF, Simonsen KL, McIntyre LM (2005) Implementing false discovery rate control: increasing your power. Oikos 108: 643–647
13. Benjamini Y, Yekutieli D (2001) The control of the false discovery rate in multiple testing under dependency. Ann Stat 29:1165–1188
14. Benjamini Y, Hochberg Y (1995) Controlling the false discovery rate: a practical and powerful approach to multiple testing. J R Stat Soc 57:289–300
15. Zeng ISL, Lumley T, Ruggiero K et al (2013) Two optimization strategies of multi-stage design in clinical proteomic studies. Stat Appl Genet Mol Biol 12:263–283
16. von Elm E, Altman D, Egger M et al (2007) The Strengthening the Reporting of Observational Studies in Epidemiology (STROBE) statement: guidelines for reporting observational studies. PLoS Med 4, e297
17. Tzoulaki I, Ebbels TMD, Valdes A et al (2014) Design and analysis of metabolomics studies in epidemiologic research: a primer on -omic technologies. Am J Epidemiol 180: 129–139
18. McShane LM, Altman DG, Sauerbrei W et al (2005) REporting recommendations for tumor MARKer prognostic studies (REMARK). Nat Clin Pract Oncol 2
19. Kuller LH, Bracken MB, Ogino S et al (2013) The role of epidemiology in the era of molecular epidemiology and genomics: summary of the 2013 AJE-sponsored Society of Epidemiologic Research Symposium. Am J Epidemiol 178:1350–1354
20. Malats N, Castaño-Vinyals G (2007) Cancer epidemiology: study designs and data analysis. Clin Transl Oncol 9:290–297
21. Zeng ISL, Browning S, Gladding P et al (2009) A multi-feature reproducibility assessment of

mass spectral data in clinical proteomic studies. Clin Proteomics 5:170–177

22. Bossuyt PM, Reitsma JB, Bruns DE et al (2003) Towards complete and accurate reporting of studies of diagnostic accuracy: the STARD initiative. BMJ 326:41–44
23. Bonassi S, Au WW (2002) Biomarkers in molecular epidemiology studies for health risk prediction. Mutat Res 511:73–86
24. Thomas D, Conti D (2004) The concept of 'Mendelian Randomization'. Int J Epidemiol 33:21–25
25. Lawlor DA, Harbord RM, Sterne JA et al (2008) Mendelian randomization: using genes as instruments for making causal inferences in epidemiology. Stat Med 27:1133–1163
26. Applied Biosystems/MDS SCIEX, ProteinPilot™ Software (2007) Getting started guide
27. Breitwieser FP, Muller A, Dayon L et al (2001) General Statistical Modeling of Data from Protein Relative Expression Isobaric Tags. J Proteome Res 10:2758–2766
28. Oberg AL, Mahoney DW (2012) Statistical methods for quantitative mass spectrometry proteomic experiments with labeling. BMC Bioinformatics 13:S7
29. Zeng ISL (2014) Doctorate thesis: Statistical methods in clinical proteomic studies 2007-2014—a protein concerto. In: Statistics2014, University of Auckland: ResearchSpace@ Auckland. p 218
30. Goldstain H (1999) Multivariate multilevel model. In: Multilevel statistics models 1999. London, Institute of Education, p 5–6
31. Li X (2005) PROcess: Ciphergen SELDI-TOF processing. R package
32. Boehm AM, Putz S, Altenhofer D et al (2007) Precise protein quantification based on peptide quantification using iTRAQ™. BMC Bioinformatics 8:214
33. Clinical proteomic program data portal: OvarianDataset4-3-02.zip. A.2.1. http://home.ccr.cancer.gov/ncifdaproteomics/ppatterns.asp
34. Hochberg Y (1988) A sharper Bonferroni procedure for multiple tests of significance. Biometrika 75:800–803
35. CPTAC, TCGA Cancer proteome study of colorectal tissue. https://cptac-data-portal.georgetown.edu/cptac/s/S016
36. Zhang B, Wang J, Wang X et al (2013) Proteogenomic characterization of human colon and rectal cancer. Nature 513:382–387

Chapter 3

Preprocessing and Analysis of LC-MS-Based Proteomic Data

Tsung-Heng Tsai, Minkun Wang, and Habtom W. Ressom

Abstract

Liquid chromatography coupled with mass spectrometry (LC-MS) has been widely used for profiling protein expression levels. This chapter is focused on LC-MS data preprocessing, which is a crucial step in the analysis of LC-MS based proteomics. We provide a high-level overview, highlight associated challenges, and present a step-by-step example for analysis of data from LC-MS based untargeted proteomic study. Furthermore, key procedures and relevant issues with the subsequent analysis by multiple reaction monitoring (MRM) are discussed.

Key words Data preprocessing, Label-free, Liquid chromatography-mass spectrometry (LC-MS), Multiple reaction monitoring (MRM), Proteomics

1 Introduction

With recent advances of mass spectrometry and separation methods, liquid chromatography coupled with mass spectrometry (LC-MS) has become an essential analytical tool in biomedical research. LC-MS provides qualitative and quantitative analyses of a variety of biomolecules in a high-throughput fashion, and there has been significant progress in systems biology research and biomarker discovery using LC-MS based proteomics [1–3].

LC-MS methods can be used for extraction of quantitative information and detection of differential abundance [4–6]. This requires that a rigorous analysis workflow be implemented. In addition to analytical considerations, crucial steps include: (1) experimental design that avoids introducing bias during data acquisition and enables effective utilization of available resource [7], (2) data preprocessing pipeline that extracts meaningful features [8], and (3) statistical test that identifies significant changes based on the experimental design [9]. Conducting these three steps in a coherent manner is key to a successful LC-MS based proteomic analysis. Good experimental design helps effectively identify true differences

Klaus Jung (ed.), *Statistical Analysis in Proteomics*, Methods in Molecular Biology, vol. 1362,
DOI 10.1007/978-1-4939-3106-4_3, © Springer Science+Business Media New York 2016

in the presence of variability from various sources. This benefit can diminish if the data analysts fail to appropriately analyze the LC-MS data and conduct the subsequent statistical tests in accordance with the experimental design. This chapter introduces data preprocessing pipelines for LC-MS based proteomics, with a focus on untargeted and label-free proteomic analysis. We provide a high-level overview of LC-MS data preprocessing and highlight associated challenges. Furthermore, we present a step-by-step example for analysis of LC-MS data from untargeted proteomic study, and how this could be utilized in subsequent evaluation using targeted quantitative approaches such as multiple reaction monitoring (MRM).

2 LC-MS Data Preprocessing

In a typical untargeted proteomic analysis, proteins are first enzymatically digested into smaller peptides, and these thousands of peptides can be profiled in a single LC-MS run. The profiling procedure involves chromatographic separation and MS based analysis. Due to the difference in hydrophobicity and polarity among other properties, each peptide elutes from the LC column at distinct retention time (RT). The eluted peptide is then analyzed by MS or tandem MS (MS/MS). An LC-MS run contains RT information in chromatogram, mass-over-charge ratio (m/z) in MS spectrum, and relative ion abundance for each particular ion. MS signals detected throughout the range of chromatographic separation are formatted in a three-dimensional map, which defines the data from a single LC-MS run, as shown in Fig. 1. The LC-MS

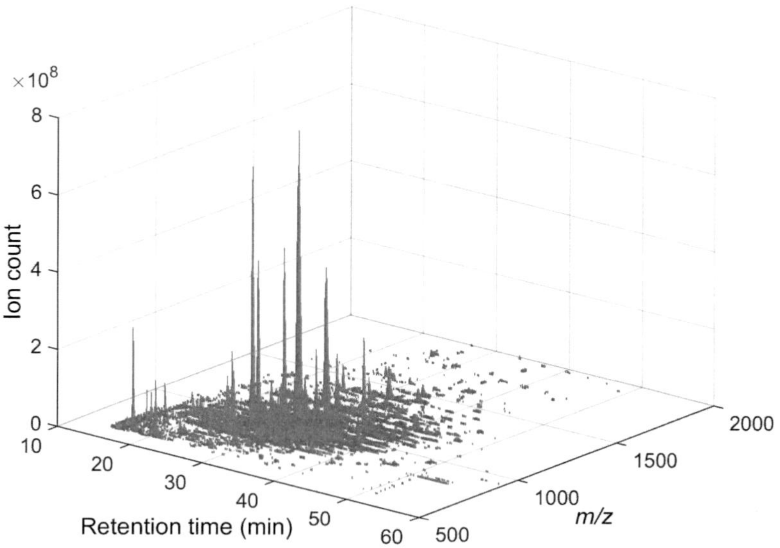

Fig. 1 An LC-MS run contains RT information in chromatogram, mass-over-charge ratio (m/z) in MS spectrum, and relative ion abundance for each particular ion

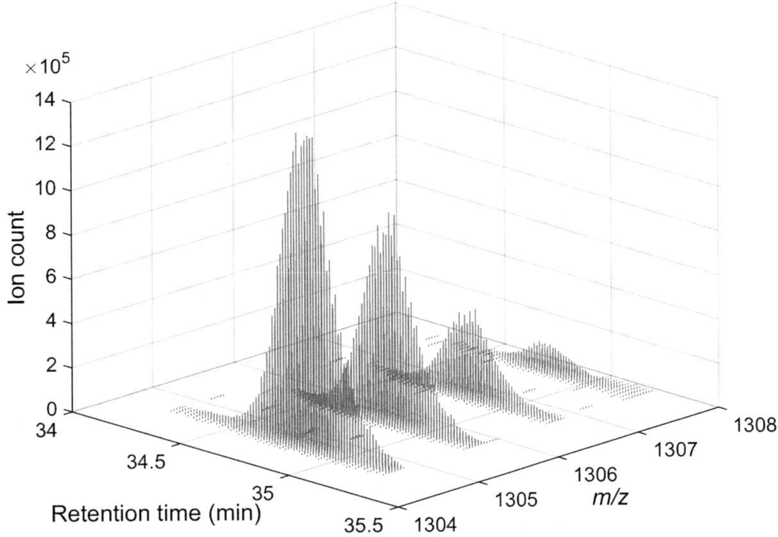

Fig. 2 A typical feature in LC-MS data

data contain quantitative information of detected peptides and their associated proteins, which are identified by de novo sequencing or database searching using MS/MS spectra [10]. A reliable preprocessing pipeline is needed to extract features (usually referred to as peaks) from LC-MS data, in which each peptide is characterized by its isotopic pattern resulting from common isotopes such as ^{12}C and ^{13}C in a set of MS spectra within its elution duration, in superposition of noise signals (Fig. 2). Adequate consideration of such characteristics is crucial for LC-MS data preprocessing, including steps of noise filtering, deisotoping, peak detection, RT alignment, peak matching and normalization. Typically, these data preprocessing steps generate a list of detected peaks characterized by their RTs, m/z values and intensities. The preprocessed data can be used in subsequent analysis, e.g., identification of significant differences between groups. Association of these peaks with peptides/proteins is achieved through MS/MS identification, which is out of scope of this chapter, and we refer to interested readers to the literature [10]. In this section, critical preprocessing steps are introduced and discussed.

2.1 Noise Filtering

LC-MS data are subject to electronic/chemical noises due to contaminants present in the column solvent or instrumental interference. Appropriate noise filtering can increase the signal-to-noise ratio (SNR) and facilitate the subsequent peak detection step. Some software tools, e.g., MZmine 2 [11], integrate the noise filtering into the peak detection step to ensure coherence. Smoothing filters such as Gaussian filter and Savitzky-Golay filter [12] are commonly applied to eliminate the effects of noises. Due to the differences in terms of resolution and detection limit among various LC-MS platforms, parameters for the smoothing filters need to be adaptively selected, preferably through a pilot experiment with similar experimental settings.

2.2 Deisotoping

Most chemical elements have naturally occurring isotopes, e.g., ^{12}C and ^{13}C are two stable isotopes of the element carbon with mass numbers 12 and 13, respectively. Consequently, each analyte gives rise to more than one ion peaks in an MS spectrum, where the peak arising solely from the most common isotope is called the monoisotopic peak. In LC-MS based proteomics, each peptide is characterized by an envelope of ion peaks due to its constituent amino acids. ^{13}C constitutes about 1.11 % of the carbon species and the approximately one dalton (Da) mass difference between ^{13}C and ^{12}C results in $1/z$ difference between adjacent ion peaks in the isotopic envelope, where z is the state of a charged peptide. The deisotoping step integrates siblings of ion peaks originating from the same peptide and summarizes by its monoisotopic mass. This facilitates the interpretation of LC-MS data and reduces the complexity in subsequent analysis. DeconTools [13] is widely used to deisotope MS spectra, which involves: (1) identification of isotopic pattern, (2) prediction of the charge state based on the distance between the ion peaks, and (3) comparison between the observed isotopic pattern and a theoretical distribution generated based on an average residue.

2.3 Peak Detection

Peak detection is a procedure to determine the existence of a peak in a specific range of RT and m/z value, and to quantify its intensity. Many LC-MS peak detection approaches [11, 14, 15] are adapted from previously established methods such as those for analysis of matrix-assisted laser desorption/ionization time of flight (MALDI-TOF) MS data [16, 17]. In consideration of the isotopic pattern naturally present in LC-MS data, alternative strategies have also been exploited (e.g., as in the MaxQuant platform [18]). Most existing methods perform peak detection via a pattern matching process, followed by a filtering step based on quantified peak characteristics. A critical issue is that the elution profiles may vary across different RTs [19]. As a result, the use of a single pattern throughout the whole RT range in the current approaches may lead to inaccurate estimates of peak characteristics and SNR, where the latter is often employed as a filtering criterion. Also, peak detection is usually performed for each LC-MS run individually, without leveraging the information from other runs in the same experiment. Utilization of multi-scale information from multiple runs has been proposed for analysis of MALDI-TOF data [20]. This idea could potentially be applied to LC-MS data and lead to a more reliable peak detection result, where the peak matching step to be introduced later plays an important role.

2.4 Normalization

Due to the presence of various analytical and technical variability in LC-MS data, it requires appropriate normalization of intensity measurements to remove systematic biases and eliminate the effect of obscuring variability. One of the typical normalization approaches carries out the task through identifying a reference for ion intensities and making adjustment based on the reference. Apparently, identification of reliable reference is crucial for the normalization process. Most existing methods assume that each of the LC-MS runs in the same experiment should have an equal concentration of molecules on average [21]. With this assumption, measures including summation, median, and quantile of the ion intensities are used as the reference for normalization. Unfortunately, the validity of this assumption is questionable as an increase of concentration in a specific group of molecules is not necessarily compensated by a decrease in other groups [22]. More rigorous approaches using regression methods based on a set of matched peaks [23] or spiked-in internal standards [22] have been proposed. However, it is unclear that if neighboring ions (in terms of RT, m/z value, or intensity) would necessarily share a similar drifting trend along the analysis order. At present, the use of quality control (QC) runs to assess and correct variability in LC-MS data appears to be the most reliable approach [24], in which QC runs can be collected using a reference sample or a mixture pooled from the analyzed samples. This idea has been successfully implemented for large-scale metabolomic studies, where variability along the analysis order is estimated for each of the detected peaks through assessment of the QC runs [24]. This circumvents the need to select an arbitrary reference, with additional experimental challenges to assure appropriate coverage and reproducible detection of ions in the QC runs. Alternatively, a recently published method called MaxLFQ [25] leverages information from every pair of peptides between samples to account for the reproducibility issue and exploits such information to accomplish normalization at protein level.

2.5 RT Alignment and Peak Matching

The peak matching step groups consensus peaks across multiple LC-MS runs prior to subsequent analysis, e.g., identification of significant differences between samples, to ensure a valid comparison of the LC-MS runs. Also, it is crucial for potential extensions of peak detection and normalization steps, by leveraging information from multiple runs. The main challenge in peak matching results from the presence of RT variability among LC-MS runs. Recent advances in MS technology have made highly precise and accurate mass measurement (low- to sub-ppm) achievable [26]. However, controlling the chromatographic variability remains challenging. Most LC-MS preprocessing pipelines, (e.g., OpenMS [14], msInspect [27], MZmine 2 [11]) integrate the estimation of RT variability into the peak matching step, in order to perform RT alignment and achieve reliable identification of consensus peaks.

RT alignment approaches can be categorized as: (1) feature-based approaches and (2) profile-based approaches [28]. The feature-based approaches perform the alignment task based on detected peaks and rely on the correct identification of a set of consensus peaks among LC-MS runs. On the other hand, the profile-based approaches utilize chromatograms of the LC-MS runs to estimate the variability along RT and then make an adjustment accordingly [29–31].

Incorporation of information from peptide identification can reduce the matching ambiguity and improve the alignment result [32, 33]. For example, the PEPPeR platform [33] integrates peak lists and MS/MS identification for RT alignment. More sophisticated approach has been implemented in MaxQuant [18], which leverages each preprocessing step to enhance the overall performance. In profile-based alignment, utilization of complementary information from various sources has also been shown to yield better alignment performance [30].

3 Pipeline for LC-MS Data Preprocessing

Several preprocessing pipelines have been made available in various software tools including OpenMS [14], msInspect [27], MZmine 2 [11], and MaxQuant [18]; however, very few studies have systematically evaluated and compared their performance [34]. As a result, determination of the most appropriate pipeline is still challenging. As a starting point, we present a step-by-step example using MaxQuant in this section. This software tool is chosen for demonstration due to (1) its ease of use, (2) its capability to handle data from large-scale LC-MS experiments, and (3) its active discussion forum.

MaxQuant can be downloaded from http://www.maxquant.org after registration. A personal computer with CPU frequency at 800 MHz and RAM at 2GB per thread is the minimum requirement for installation. Multicore processor is recommended for parallel computation. Prerequisite software/plug-ins include Xcalibur, MSFileReader, and .NET Framework 4.5. A peptide search engine, Andromeda [35], is integrated as part of MaxQuant and downstream bioinformatics and statistical analyses on the outputs of MaxQuant can be performed using Perseus, if needed. Users are referred to the forum (https://groups.google.com/forum/#!forum/maxquant-list) for related discussions and possible solutions. For comparative analysis by label-free LC-MS methods, detailed preprocessing steps using MaxQuant (version 1.4.1.2) are described in the following.

3.1 Importing Files

1. Launch the MaxQuant graphical interface (Fig. 3) and load the .raw files (from Thermo instruments) to be processed. The basic information (file name, size, etc.) of the imported data will be displayed on the interface. Specify additional informa-

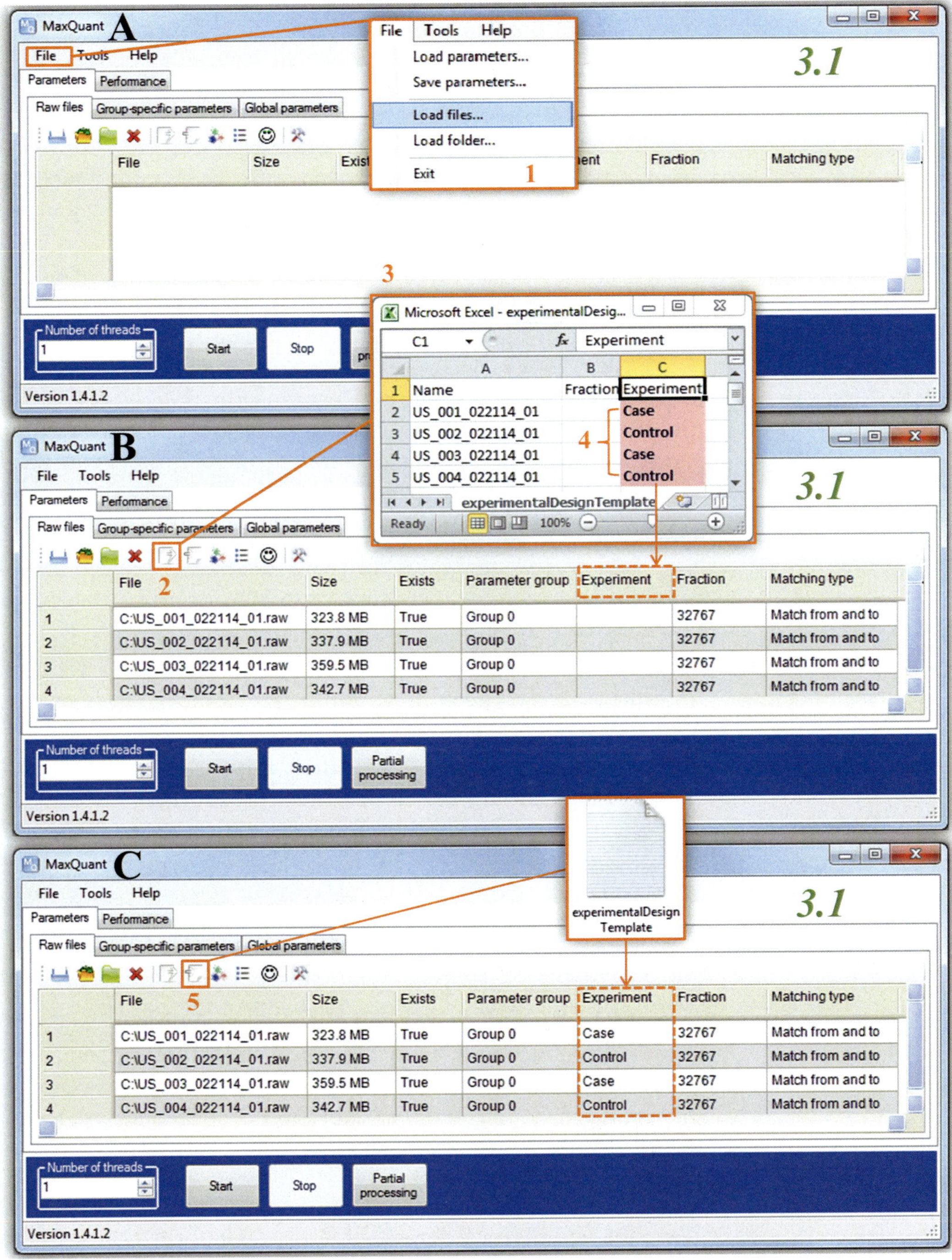

Fig. 3 Procedure of MaxQuant (Subheading 3.1): loading files (panel a) and setting up experimental design template (panels b–c)

tion (e.g., fraction labels) for the MaxQuant analysis using the experimental design template.

2. Click the icon of "Write template" to generate a "combined" folder in the same location of the .raw files.

3. Under the newly generated "combined" folder, open the template file "experimentalDesignTemplate.txt" using appropriate text editor (e.g., Microsoft Excel).
4. The template file presents a table with three columns, where the "Name" column should have been filled in with the .raw file names. Complete the table with distinct numbers in the "Fraction" column and group information in the "Experiment" column (*see* **Note 1**). Save these changes.
5. Click the "Read from file" icon and select the modified template file to import the specified information.

3.2 Setting Group-Specific Parameters

Click tab "Group-specific parameters" (Fig. 4a), where default values are given for general experiment information, label-free quantification, first search, and advanced settings. If data with different experimental protocols are processed together, users can set specific parameters for each group. Modify settings within each parameter group according to specific experiments.

1. The "Type" setting is machine dependent. Select "All Ion Fragmentation" if an Exactive is used. "Standard" (default) should be selected for other Thermo instruments (XL, Velos, etc.).
2. Specify labels, if a labelling strategy is used. For a label-free analysis, select "Multiplicity" as "1".
3. "Variable modifications" settings describe the chemical reactions on the proteins. This does not include fixed modifications that should be selected under "Global parameters".
4. Select the enzyme used to digest the proteins in "Digestion mode". Trypsin is used in most cases.
5. Indicate maximum allowable missed cleavages during enzymatic digestion. Default allowable value is "2".
6. Specify the instrument type.
7. Select "LFQ" for label-free analysis.
8. The "First Search" and "Main Search" (under "Advanced") specify a two-step search in MaxQuant, where a number of peptides are selected for calibration of mass and RT, followed by a refined search.

3.3 Setting Global Parameters

Click tab "Global parameters" (Fig. 4b), where default values are given for settings including general analysis information, sequences, identification, protein quantification, site quantification, label-free quantification, isobaric label quantification, etc. These settings apply for all data files. We describe critical settings to modify parameters according to specific experimental designs in the following steps.

1. Click "Add file" to load the .fasta files for the database against which the processed spectra are searched. The files are parsed through Andromeda configuration (*see* **Note 2**).

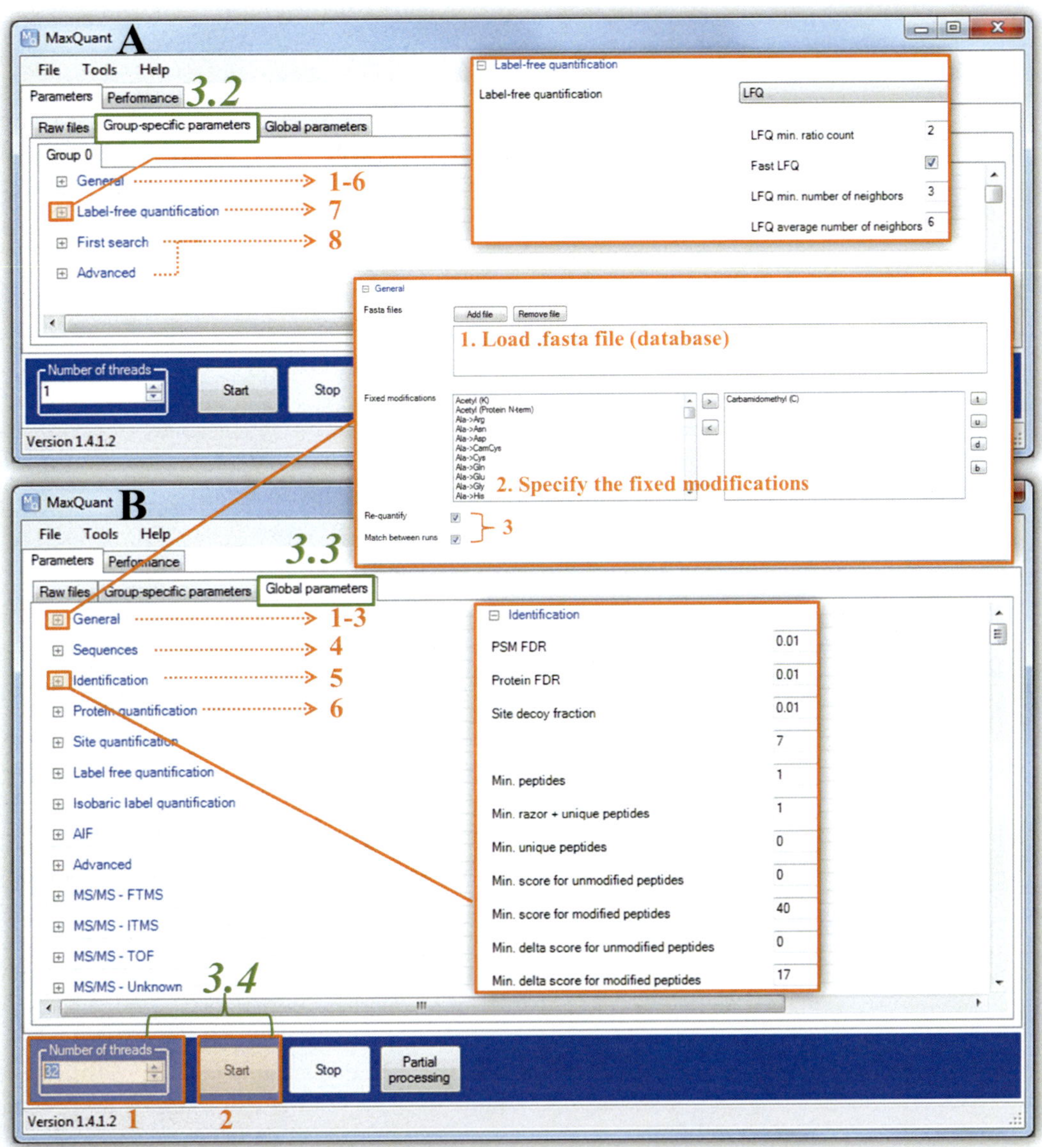

Fig. 4 Procedure of MaxQuant (Subheadings 3.2–3.4): setting group-specific parameters (panel **a**); setting global parameters and starting analysis (panel **b**)

2. Specify the fixed modifications such as carbamidomethylation of cysteine.
3. "Re-quantify" allows the first search as calibration steps prior to the more exact main search and re-calibration steps. "Match between runs" enables association of spectral identification across LC-MS/MS runs based on RT and accurate mass. These two boxes are recommended to be selected.
4. In "Sequences" section, set "Decoy mode" and "Special AAs", select "Include contaminants", and load other .fasta files if the

database used for first search is different from the one loaded in "Fasta files".

5. Set the searching parameters in "Identification", such as false discovery rate (FDR), number of peptides required for a valid identification, minimum peptide length, minimum number of unique (*see* **Note 3**) and razor peptides, posterior error probability (PEP), and score cutoff. Deselect the "Filter labelled amino acids" box for label-free analysis.
6. Specify the quantification methods in "Protein quantification," including minimum ratio count, peptide type for quantification, and whether modified peptides are considered.

3.4 Starting Analysis

1. Set "Number of threads" available to the analysis on the bottom of the setting window for global parameters (Fig. 4b). Using more threads yields faster computation times.
2. Start the analysis with the above settings. The progress can be monitored in the "Performance" tab.

4 Analysis of Targeted Quantitative Proteomic Data

Untargeted LC-MS based proteomics is generally biased towards analysis of the most abundant and observable proteins. Biologically relevant molecular responses, however, are often less discernible in that analysis. Targeted quantification by multiple reaction monitoring (MRM) using triple quadrupole (QqQ) mass spectrometers has been introduced to overcome the limitations of untargeted analysis [36]. Briefly, the MRM method organizes the analysis of a specific list of peptides associated with targeted proteins, characterized by the m/z values of their precursor and fragment ions. The precursor-fragment ion pairs are called transitions, which are highly specific and unique for the targeted peptides. A specific ion is selected in the first quadrupole (Q1) on the basis of its precursor m/z value. The ion gets fragmented by collision-induced dissociation (CID) in the second quadrupole. Only the relevant ions produced by the fragmentation are selected in the third quadrupole (Q3). The resulting transitions are then used for quantification. As the data acquisition is highly specific with less interference from irrelevant ions, the MRM analysis can yield more sensitive and accurate quantification results.

Most bioinformatics tools developed for targeted proteomic data analysis have been either limited in their functions or restricted to specific instrument vendors [37]. Freely available software, such as MaRiMba [38], MRMaid [39], and TIQAM [40], are only designed to aid creation of transition list. Other proprietary software, such as Agilent Mass Hunter Workstation, Applied Biosystems MRMPilot, Thermo-Fisher Pinpoint, and Waters TargetLynx, are limited to specific instrument vendor and not freely accessible. MRMer [41] and Skyline [37] are two instrument-independent and freely available platforms used for MRM analysis. In this following, we briefly present major steps for targeted quantification using Skyline, including design of transition list and analysis of acquired MRM data. This software can be downloaded from https://proteome.gs.washington.edu/software/skyline.

To design a transition list using Skyline, users should import spectral libraries (e.g., public spectral libraries or results from search engines such as Andromeda applied in untargeted analysis) and background proteome files (e.g., human proteome database) to provide background information of the targeted proteomic experiments, upon which, the Skyline can read and match the inserted targeted protein list (in fasta sequences or protein IDs, typically from untargeted proteomic data analysis). Skyline allows the users to customize the parameters of generated transitions (e.g., precursor charges, ion types, and product ions). The selected transitions and corresponding spectra are well visualized in Skyline windows. This facilitates further refinement such as removing poor matches in the spectral library before exporting the list. To analyze MRM data acquired with transition lists already designed (unnecessarily by Skyline), we set up the background proteome information and insert the transition list with associated proteins into Skyline. The data collected on a QqQ MS instrument using this transition list are then imported. Skyline begins loading the files into their high-performance data caches, where the relevant information can be retrieved efficiently. Meanwhile, peak detection is automatically performed and detected peaks are assigned to their corresponding transitions. Once completed, Skyline highlights the transitions with their integration boundaries and measured signals. The users can inspect the data by comparing replicates (across samples) in terms of their RT and intensity ratios. Manual curations are allowed to correct erroneous assignment and adjust the integration boundaries (*see* **Note 4**). Finally, the quantification results can be customized and exported into a .csv file, on which, the downstream statistical analysis can be performed.

5 Notes

1. If the user specifies an identical name for several LC-MS runs in the experiment column, their information will be combined and these individual runs will not be compared. This is, however, an ideal setting if they are all fractions of the same sample.
2. Andromeda configuration is required before starting MaxQuant to correctly retrieve protein sequence information from the .fasta files, as different databases may be delimited in

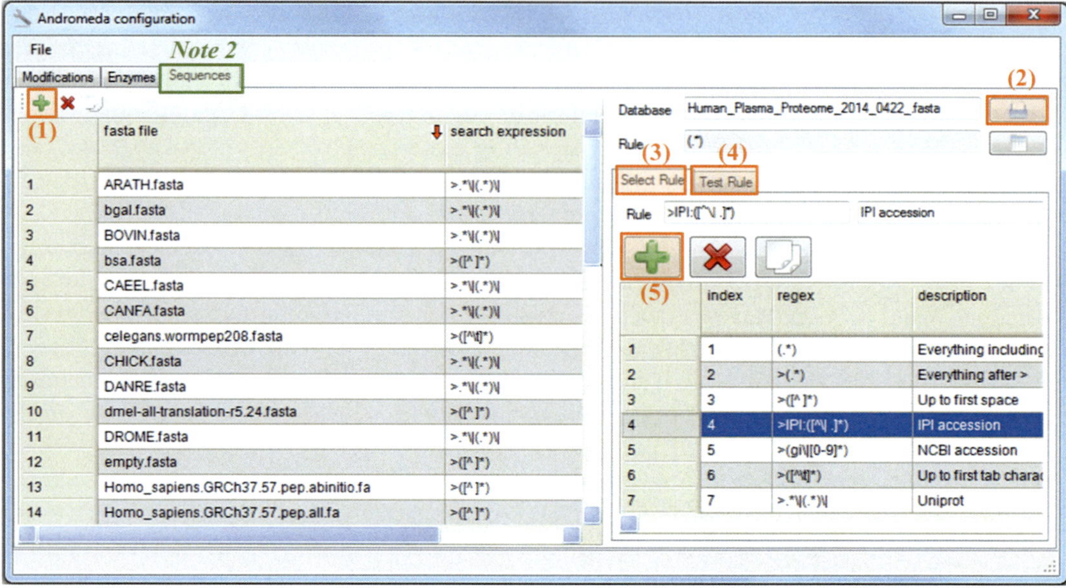

Fig. 5 Configuration of Andromeda

distinct ways. Figure 5 illustrates the main configuration steps including (1) loading a new database entry by clicking the green plus button ("+") in tab "Sequence"; (2) importing user-defined .fasta file; (3) specifying a parsing rule form the list in the "Select Rule" tab; (4) checking if Andromeda is able to retrieve the information from the .fasta file correctly in the "Test Rule" tab; (5) clicking the green plus button ("+") in the top-left corner of the "Select Rule" panel if users need to write specific rules.

3. The uniqueness of peptide is related to the proteome database. In MaxQuant, a peptide is recognized as unique to a group of proteins (termed protein group) if on the entire proteome its sequence only occurs in this group.
4. The inspection is crucial for cases where multiple peaks are detected, and consequently selection of the best peak may not be consistent across samples. To improve the performance of peak selection, Skyline also allows users to create custom advanced selection models and to utilize information from iRT retention time prediction of peptides.

Acknowledgements

This work was supported by the NIH Grants R01CA143420 and R01GM086746.

References

1. Diamandis EP (2004) Mass spectrometry as a diagnostic and a cancer biomarker discovery tool: opportunities and potential limitations. Mol Cell Proteomics 3:367–378
2. Gstaiger M, Aebersold R (2009) Applying mass spectrometry-based proteomics to genetics, genomics and network biology. Nat Rev Genet 10:617–627
3. Ahrens CH, Brunner E, Qeli E et al (2010) Generating and navigating proteome maps using mass spectrometry. Nat Rev Mol Cell Biol 11:789–801
4. Aebersold R, Mann M (2003) Mass spectrometry-based proteomics. Nature 422: 198–207
5. Domon B, Aebersold R (2006) Mass spectrometry and protein analysis. Science 312: 212–217
6. Elias JE, Haas W, Faherty BK et al (2005) Comparative evaluation of mass spectrometry platforms used in large-scale proteomics investigations. Nat Methods 2:667–675
7. Oberg AL, Vitek O (2009) Statistical design of quantitative mass spectrometry-based proteomic experiments. J Proteome Res 8:2144–2156
8. Karpievitch YV, Polpitiya AD et al (2010) Liquid chromatography mass spectrometry-based proteomics: biological and technological aspects. Ann Appl Stat 4:1797–1823
9. Karpievitch Y, Stanley J, Taverner T et al (2009) A statistical framework for protein quantitation in bottom-up MS-based proteomics. Bioinformatics 25:2028–2034
10. Eng JK, Searle BC, Clauser KR et al (2011) A face in the crowd: recognizing peptides through database search. Mol Cell Proteomics 10:R111.009522
11. Pluskal T, Castillo S, Villar-Briones A et al (2010) MZmine 2: modular framework for processing, visualizing, and analyzing mass spectrometry-based molecular profile data. BMC Bioinformatics 11:395
12. Savitzky A, Golay MJE (1964) Smoothing and differentiation of data by simplified least squares procedures. Anal Chem 36:1627–1639
13. Jaitly N, Mayampurath A, Littlefield K et al (2009) Decon2LS: an open-source software package for automated processing and visualization of high resolution mass spectrometry data. BMC Bioinformatics 10:87
14. Sturm M, Bertsch A, Gropl C et al (2008) OpenMS—an open-source software framework for mass spectrometry. BMC Bioinformatics 9:163
15. Yu T, Park Y, Johnson JM et al (2009) apLCMS—adaptive processing of high-resolution LC/MS data. Bioinformatics 25:1930–1936
16. Coombes KR, Tsavachidis S, Morris JS et al (2005) Improved peak detection and quantification of mass spectrometry data acquired from surface-enhanced laser desorption and ionization by denoising spectra with the undecimated discrete wavelet transform. Proteomics 5:4107–4117
17. Du P, Kibbe WA, Lin SM (2006) Improved peak detection in mass spectrum by incorporating continuous wavelet transform-based pattern matching. Bioinformatics 22:2059–2065
18. Cox J, Mann M (2008) MaxQuant enables high peptide identification rates, individualized p.p.b.-range mass accuracies and proteome-wide protein quantification. Nat Biotechnol 26:1367–1372
19. Steen H, Mann M (2004) The abc's (and xyz's) of peptide sequencing. Nat Rev Mol Cell Biol 5:699–711
20. Zhang P, Li H, Wang H et al (2011) Peak tree: a new tool for multiscale hierarchical representation and peak detection of mass spectrometry data. IEEE/ACM Trans Comput Biol Bioinform 8:1054–1066
21. Kultima K, Nilsson A, Scholz B et al (2009) Development and evaluation of normalization methods for label-free relative quantification of endogenous peptides. Mol Cell Proteomics 8:2285–2295
22. Sysi-Aho M, Katajamaa M, Yetukuri L et al (2007) Normalization method for metabolomics data using optimal selection of multiple internal standards. BMC Bioinformatics 8:93
23. Callister SJ, Barry RC, Adkins JN et al (2006) Normalization approaches for removing systematic biases associated with mass spectrometry and label-free proteomics. J Proteome Res 5:277–286
24. Dunn WB, Broadhurst D, Begley P et al (2011) Procedures for large-scale metabolic profiling of serum and plasma using gas chromatography and liquid chromatography coupled to mass spectrometry. Nat Protoc 6:1060–1083
25. Cox J, Hein MY, Luber CA et al (2014) Accurate proteome-wide label-free quantification by delayed normalization and maximal peptide ratio extraction, termed MaxLFQ. Mol Cell Proteomics 13:2513–2526
26. Mann M, Kelleher NL (2008) Precision proteomics: the case for high resolution and high mass accuracy. Proc Natl Acad Sci U S A 105:18132–18138

27. Bellew M, Coram M, Fitzgibbon M et al (2006) A suite of algorithms for the comprehensive analysis of complex protein mixtures using high-resolution LC-MS. Bioinformatics 22:1902–1909
28. Mathias V, Sebastien Li-Thiao T, Hans-Michael K et al (2008) Alignment of LC-MS images, with applications to biomarker discovery and protein identification. Proteomics 8:650–672
29. Listgarten J, Emili A (2005) Statistical and computational methods for comparative proteomic profiling using liquid chromatography-tandem mass spectrometry. Mol Cell Proteomics 4:419–434
30. Tsai TH, Tadesse MG, Di Poto C et al (2013) Multi-profile Bayesian alignment model for LC-MS data analysis with integration of internal standards. Bioinformatics 29:2774–2780
31. Tsai TH, Tadesse MG, Wang Y et al (2013) Profile-based LC-MS data alignment—a Bayesian approach. IEEE/ACM Trans Comput Biol Bioinform 10:494–503
32. Fischer B, Grossmann J, Roth V et al (2006) Semi-supervised LC/MS alignment for differential proteomics. Bioinformatics 22:e132–e140
33. Jaffe JD, Mani DR, Leptos KC et al (2006) PEPPeR, a platform for experimental proteomic pattern recognition. Mol Cell Proteomics 5:1927–1941
34. Tuli L, Tsai TH, Varghese RS et al (2012) Using a spike-in experiment to evaluate analysis of LC-MS data. Proteome Sci 10:13
35. Cox J, Neuhauser N, Michalski A et al (2011) Andromeda: a peptide search engine integrated into the MaxQuant environment. J Proteome Res 10:1794–1805
36. Picotti P, Aebersold R (2012) Selected reaction monitoring-based proteomics: workflows, potential, pitfalls and future directions. Nat Methods 9:555–566
37. MacLean B, Tomazela DM, Shulman N et al (2010) Skyline: an open source document editor for creating and analyzing targeted proteomics experiments. Bioinformatics 26: 966–968
38. Sherwood CA, Eastham A, Lee LW et al (2009) MaRiMba: a software application for spectral library-based MRM transition list assembly. J Proteome Res 8:4396–4405
39. Mead JA, Bianco L, Ottone V et al (2009) MRMaid, the web-based tool for designing multiple reaction monitoring (MRM) transitions. Mol Cell Proteomics 8:696–705
40. Lange V, Malmstrom JA, Didion J et al (2008) Targeted quantitative analysis of Streptococcus pyogenes virulence factors by multiple reaction monitoring. Mol Cell Proteomics 7: 1489–1500
41. Martin DB, Holzman T, May D et al (2008) MRMer, an interactive open source and cross-platform system for data extraction and visualization of multiple reaction monitoring experiments. Mol Cell Proteomics 7: 2270–2278

Chapter 4

Normalization of Reverse Phase Protein Microarray Data: Choosing the Best Normalization Analyte

Antonella Chiechi

Abstract

Reverse phase protein microarray (RPMA) are a relatively recent but widely used approach to measure a large number of proteins, in their original and posttranslational modified forms, in a small clinical sample. Data normalization is fundamental for this technology, to correct for the sample-to-sample variability in the many possible confounding factors: extracellular proteins, red blood cells, different number of cells in the sample. To address this need, we adopted gene microarray algorithms to tailor the RPMA processing and analysis to the specific study set. Using geNorm and NormFinder algorithms, we screened seven normalization analytes (ssDNA, glyceraldehyde 3-phosphate dehydrogenase (GAPDH), α/β-tubulin, mitochondrial ribosomal protein L11 (MRPL11), ribosomal protein L13a (RPL13a), β-actin, and total protein) across different sample sets, including cell lines, blood contaminated tissues, and tissues subjected to laser capture microdissection (LCM), to identify the analyte with the lowest variability. Specific normalization analytes were found to be advantageous for different classes of samples, with ssDNA being the optimal analyte to normalize blood contaminated samples.

Key words Reverse phase protein microarray, Data normalization, geNorm, NormFinder, Reverse phase protein microarray analysis suite, Proteomics, Single stranded DNA

1 Introduction

Reverse phase protein microarray is a quantitative, multiplexed array of heterogeneous mixtures of cellular proteins derived from cells, serum, or body fluids [1–3]. Arrays are probed with specific validated antibodies against target proteins and successively scanned for imaging [4]. Image analysis software are used to measure the pixel intensity for each array spot, which is directly proportional to the amount of protein in the spot [3].

Tissue samples can be extremely heterogeneous. Data normalization allows to correct for differences in disease state, organ of origin, cellular and extracellular contribution, and blood contamination [5, 6]. Commonly, RPMA data are normalized by subtracting the nonspecific staining first, and then dividing each spot value

Klaus Jung (ed.), *Statistical Analysis in Proteomics*, Methods in Molecular Biology, vol. 1362,
DOI 10.1007/978-1-4939-3106-4_4, © Springer Science+Business Media New York 2016

by the total protein staining value for the correspondent spot. Following normalization, RPMA data can be compared both between and within data sets.

While total protein staining might be a reliable normalization factor for cell culture samples, it has many limits when used in data sets including tissue samples, blood, and bone marrow aspirates. These samples can have equal total protein, but contain different amounts of target cells mixed with different amounts of extracellular components and contaminants [7, 8].

Based on this need, we recently developed a more versatile approach to RPMA data normalization, which utilizes several analytes deriving from different cellular compartments [8]. We adapted algorithms used for gene microarrays (geNorm and NormFinder) to RPMA normalization analytes in order to identify the best analyte with the greatest reduction in sample-to-sample variability [9, 10]. In this approach, we use seven analytes: total protein, single stranded DNA (ssDNA), β-actin, glyceraldehyde 3-phosphate dehydrogenase (GAPDH), α/β-tubulin, mitochondrial ribosomal protein L11 (MRPL11), and ribosomal protein L13a (RPL13a). Staining for ssDNA requires a few modifications to the array preparation and staining, as will be described in Subheading 3 of this chapter [7, 8].

When tested across several data set types, different protein analytes were identified as best analytes in different types of data sets, with ssDNA being the best normalization factor for blood contaminated samples, as expected, because able to eliminate red blood cell (RBC) contribution (RBCs are devoid of a nucleus). Normalizing RPMA data in a specific data set using different normalization analytes results in often significantly different results. Data normalization performed alternatively using analytes of similar stability produces similar results [2, 3].

This method is a fast and easy way to determine reliable normalization factors for specific data sets, allowing for a more reliable and precise data analysis.

2 Materials

Ultrapure water must be used every time the procedure calls for water. Solutions must be prepared fresh each time, unless indicated differently.

2.1 RPMA Preparation

Use glass backed nitrocellulose array slides (Grace Bio-Labs, Bend, OR, USA) (*see* **Note 1**) as microarray support. Arrays are printed with Aushon 2470 arrayer equipped with 350 μm pins (Aushon Biosystems, Billerica, MA, USA).

Prepare protein lysates from all samples in the data set following routine protocol for western blot sample preparation (*see* **Note 2**). Protein lysates are diluted in sample buffer—45 % T-Per (Pierce, Thermo Fisher Scientific, Rockford, IL, USA), 45 % Novex

Tris-Glycine SDS Sample Buffer (2×) (Invitrogen, Life Technologies, Thermo Fisher Scientific, Waltham, MA, USA), 10 % TCEP Bond Breaker (Pierce, Thermo Fisher Scientific, Rockford, IL, USA)—prior to array printing. (1) When preparing the sample buffer, quickly warm up Novex Tris-Glycine SDS Sample Buffer (2×) on a heating block or in a microwave oven and make sure that the reagent solution is uniform and all the clumps are dissolved before pipetting.

Dessiccant Drierite (W.A. Hammond, Xenia, OH, USA) is used to keep reverse phase protein microarrays dry during storage at −20 °C.

2.2 Staining

Prepare 1× antibody stripping solution as a 1:10 working dilution of 10× ReBlot Plus Mild antibody stripping solution (EMD Millipore, Merck, Darmstadt, Germany), in deionized water. Mix well. Prepare fresh each time. Approximately 30 ml of 1× antibody stripping solution are needed for ten slides.

Use 1× Sterile Phosphate Buffered Saline (PBS) without calcium or magnesium (ATCC, Manassas, VA, USA) where required for washing and solution preparation. You can prepare your own PBS from scratch, if you prefer; however, using sterile ready-to-use PBS improves results consistency.

To prepare the blocking solution, pour 500 ml of 1× PBS without calcium or magnesium (ATCC) into a beaker, add 1 g of I-Block powder (Life Technologies, Thermo Fisher Scientific, Waltham, MA, USA), place on a hot stirring plate and heat (low) and stir until the solution is clear. Allow to cool at room temperature, then add 500 μl of Tween 20 (Sigma-Aldrich, St. Louis, MO, USA) and mix well. Make sure you do not boil or excessively heat the solution. The beaker should feel warm at the touch, but not hot. Avoid leaving the solution on the heating plate for longer than necessary. Cool before use. Blocking solution can be prepared in advance the previous day to allow enough time to cool down. Store at 4 °C for up to 7 days.

An automated slide stainer (Dako, Carpinteria, CA, USA) is used for RPMA immunostaining. Perform immunostaining using the following reagents: Catalyzed Signal Amplification (CSA) System kit, Biotin Blocking System, Antibody Diluent with Background Reducing Components, and TBST Tris Buffered Saline with Tween 20 10× Concentrate (DAKO).

The following western blot validated primary antibodies are used for normalization analytes immunostaining: rabbit anti-MRPL11, rabbit anti-RPL13a, mouse anti-GAPDH, rabbit anti-α/β-tubulin, rabbit anti-β-actin (Cell Signaling Technology, Danvers, MA, USA), rabbit anti-ssDNA (Immuno-Biological Laboratories Co (IBL), Gunma, Japan) (*see* **Note 3**). Goat anti-rabbit IgG H+L (Vector Laboratories, Burlingame, CA, USA) and rabbit anti-mouse IgG (part of the CSA System kit) (DAKO) are used.

SYPRO Ruby Protein Blot Stain (Life Technologies) is used to perform total protein staining. To prepare the total protein fixative solution, mix 7 % acetic acid and 10 % methanol in deionized water.

2.3 Imaging

NovaRay CCD imager (Alpha Innotech, San Leonardo, CA, USA) equipped with a Cy3 filter is used for total protein RPMAs imaging. Any comparable CCD imager equipped with Cy3 filter can be used. For immunostained RPMAs imaging a UMAX 2100XL flatbed scanner is used. Any comparable flatbed scanner can be used.

2.4 Data Analysis

Image Quant (GE Healthcare, Pittsburgh, PA, USA) image analysis software is used for RPMAs image analysis. Any other comparable software can be used for image analysis. Data processing is carried out using Reverse-Phase Protein Microarray Analysis Suite (RAS) (available for free download at http://capmm.gmu.edu/rpma-analysis-suite). geNorm (Biogazelle, Zwijnaarde, Belgium) and/or NormFinder (available for free download at http://moma.dk/normfinder-software) are used for normalization analyte selection.

3 Methods

3.1 RPMA Construction

Dilute protein lysates in sample buffer to a 0.5 mg/ml starting concentration. Heat protein lysates for 5 min at 100 °C. Set up arrayer printing options according to Aushon's instructions, choosing three depositions per spot and two duplicate spots for each sample (so that each sample is printed in triplicate). Prepare a 3-point 2-fold dilution curve for each sample. Load sample dilutions in multiwell plates following Aushon's instructions for array construction (*see* **Note 4**).

When array printing is finished, proceed to DNA fixation immediately.

3.2 DNA Fixation on Nitrocellulose

Set a laboratory oven to 80 °C. When the oven has reached the temperature, place array slides in the oven for 2 h. At the end of the 2 h, extract array slides from the oven, place them on the bench and allow to cool at room temperature. Cover slides to avoid contamination. When cool, array slides can be stored at −20 °C with Drierite. In this case, the staining procedure can be performed on a different day. Array slides can be kept at −20 °C for several months.

3.3 Total Protein Staining and Imaging

Use one array slide for total protein staining. If array slides were stored at −20 °C, allow the slide to thaw at room temperature before proceeding with the staining. Perform the staining in a chemical hood or use a container with hermetic lid to avoid inhaling toxic vapors.

Wash the slide in total protein fixative solution for 15 min with gentle agitation (e.g., on an orbital shaker). After fixation, discard the fixative solution and wash the slide in deionized water for 5 min, 4 times, with gentle agitation. After the forth wash, incubate the slide in SYPRO Ruby Blot Stain solution for 15 min with gentle agitation. During this step, protect array slide from light using a dark container or covering the container with aluminum foil. The slide must be protected from light until the end of the staining. At the end of the 15 min, discard the SYPRO Ruby Blot Stain solution (*see* **Note 5**) and wash the slide in deionized water for 1 min, 3 times, with gentle agitation. Allow the slide to air-dry in the dark.

Proceed to imaging using NovaRay CCD imager with a Cy3 filter. Export the image as a .tiff file.

3.4 Immunostaining and Imaging

Use nine array slides for normalization analyte immunostaining. Place seven of the nine array slides in a shallow container and add 1× Antibody stripping solution to cover the slides. The remaining array slides will be used to stain for ssDNA and must not be incubated in antibody stripping solution (*see* **Note 6**). Incubate in the solution for 15 min with gentle agitation. Do not exceed 15 min incubation.

Discard antibody stripping solution and rinse briefly in 1× PBS. Clearly mark the remaining two slides for ssDNA staining so that they are easily identifiable and add them into the container with all other slides. Wash all slides in 1× PBS twice for 5 min with gentle agitation. Successively, incubate array slides with blocking solution for at least 1 h and up to 4 h at room temperature, or overnight at 4 °C, with gentle agitation. Do not exceed 4 h incubation in blocking solution at room temperature. If you do not have time to perform the staining the same day, you can choose to block the slides overnight and proceed to immunostaining the next day. Once blocked, array slides must be stained without further storage.

Use the seven array slides treated with 1× antibody stripping solution for protein staining: use one slide for each anti-protein antibody and two slides as negative control, one to be probed with anti-rabbit secondary antibody and one with anti-mouse secondary antibody. Use array slides that were not treated with 1× antibody stripping solution for ssDNA staining: use one slide for anti-ssDNA antibody and one slide as negative control, to be probed with anti-rabbit secondary antibody. Negative control slides will be probed with antibody diluent only in place of primary antibody and will be regularly probed with secondary antibody (*see* **Note 7**).

Program DAKO autostainer following manufacturer instructions for default staining program (*see* **Note 8**). Add an incubation step with avidin followed by an incubation step with biotin, each of 10 min, to the default program. Insert these two steps right after the peroxidase incubation step.

Prepare all reagents following CSA kit manufacturer instructions. Dilute primary antibodies in Antibody Diluent as follows: anti-β-actin 1:500, anti-α/β-tubulin 1:500, anti-MRPL11 1:2000, anti-RPL13a 1:1000, anti-GAPDH 1:10000, and anti-ssDNA 1:15000. Dilute secondary antibodies in Antibody Diluent as follows: anti-rabbit 1:7500; anti-mouse 1:10. When all reagents are ready, load array slides into the autostainer and immediately proceed to immunostaining. It is important to not allow slides to dry before immunostaining.

When immunostaining is completed, let array slides air-dry covered and proceed to imaging. Scan the slides on the flatbed scanner using the following settings: white balance 255, black 0, middle tone1.37, 600dpi, 14 bit. Save images as .tiff files.

4 Data Analysis

4.1 Image Analysis

Use ImageQuant to measure staining intensity in pixels (*see* **Note 9**). The software inverts all images prior to analysis, assuming we are using fluorescence or chemiluminescence. For this reason, images of colorimetric arrays or blots need to be inverted before being utilized in ImageQuant. Invert all images in Photoshop, so that a positive image (black spots on white background) becomes a negative one (white spots on black background) (Fig. 1). This way, when you open files in ImageQuant you will see a positive image, with black spots on white background. Do not invert total protein slide image: it already has white spots on black background.

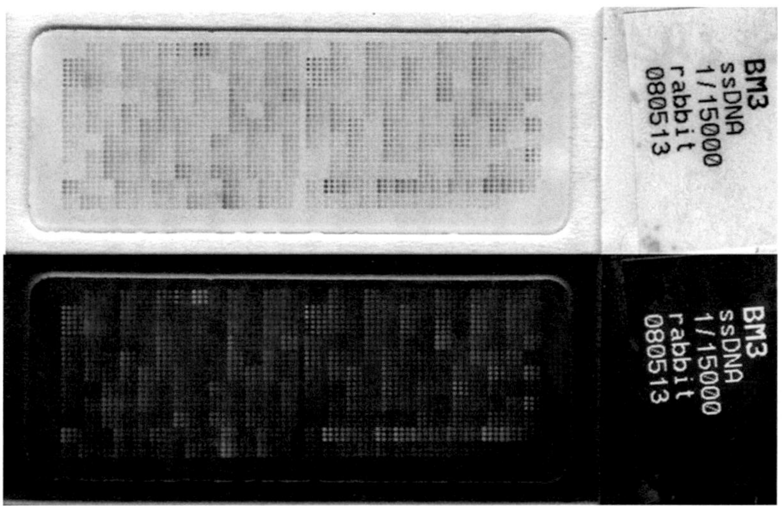

Fig. 1 Example of reverse phase protein microarray slide. Original scanned image at the *top* and inverted image at the *bottom*

Following ImageQuant user's instructions, analyze all images. Draw boxes around each spot on the array. A convenient way to do this is to draw a 3×3 grid that includes all three dilutions and all three replicates for the first sample. Copy and paste the grid onto all remaining samples on the array. This assures that boxes are all the same size, reducing possible variability in this step. Make the boxes/grids big enough to contain the spots, paying attention that the borders do not touch the spot. Set the box perimeter as local background. Check the local background box in the output setting window. ImageQuant will produce a table listing the measured staining intensity and background for each spot in pixels. After analyzing the first image, copy and paste all grids from one slide image to the next one and adjust their position if necessary. It is not possible to copy and paste grids from and onto the total protein slide image because of different resolution of the image.

After analyzing each image, check for artifacts, saturation, dust, and spots that touch the box borders and take note of them, so that they can be excluded in following steps of data analysis.

4.2 Data Reduction

Reverse phase protein microarray data processing can be intricate and time consuming when analyzing a large number of samples and endpoints. Reverse-Phase Protein Microarray Analysis Suite (RAS) excel macro allows us to input intensity values measured with image analysis software and obtain processed and normalized data, eliminating operator related errors. RAS will process data and display every step of data reduction in a separate sheet in the excel spreadsheet, so you will still be able to check and use data from each step of the process.

Before proceeding to data reduction, order your raw data so that replicate spot values come one after the other in the output table produced by ImageQuant. If you performed image analysis drawing 3×3 grids, your raw data will be already in the correct order in the output table.

Open RAS macro and set up the data reduction parameters in the setup sheet. Indicate number of endpoints, including negative controls and normalizers: this number will depend on the number of target proteins you are analyzing in your study. This will create four new sheets: "Enter_Raw_Data", "Enter_Flags", "Enter_Neg", and "Enter_Norm". In "Enter_Raw_Data", input spot intensity and background values for each spot for each endpoint, including negative controls and normalizers, as obtained from ImageQuant. Label each endpoint column, grid ID, sample ID, and Dilution; these labels will be automatically copied in all sheets that will be created by RAS during data reduction. In "Enter_Flags", flag all those spots that must be excluded from the analysis due to artifacts, dust, saturation, or box overlapping, as noted during image analysis. In "Enter_Neg", match the correct negative control (in columns) with each endpoint (in rows) checking the correspondent cell. The sheet "Enter_Norm" is where we choose the normalizer(s) for the analysis. No normalizer must be chosen at this stage and this sheet must be left empty.

Go back to the "Setup" sheet to set all remaining parameters. Indicate number of replicates, which is 3 in our case. Set background cutoff to 2: this means that any intensity value lower than 2 times the background standard deviation will be set to 0 (value not above background). Set multiplier to 100: this will multiply your final normalized data values by 100, making your data more manageable. Check the "Show Intermediate Results" box: this will create a new sheet for each step of data reduction and allow you to access and use any partially processed data at any intermediate step. Set low value threshold to 0. Set CV cutoff above low value threshold to 20: this means that data values with a replicate CV higher than 20 will be excluded from the analysis. The CV cutoff below low value threshold is irrelevant, since the low value threshold is set to 0. Check the "Filter Negative Control by CV" box and set CV cutoff for negative control to 20: with this setting, negative control data values with a replicate CV higher than 20 will be excluded from the analysis.

Click the "Normalize" tab. RAS will start processing the data and will produce six new sheets, one for each data reduction step. Data reduction steps are: (1) remove flagged spots from analysis; (2) remove negative values and replace them with 0; (3) calculate replicate mean, SD and CV for all data point; (4) filter data by background cutoff and CV cutoff; (5) subtract negative control value; (6) normalize data. You can perform data reduction steps manually, if you prefer. However, RAS simplifies RPMA data processing, reducing hands-on time and operator errors. Once RAS has finished reducing your data, save your analysis file. Make sure you save the file as an Excel Macro file, so that you can go back and complete your data analysis after selection of the best normalizer.

4.3 Normalizer(s) Selection

After first data reduction, retrieve data from the "Calc_SubNegCtrl" sheet in RAS. These are completely processed (filtered and negative subtracted), but not normalized data. Copy normalization analytes intensity data for all samples in the data set and paste them in a new Excel spreadsheet. Choose one dilution point (the same one for all samples), the one with the best intensity without saturation across the sample set, and use only data values correspondent to that dilution point for following analysis. We are now ready to process this group of data through the geNorm or NormFinder algorithm. Both geNorm and NormFinder are macros for Microsoft Excel and can be downloaded and run on different systems.

Organize data as requested by the specific macro you are using. For geNorm you will have to create an Excel spreadsheet with samples listed in rows and normalization analytes listed in columns.

This input file must be saved in the geNorm InputData directory. To analyze the data, open geNorm, import the input data file and start the analysis. For NormFinder, create an Excel spreadsheet with samples listed in columns and normalization analytes in rows. If your samples belong to different groups, for example normal and tumor, you should indicate the group in the bottom row with an integer (e.g., normal = 1 and tumor = 2) (*see* **Note 10**). When your list is complete, enable NormFinder checking the correspondent box in the Excel Addin list and start the analysis.

geNorm and NormFinder will calculate the analyte variability and rank normalizers based on their expression stability across the data set. geNorm will produce a more visual output, with tables and graphs showing stability rank and pairwise variation (Fig. 2). NormFinder will provide stability ranking both intergroup and intragroup for each single analyte and for pair of analytes, in the table format. Both macros will calculate the geometric mean of the most stable normalizers of your choice. After screening normalization analytes for stability, you can choose to use one single normalizer, the most stable, or a group of normalizers. If you are using geNorm, the pairwise variation graph will show you what is the best number of analytes to use for normalization in your dataset: the combination with the lowest pairwise variation is the best combination to use. If you are using NormFinder you will use the analyte with intergroup variation as close to 0 as possible and a small intra-group variation (this will be the first ranked analyte for stability in NormFinder). NormFinder will also produce a stability rank table for pairs of normalization analytes.

	a/b-Tubulin	B-Actin	L13a	MRPL11	ssDNA	Total Protein	Normalisation Factor
A	5.98E-01	5.62E-01	5.17E-01	6.66E-01	5.25E-01	2.66E-01	0.8046
B	6.41E-01	5.53E-01	5.81E-01	6.94E-01	5.23E-01	2.69E-01	0.8344
C	7.13E-01	6.18E-01	5.21E-01	6.45E-01	5.46E-01	2.96E-01	0.8591
D	8.96E-01	7.82E-01	7.91E-01	8.66E-01	9.75E-01	4.64E-01	1.2404
E	6.58E-01	6.98E-01	7.41E-01	1.00E+00	6.81E-01	3.89E-01	1.0712
F	5.10E-01	6.68E-01	5.28E-01	6.01E-01	4.51E-01	2.12E-01	0.7465
G	7.86E-01	7.28E-01	8.67E-01	8.68E-01	9.58E-01	6.39E-01	1.2814
H	6.21E-01	6.81E-01	6.08E-01	6.73E-01	6.11E-01	2.28E-01	0.8596
I	1.00E+00	1.00E+00	1.00E+00	9.53E-01	1.00E+00	1.00E+00	1.5870
	0.699577979	0.688390393	0.665270164	0.761729469	0.66753196	0.366066887	
M < 1.5	**0.264**	**0.286**	**0.228**	**0.282**	**0.263**	**0.485**	

Fig. 2 geNorm analysis output: stability ranking output screen. Samples are listed in *rows* from A through I, normalization analytes are listed in *columns*. Values in *bold black* in the *bottom row* are a measure of variability as it is calculated by geNorm algorithm. Value corresponding to the least variable (or most stable) analyte is highlighted in *green*, the most variable in *red*. Normalization factor is the geometric mean of the analytes for each sample

1.5	L13a	MRPL11	Normalisation Factor
A	1.99E+04	3.88E+04	0.8245
B	2.24E+04	4.05E+04	0.8927
C	2.00E+04	3.76E+04	0.8142
D	3.05E+04	5.05E+04	1.1627
E	2.85E+04	5.83E+04	1.2091
F	2.03E+04	3.50E+04	0.7907
G	3.34E+04	5.06E+04	1.2189
H	2.34E+04	3.92E+04	0.8982
I	3.85E+04	5.55E+04	1.3712
	25606.47037	44408.57416	
M < 1.5	0.164	0.164	

Fig. 3 geNorm analysis output: output screen showing the chosen normalization analytes to use for data processing and their geometric mean. Samples are listed in *rows* from A through I, normalization analytes are listed in *columns*. Normalization factor is the geometric mean of the analytes for each sample

Based on stability ranking, choose the one most stable analyte or a group of stable analytes to use for normalization in your dataset. Using a single analyte has the convenience of not requiring further calculations, although using a group of stable analytes makes data reduction even more robust and reliable. If you choose to use a group of normalizers in your analysis, you will use the geometric mean of the intensity values of the analytes for each spot in your array as normalization factor. To obtain the geometric mean of the group of analytes of your choice, run the algorithms using only the chosen analytes (Fig. 3).

4.4 Data Normalization

We are now ready to proceed to the last step of data reduction: data normalization. Open your RAS file where you saved your analysis previously. In the "Enter_Norm" sheet, select the normalizer you identified as the most stable and check the correspondent cell for all the protein endpoint in your dataset. If you decided to use the geometric mean of a group of analytes, enter geometric mean values in a new column in "Enter_Raw_Data" and label the column as "geometric mean" (or any label that will help you identify it in the next steps). In "Enter_Normalizer" select "geometric mean" as normalizer and check the correspondent cell for all protein endpoints. Go back to the "Setup" sheet and click the "Normalize" tab. RAS will reprocess your data and, this time, also normalize them. Final reduced and normalized data will appear in the "Calc_Norm" sheet. You can now use these data for statistical analysis.

Using the most appropriate normalizer for a specific dataset will improve data reliability and statistical analysis (Fig. 4).

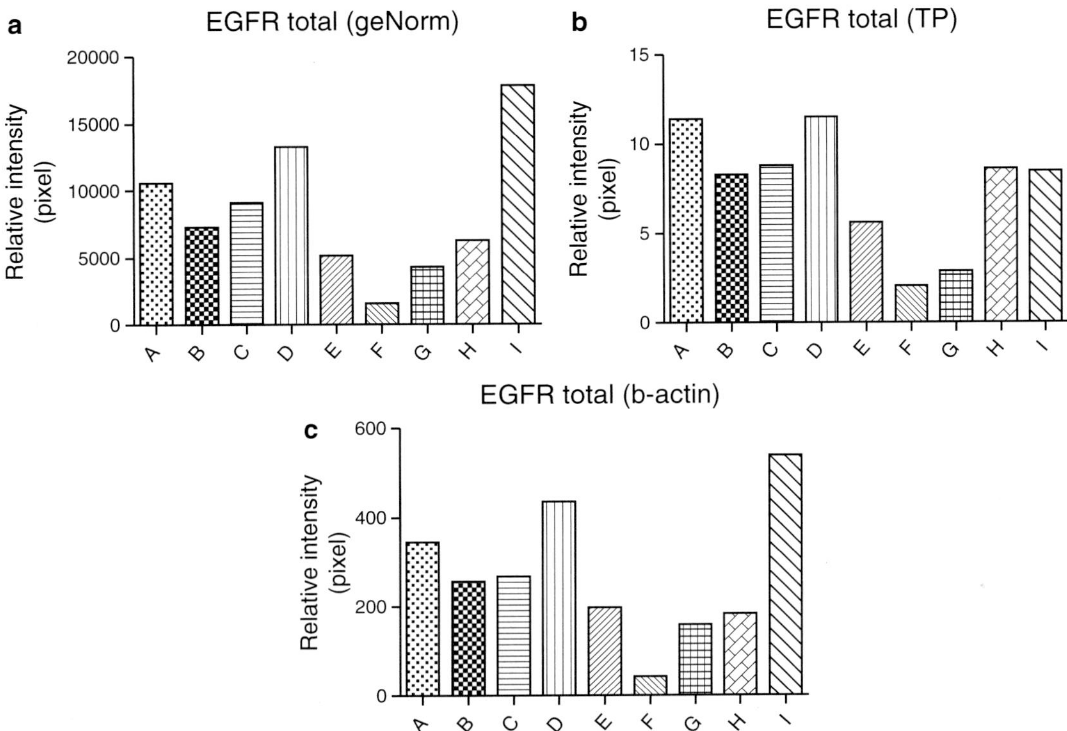

Fig. 4 EGFR protein expression data results in a tissue samples set, using different normalization analytes for data processing: (**a**) data normalized using the geometric mean of the two least variable analytes as calculated by geNorm algorithm; (**b**) data normalized using the most variable analyte, total protein, as for geNorm; (**c**) data normalized using a medium ranking analyte for stability in geNorm, β-actin. The graphs illustrate the necessity of using stable analytes for normalization in order to obtain more accurate results

5 Notes

1. Glass backed nitrocellulose array slides exist in several variations of porosity and protein binding capacity and are marketed by a number of different companies. Choose the array slides that best perform with your samples and staining system (chromogenic or fluorescence).

2. Prepare protein lysates so that they have the same protein concentration. If it is not possible to determine protein concentration, lysates should be prepared starting from a same number of cells, or a same tissue weight or volume, in order to keep the protein concentration as similar as possible. Substitution of β-mercaptoethanol and/or DTT with TCEP (Pierce, Thermo Fisher Scientific, Rockford, IL, USA) as a reducing agent in the extraction/loading buffer is suggested, as TCEP is the agent that produces the lowest background signal during total protein staining detection.

3. You can choose to use different antibodies directed against different analytes that you believe could represent better normalization analytes in your specific data set. Because we are using RPMAs, each primary antibody specificity has to be validated by western blot. For an antibody to be suitable for RPMA, specific bands on a western blot have to account for at least the 80 % of the total signal detected on the blot.
4. Keep in mind that different nitrocellulose backed slides have different binding capacity, so adjust your sample starting concentration to the slides you choose to use. You must also keep into account the number of deposition per spot that you decide to use during printing and the size of pins with which your arrayer is equipped, in case these differ from the protocol in this chapter. Moreover, since different antibodies have different reaction curves, printing a dilution curve of the samples will allow you to choose the best dilution spot for each antibody, which is the one that falls in the linear range of the reaction curve.
5. SYPRO Ruby Blot Stain solution can be reused a few times. If you wish to reuse the reagent, place it in a dark bottle and store it at room temperature in the dark.
6. While the antibody stripping solution improves protein immunostaining (improving protein linearity), it can remove ssDNA bound on nitrocellulose because of strong charge interactions due to its alkaline pH.
7. It is very important that you use the array slides that were not treated with 1× antibody stripping solution for ss-DNA staining and the treated one for protein staining. Using treated slides for ss-DNA staining could produce negative results because ss-DNA was washed off by 1× antibody stripping solution. Using untreated slides for protein staining will result in lighter staining intensity and non optimal results.
8. Immunostaining can be performed by hand instead than using an autostainer. If you wish to do so, proceed to the immunostaining as you would do for immunohistochemistry staining. Using an autostainer reduces hands-on time and overall procedure time, and improves consistency of results.
9. If you prefer, you can use alternative software for data analysis. One widely used alternative software is MicroVigene (Vigene Tech, Carlisle, MA, USA) in its specific version for reverse phase protein microarray.
10. NormFinder will calculate variability between and within groups, when groups are indicated. This allows for a more accurate evaluation of normalizer stability. However, NormFinder and geNorm analyte stability evaluation generally produces comparable results.

References

1. Paweletz CP, Charboneau L, Bichsel VE et al (2001) Reverse phase protein microarrays which capture disease progression show activation of pro-survival pathways at the cancer invasion front. Oncogene 20:1981–1989
2. Liotta LA, Espina V, Mehta AI et al (2003) Protein microarrays: meeting analytical challenges for clinical applications. Cancer Cell 3:317–325
3. Mueller C, Liotta LA, Espina V (2010) Reverse phase protein microarrays advance to use in clinical trials. Mol Oncol 4:461–481
4. Espina V, Wulfkuhle JD, Calvert VS et al (2007) Reverse phase protein microarrays for monitoring biological responses. Methods Mol Biol 383:321–336
5. VanMeter AJ, Rodriguez AS, Bowman ED et al (2008) Laser capture microdissection and protein microarray analysis of human non-small cell lung cancer: differential epidermal growth factor receptor (EGPR) phosphorylation events associated with mutated EGFR compared with wild type. Mol Cell Proteomics 7:1902–1924
6. Emmert-Buck MR, Bonner RF, Smith PD et al (1996) Laser capture microdissection. Science 274:998–1001
7. Chiechi A, Novello C, Magagnoli G et al (2013) Elevated TNFR1 and serotonin in bone metastasis are correlated with poor survival following bone metastasis diagnosis for both carcinoma and sarcoma primary tumors. Clin Cancer Res 19:2473–2485
8. Chiechi A, Mueller C, Boehm KM et al (2012) Improved data normalization methods for reverse phase protein microarray analysis of complex biological samples. BioTechniques 0:1–7
9. Vandesompele J, De Preter K, Pattyn F et al (2002) Accurate normalization of real-time quantitative RT-PCR data by geometric averaging of multiple internal control genes. Genome Biol 3:RESEARCH0034
10. Andersen CL, Jensen JL, Orntoft TF (2004) Normalization of real-time quantitative reverse transcription-PCR data: a model-based variance estimation approach to identify genes suited for normalization, applied to bladder and colon cancer data sets. Cancer Res 64:5245–5250

Chapter 5

Outlier Detection for Mass Spectrometric Data

HyungJun Cho and Soo-Heang Eo

Abstract

Mass spectrometry data are often generated from various biological or chemical experiments. However, due to technical reasons, outlying observations are often obtained, some of which may be extreme. Identifying the causes of outlying observations is important in the analysis of replicated MS data because elaborate pre-processing is essential in order to obtain successful analyses with reliable results, and because manual outlier detection is a time-consuming pre-processing step. It is natural to measure the variability of observations using standard deviation or interquartile range calculations, and in this work, these criteria for identifying outliers are presented. However, the low replicability and the heterogeneity of variability are often obstacles to outlier detection. Therefore, quantile regression methods for identifying outliers with low replication are also presented. The procedures are illustrated with artificial and real examples, while a software program is introduced to demonstrate how to apply these procedures in the **R** environment system.

Key words Outlier detection, Data preprocessing, Standard deviation, Interquartile range, Quantile regression

1 Introduction

Mass spectrometry (MS) data are often generated from various biological or chemical experiments. Such large amounts of data are usually analyzed automatically in a computing process that consists of pre-processing, significance testing, classification, and clustering. Elaborate pre-processing is essential to obtain successful analyses with reliable results. A key pre-processing step is the detection of outliers, which may have extreme values due to technical reasons [1]. Possible outlying observations need to be examined carefully, and then corrected for or eliminated if necessary. However, as the manual examination of all observations for outliers is time-consuming, possible outliers must be detected automatically.

An outlier is an observation that falls well above or well below the overall bulk of the data [2–4]. A natural approach to detect outliers is to investigate the distribution of the observations and evaluate the outlying degrees of potential outliers. The investigation can

Klaus Jung (ed.), *Statistical Analysis in Proteomics*, Methods in Molecular Biology, vol. 1362,
DOI 10.1007/978-1-4939-3106-4_5, © Springer Science+Business Media New York 2016

be conducted for each peptide because the distributions of observations of peptides may differ substantially. It is natural to measure the variability of observations for each peptide by calculating the standard deviation (SD) or interquartile range (IQR) of each sample [5].

The SD and IQR criteria may produce unreliable outcomes in the case of a few replicates. Furthermore, they are not applicable for duplicated samples. Another, perhaps naive, approach for detecting outliers statistically involves constructing lower and upper fences of differences between two samples for all peptides. A suspected outlier is then an observation whose value is either smaller than the lower fence or greater than the upper fence. However, this may generate a spurious result because variability is heterogeneous in high-throughput data generated even from MS experiments. Naive outlier detection methods such as these ignore the heterogeneity of variability, and may often miss true outliers at high levels and select false outliers at low levels. If a number of technical replicates for each peptide under the same biological condition can be obtained in MS experiments, a search for outliers can be conducted for each peptide. However, only a small number of replicates are usually subjected to MS experiments due to the high cost of experiments and the limited supply of biological samples. Instead, a more elaborate approach for detecting outliers with low false-positive and false-negative rates in MS data is to utilize quantile regression, which is especially useful when the number of technical replicates is small. The outlier detection procedures are illustrated in the next section, with artificial and real datasets in the **R** environment system.

2 Outlier Detection Methods

Suppose that there are n replicated samples and p peptides in an MS dataset. Then let xij be the i-th replicated observation for the j-th peptide from experiments under the same biological or experimental condition, where $i = 1, \ldots, n$ and $j = 1, \ldots, p$ and let $y_{ij} = \log_2\left(x_{ij}\right)$. Typically, n is small and p is very large in high-throughput data, i.e., $p \gg n$. We introduce the standard deviation, interquartile range, and quantile regression approaches for identifying outliers in this section.

2.1 Standard Deviation Criteria

The standard deviation describes the distance between the data and the mean, thus providing a measure of the variability of the data. The standard deviation s is defined as the square root of the sum of squared deviations divided by the sample size minus 1, i.e., $s = \sum_i \left(y_i - \overline{y}\right)^2 / \left(n-1\right)$. The z-score for an observation is the number of standard deviations that it falls away from the mean. A positive z-score indicates the observation is above the mean, while a negative z-score indicates that the observation is below the mean. For sample data, an observation from a bell-shaped distribution is a potential outlier if its z-score < −3 or > +3. The z-score criterion for identifying outliers is summarized below:

1. Compute the standard deviation, sj for each peptide j, and then z-score $z_{ij} = \left(y_{ij} - \overline{y}_j\right) / s_j$, where $\overline{y}_j$ and sj are the sample mean and standard deviation, respectively.
2. For each peptide j, observation yij is flagged as an outlier if $z_{ij} < -k$ or $z_{ij} > k$, where $k = 2$ or 3.
3. This z-score criterion works well when the data follows a bell-shaped, normal distribution. Thus, the thresholds $k = 2$ and 3 indicate that 95 and 99.7 % of the observations fall within 2 and 3 SDs of the mean, respectively.

Grubbs et al. [6] developed a more elaborate procedure, where the threshold is more precise, and outliers are removed recursively. This is the Grubbs' test, and its method for identifying outliers is summarized below:

1. Compute the test statistic $G_{ij} = \max_{i=1,\ldots,n} | y_{ij} - \overline{y}_j | / s_j$, where the sample mean is $\overline{y}_j$ and standard deviation is sj for peptide j.
2. For each peptide j, observation yij is flagged as an outlier if $G_{ij} > c$, where c is the critical value (*see* **Note 1**).
3. Remove the detected outlier, and then repeat steps 1–3 until no further outliers are detected.

If $n = 2$, the statistic is always $1 / \sqrt{n}$; thus, this test is applicable for $n > 2$. Grubbs' test is based on the assumption of normality; therefore, one should first verify that the data could be reasonably approximated by a normal distribution before applying the test. Grubbs' test detects one outlier at a time. This outlier is expunged from the dataset and the test is reiterated until no further outliers are detected. However, multiple iterations change the probabilities of detection, and the test should not be used for sample sizes of six or less since it frequently tags most of the points as outliers [7].

2.2 Interquartile Range Criteria

The p-th percentile is a value such that p percentages of the observations fall at, or below, a certain value. Three useful percentiles are the quartiles. The first quartile Q_1 is the 25th percentile, where the lowest 25 % of the data fall below it. The second quartile Q_2 is the 50th percentile, which is the median. The third quartile Q_3 is the 75th percentile, and the highest 25 % of the data exists above it. The quartiles split the data into four parts, each containing quarter (25 %) of the observations. The interquartile range (IQR) is the distance between the third and first quartiles, i.e., IQR $= Q_3 - Q_1$. An observation is declared an outlier if it is greater than 1.5 IQR below the first quartile or more than 1.5 IQR above the third quartile. Thus, the lower and upper fences for outliers are $Q_1 - 1.5\,\text{IQR}$ and $Q_3 + 1.5\ \text{IQR}$ [8]. This IQR criterion for identifying outliers is summarized as follows:

1. Compute the first and third quartiles, Q_1j and Q_3j, for each peptide j, and then its IQR: $\text{IQR}_j = Q_{3j} - Q_{1j}$.
2. For each peptide j, observation yij is flagged as an outlier if $y_{ij} < Q_{1j} - k\ \text{IQR}_j$ or $y_{ij} > Q_{3j} + k\ \text{IQR}_j$, where $k = 1.5$ or 3.

In this IQR criterion, a coefficient k determines the strictness of capturing outlying observations. Values of $k = 1.5$ or 3 are often used. A larger value of k selects outlying observations more conservatively.

The distribution of observations may not be symmetric about the median, but instead may be skewed to the left or the right, implying that the middle of the first and third quartiles is in fact not the median. Thus, the distance from the first quartile to the median is significantly different of that from the third quartile to the median. In this situation, IQR can be too large for one side and too small for the other.

As an alternative, the semi-interquartile range (SIQR) can be more effective. That is, the left and right SIQRs are used rather than IQR. This SIQR criterion for identifying outliers is summarized as follows:

1. Compute the first, second, and third quartiles, Q_1j, Q_2j, and Q_3j, for each peptide j, and then its SIQR: $\text{SIQR}j^L = Q_2j - Q_1j$ and $\text{SIQR}j^U = Q_3j - Q_2j$.
2. For each peptide j, observation yij is flagged as an outlier if $y_{ij} < Q_{1j} - 2k\ \text{SIQR}_j^L$ or $y_{ij} > Q_{3j} + 2k\ \text{SIQR}_j^U$, where k=1.5 or 3.

2.3 Quantile Regression Approaches

The above IQR and SD criteria require for the data to follow a normal distribution, and for the sample sizes to be large enough (*see* **Note 2**). However, the assumptions may not be satisfied for some MS analyses, and in particular, the sample size is often small (*see* **Note 3**).

In duplicated experiments ($n = 2$), two observed values for each peptide should be theoretically identical, but are not identical in practice due to their variability; however, they should not differ substantially. The tolerance of the difference between the two observed values from the same condition is not constant because their variability is heterogeneous. The variability of high-throughput data depends on the intensity levels.

Lower and upper fences can be constructed for detecting outliers using quantile regression in an M–A plot with M and A values in vertical and horizontal axes, respectively, where Mj is the difference between replicated samples for j and Aj is the average, i.e., $M_j = y_{1j} - y_{2j} = \log_2\left(x_{1j} / x_{2j}\right)$ and

$A_j = (y_{1j} + y_{2j})/2 = (1/2)\log_2(x_{1j}x_{2j})$ to detect the outliers accounting for the heterogeneity of variability [9]. By applying the regression, we compute the 0.25 and 0.75 quantile estimates, $Q_1(A)$ and $Q_3(A)$, of the differences, M, depending on the levels, A. Then we construct the lower and upper fences: $Q_1(A) - 1.5\ \mathrm{IQR}(A)$ and $Q_3(A) + 1.5\ \mathrm{IQR}(A)$, where $\mathrm{IQR}(A) = Q_3(A) - Q_1(A)$. To obtain quantile estimates that depend on the levels more flexibly, nonlinear or nonparametric quantile regression can be utilized [10]. This quantile regression approach [1], called the *OutlierD* algorithm, is summarized as follows:

1. Generate an M–A plot with M and A values in vertical and horizontal axes, respectively, where Mj is the difference between replicated samples for j and Aj is the average.
2. Apply linear, nonlinear, or nonparametric regression and then compute the 0.25 and 0.75 quantile estimates, $Q_1(A)$ and $Q_3(A)$, of the differences, M, depending on the levels, A.
3. Construct the lower and upper fences: $Q_1(A) - k\ \mathrm{IQR}(A)$ and $Q_3(A) + k\ \mathrm{IQR}(A)$, where $\mathrm{IQR}(A) = Q_3(A) - Q_1(A)$ and $k = 1.5$ or 3.
4. Peptide j is claimed as containing an outlying observation if $M_j < Q_1(A_j) - k\ \mathrm{IQR}(A_j)$ or $M_j > Q_3(A_j) + k\ \mathrm{IQR}(A_j)$, where $k = 1.5$ or 3.

A larger value of k selects outliers more conservatively. In this approach, one of the two samples is outlying, but which one is not known.

In multiple experiments $(n \geq 2)$, it is natural to search for outliers based on all observed values in a high-dimensional space. An outlier will be at a very large distance from the center of the distribution of a peptide. The cutoffs of distances for classification of outliers depend on the degree of variability from the center. The degree of variability is dependent on intensity levels, and the center can be defined as a 45° line from the origin. More flexibly, the center can be obtained by principal component analysis (PCA) [11]. The first principal component (PC) becomes the center of each intensity level, i.e., a new axis for intensity levels. The experiments are replicated under the same biological and technical condition; hence, the first PC can explain most variations. It implies that it is enough to use the first PC practically. An outlier will be at a large distance from its projection. Following the notations for applying quantile regression, we can define the distance of peptide j to the projection as Mj and the length of the projection on the new axis as Aj. Then the first and third quantiles can be obtained by applying quantile regression on an M–A plot with M and A on the vertical and horizontal axes, respectively. The quantile regression algorithm that uses this projection [7] is called the *OutlierDM* algorithm, and is summarized as follows:

1. Shift the sample means to the origin (0,…,0), i.e., $y_{ij}^{*} = y_{ij} - \overline{y}_{i}$.
2. Find the first PC vector v using principal component analysis (PCA) on the space of $y_1^*, \ldots, y_n^*$.
3. Obtain the projection of a vector $y_j^* = (y_{1j}^*, \ldots, y_{nj}^*)$ of each peptide j on v, where $j = 1, \ldots, p$.
4. Compute the signed length, Aj, of the projection and the length, Mj of the difference between a vector of peptide j and the projection, where $j = 1, \ldots, p$.
5. Obtain the first and third quantile values $Q_1(A)$ and $Q_3(A)$, on an M–A plot using a quantile regression approach. Then calculate $\text{IQR}(A) = Q_3(A) - Q_1(A)$.
6. Construct the lower and upper fences, $\text{LB}(A) = Q_1(A) - k\ \text{IQR}(A)$ and $\text{UB}(A) = Q_3(A) + k\ \text{IQR}(A)$, where $k = 1.5$ or 3.
7. Peptide j is claimed as containing one or more outlying observations if it is located above the upper fence or under the lower fence.

This projection quantile regression approach utilizes all of the multiple replicates simultaneously, and a high-dimensional problem reduces to two-dimensional one that can easily be solved. Note that the quantile regression approaches only determines whether each peptide contains one or more outliers, but not which observation is an outlier. A visual approach (*see* **Note 4**) is useful to identify which observation(s) of the selected peptide is (are) outlying, is illustrated in the next section.

3 Illustrations

In this section, we illustrate how to detect outliers in two cases with artificial and real examples by using an analysis written in **R** package *OutlierDM* [12] (*see* **Note 5**). The first case is illustrated with an artificial dataset to detect outlying samples for each peptide, while the second case uses a real dataset to detect the peptides containing at least one outlying observation when the number of replicates is small.

3.1 When the Number of Replicates Is Sufficiently Large

The primary purpose of outlier detection in MS data is to determine which observations for each peptide are outlying. If the number of replicates is large enough (*see* **Note 2**), one of the SD and IQR criteria can seek out the outliers within each peptide. For illustration, an artificial data set with 200 peptides and 15 samples is generated [7]. This dataset (called "toy") contains ten peptides,

```
 Head of the Output Results
  Outlier G1 G2 G3 G4     G5    G6 G7 G8    G9 G10 G11 G12 G13    G14    G15
1    TRUE  .  .  .  .  3.491     .  .  .     .   .   .   .   .      .      .
2    TRUE  .  .  .  . -3.581     .  .  .     .   .   .   .   .      .      .
3    TRUE  .  .  .  .      .     .  .  .     .   .   .   .   .  2.434 -3.505
4    TRUE  .  .  .  .      .     .  .  .     .   .   .   .   .  3.477      .
5    TRUE  .  .  .  .      . 3.499  .  .     .   .   .   .   .      .      .
6    TRUE  .  .  .  .      .     .  .  . 2.378   .   .   .   .      . -3.582
To see the full information for the result, use a command, 'output(your_object_name)'.
```

Fig. 1 Outlier detection using the Grubbs' test for the first six peptides of the toy dataset; the test statistics are given for the detected outliers and the *dots* are given for non-outliers

and each of them has one outlying observation. This toy dataset can be called within the **R** package *OutlierDM* by using the following commands:

```
> library(OutlierDM)
> data(toy)
```

To detect outlying observations using the Grubbs' test with significance level 0.01, the function *odm()* of *OutlierDM* can be called as follows:

```
> fit = odm(x = toy, method = "grubbs", alpha = 0.01)
> fit
```

These **R** commands create an object *fit* using the three input arguments, dataset used (**x** = toy), outlier detection method (**method** = "grubbs"), and significance level (**alpha** = 0.01), and then display a table consisting of dots (test statistics for the detected outliers) for the first six peptides as an output (Fig. 1).

In the output, the first column is the row number and the second column indicates whether each peptide contains one or more outlying observations, shown as TRUE. Columns G1–G15 give the test statistics for the detected outliers, while the dots for non-outliers. To see all the peptides, the function *output(fit)* can be conducted in the **R** environment. In this example, 12 peptides were flagged as containing one or more outlying observations, two of which were flagged falsely. In the first six peptides shown, peptide 3 found two outlying observations, but one of them was flagged falsely. The other five peptides detected all the outlying samples correctly. The detected outlier for each peptide can be shown graphically by the function *oneplot()*:

```
> oneplot(fit, i = 1)
```

The object *fit* was generated from the function *odm()* and index "i" indicates the row number corresponding to a peptide. Figure 2 shows the dot plot of log2-transformed data points with one outlier (marked by an asterisk) detected by the Grubbs' test with significance level 0.01 (*see* **Note 4**).

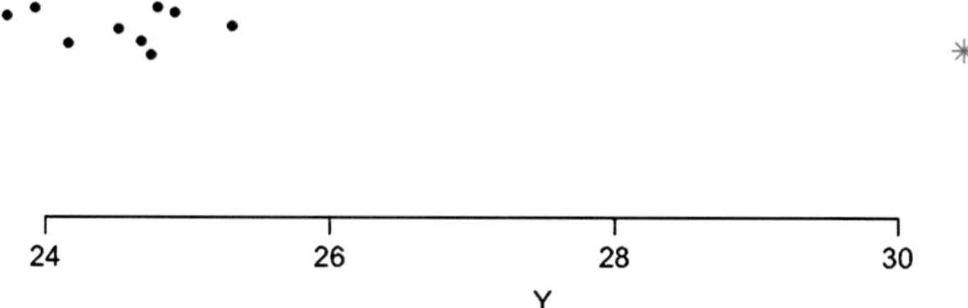

Fig. 2 Outlier detection using the Grubbs' test for the first peptide of the toy dataset; the outlier is indicated as an *asterisk*

3.2 When the Number of Replicates Is Small

We would like to know which observations for each peptide are outlying, but for cases where the number of replicates is small (*see* **Note 3**). In these events, a quantile regression approach can be utilized to detect the peptides having at least one outlying observation. For illustration, we consider a real-life dataset obtained from three replicated LC/MS/MS experiments with 922 peptides ($n = 3$ and $p = 922$). The details regarding the experiment can be found in refs. 1 and 7. This dataset can be called up by the following command:

```
> data(lcms3)
```

We first illustrate how to detect outliers under the duplicated experiment ($n = 2$). For instance, consider the first two replicates of the "lcms3" dataset and apply the *OutlierD* algorithm to the duplicated data set:

```
> fit2 = odm(x = lcms3[,1:2], method = "pair", k = 3)
> outliers(fit2)
> plot(fit2)
```

The argument method = "pair" is for the *OutlierD* algorithm and $k = 3$ is a threshold (i.e., a coefficient) used within IQR. Using the function *outliers(fit2)* generates the output shown in Fig. 3. In this output, the first column indicates the row numbers of the peptides containing an outlier observation. The next columns consist of log2-transformed values (N_1 and N_2), A and M values, the first and third quartiles (Q_1 and Q_3), and lower and upper bounds (LB and UB), respectively. Figure 4 shows the M–A plot from the object *fit2* and the superimposed lines separate outlying peptides from normally observed peptides.

Next, we use all the three replicates simultaneously to detect outliers ($n = 3$). The number of replicates is still small, so the SD and IQR criteria are not applicable. In this case, the *OutlierDM* algorithm is applied to the lcms3 dataset:

```
> fit3 = odm(lcms3, method = "proj", k = 3)
> outliers(fit3)
> plot(fit3)
```

```
    Outlier   N1   N2    A      M      Q1     Q3     LB    UB
66     TRUE 32.3 32.8 32.6 -0.600 -0.0317 0.0317 -0.222 0.222
94     TRUE 23.7 18.2 21.0  5.220 -0.6227 0.6227 -4.359 4.359
145    TRUE 23.9 19.1 21.5  4.635 -0.5960 0.5960 -4.172 4.172
236    TRUE 24.9 19.8 22.4  4.878 -0.5505 0.5505 -3.854 3.854
319    TRUE 15.0 22.2 18.6 -7.473 -0.7407 0.7407 -5.185 5.185
324    TRUE 26.9 21.7 24.3  4.946 -0.4531 0.4531 -3.172 3.172
413    TRUE 19.7 24.2 21.9 -4.742 -0.5740 0.5740 -4.018 4.018
448    TRUE 31.7 32.1 31.9 -0.542 -0.0679 0.0679 -0.475 0.475
458    TRUE 25.3 28.6 26.9 -3.522 -0.3186 0.3186 -2.230 2.230
460    TRUE 28.7 17.5 23.1 10.955 -0.5122 0.5122 -3.586 3.586
541    TRUE 25.9 20.4 23.2  5.319 -0.5111 0.5111 -3.578 3.578
661    TRUE 15.5 22.7 19.1 -7.436 -0.7188 0.7188 -5.032 5.032
751    TRUE 18.0 23.5 20.7 -5.773 -0.6343 0.6343 -4.440 4.440
782    TRUE 28.1 24.9 26.5  3.029 -0.3419 0.3419 -2.393 2.393
796    TRUE 25.3 17.0 21.1  8.116 -0.6134 0.6134 -4.294 4.294
906    TRUE 30.6 20.2 25.4 10.297 -0.3975 0.3975 -2.783 2.783
```

Fig. 3 A list of the outliers detected by the *OutlierD* algorithm for the lcms3 dataset

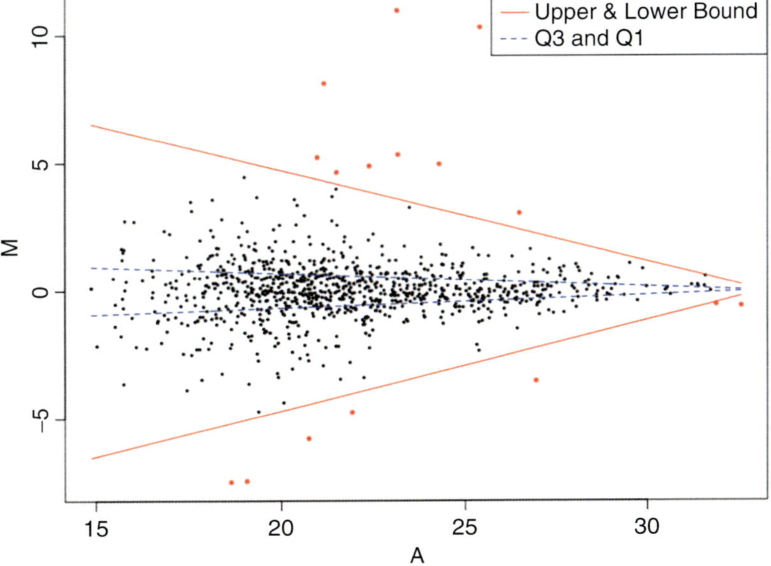

Fig. 4 Outlier detection using the *OutlierD* algorithm in a linear quantile regression analysis for the first two replicates of the lcms3 dataset; the outliers are shown as *red asterisks*

The argument method = "proj" is for *OutlierDM* and $k = 3$ is again a threshold used by IQR. Using the function *outliers(fit3)* generates the output Fig. 5. In this output, the first column indicates the row numbers of the peptides containing an outlier observation. The next columns consist of log2-transformed values (N_1, N_2, and N_3), A and M values, the first and third quartiles (Q_1 and Q_3), and lower and upper bounds (LB and UB), respectively. Figure 6 shows the M–A plot from the object *fit3* and the superimposed lines separate outlying peptides from normally observed peptides.

	Outlier	N1	N2	N3	A	M	Q1	Q3	LB	UB
18	TRUE	-3.111	-1.6157	8.619	2.2484	9.029	0.3055	0.9273	-1.56008	2.7928
50	TRUE	8.614	8.6603	1.027	10.5631	6.219	0.1588	0.4635	-0.75551	1.3778
66	TRUE	10.293	11.0763	10.426	18.3552	0.647	0.0213	0.0289	-0.00151	0.0518
94	TRUE	1.658	-3.5350	1.733	-0.0713	4.271	0.3464	1.0567	-1.78455	3.1876
120	TRUE	-2.042	1.4419	2.160	0.8942	3.180	0.3293	1.0028	-1.69113	3.0233
145	TRUE	1.887	-2.7151	1.595	0.4533	3.643	0.3371	1.0274	-1.73379	3.0983
211	TRUE	7.261	7.1830	5.221	11.3525	1.639	0.1449	0.4195	-0.67913	1.2435
236	TRUE	2.896	-1.9359	2.815	2.1906	3.906	0.3065	0.9305	-1.56568	2.8027
319	TRUE	-6.994	0.4746	-2.498	-5.2217	5.302	0.4372	1.3439	-2.28293	4.0641
324	TRUE	4.833	-0.0407	4.944	5.6329	4.010	0.2458	0.7385	-1.23259	2.2169
380	TRUE	3.174	-0.0174	3.841	4.0475	2.906	0.2737	0.8270	-1.38600	2.4867
413	TRUE	-2.371	2.4102	-0.352	-0.1910	3.394	0.3485	1.0634	-1.79614	3.2080
440	TRUE	1.556	-2.3869	1.491	0.3905	3.192	0.3382	1.0309	-1.73986	3.1090
448	TRUE	9.614	10.3301	9.703	17.1155	0.602	0.0432	0.0981	-0.12147	0.2627
458	TRUE	3.227	6.8557	3.069	7.5848	3.050	0.2113	0.6297	-1.04371	1.8847
460	TRUE	6.683	-4.2165	5.870	4.8380	8.573	0.2598	0.7829	-1.30951	2.3522
470	TRUE	-1.971	-5.5501	0.641	-3.9630	4.404	0.4150	1.2737	-2.16114	3.8499
477	TRUE	-4.025	0.3565	0.231	-1.9931	3.523	0.3803	1.1639	-1.97051	3.5147
541	TRUE	3.887	-1.3753	0.346	1.6610	3.790	0.3158	0.9601	-1.61693	2.8928
661	TRUE	-6.547	0.8896	1.591	-2.3615	6.373	0.3868	1.1844	-2.00617	3.5774
747	TRUE	-0.665	-5.1162	-0.190	-3.4368	3.853	0.4057	1.2444	-2.11022	3.7604
751	TRUE	-4.066	1.7310	-2.152	-2.6033	4.169	0.3910	1.1979	-2.02956	3.6185
782	TRUE	6.047	3.1195	5.993	8.7592	2.344	0.1906	0.5642	-0.93007	1.6848
796	TRUE	3.288	-4.7988	3.450	1.1383	6.667	0.3250	0.9892	-1.66750	2.9818
844	TRUE	4.788	5.3965	-4.599	3.2194	7.927	0.2883	0.8731	-1.46613	2.6276
906	TRUE	8.595	-1.6157	-3.343	2.1187	9.120	0.3077	0.9345	-1.57263	2.8149

Fig. 5 A list of the outliers detected by the *OutlierDM* algorithm for the lcms3 dataset

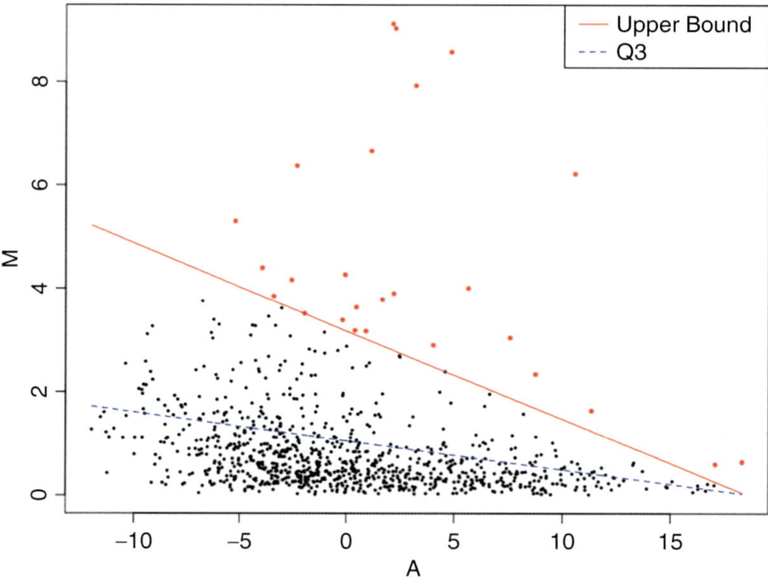

Fig. 6 Outlier detection using the *OutlierDM* algorithm in a linear quantile regression analysis on the lcms3 dataset; the outliers are shown as *red asterisks*

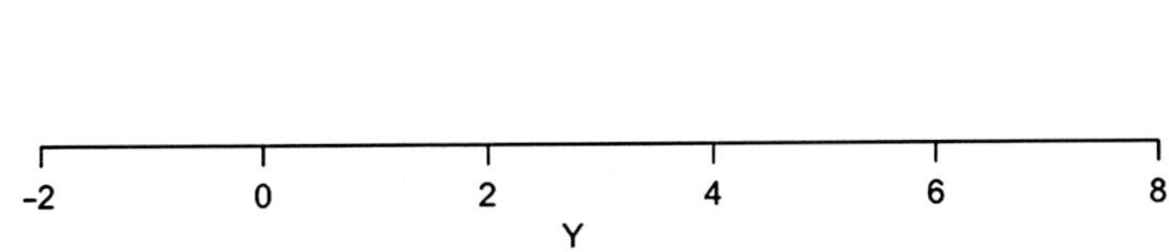

Fig. 7 A dot plot for the 18th peptide of the lcms3 dataset

After detecting the outlying peptides, their raw data points can be plotted to see which observations are furthest from the others using (*see* **Note 4**):

```
> oneplot(fit3, i = 18)
```

This generates the dot plot of the log2-transformed values for the 18th peptide, as shown in Fig. 7. It is seen that one observation is far from the other two for the 18th peptide.

4 Notes

1. In Grubbs' test, the critical value *c* is $\frac{n-1}{\sqrt{n}}\sqrt{t^2_{\frac{\alpha}{2n},n-2} \Big/ \left(n-2+t^2_{\frac{\alpha}{2n},n-2}\right)}$, where $t^2_{\frac{\alpha}{2n},n-2}$ is a *t*-distribution with a degree of freedom $n-2$ and significance level $\alpha/2n$.
2. The standard deviation (SD) and IQR criteria are used to detect outliers for each peptide. These require a sample size greater than six: $n > 6$.
3. The quantile regression approaches are used to detect peptides containing one or more outliers when a sample size is small, usually, $n \leq 6$. They also work for a sample size of two ($n = 2$).
4. After detecting peptides containing one or more outliers using a quantile regression approach, a visual analysis such as a dot plot can be used to reveal which observations are outlying for a selected peptide.
5. A software program *OutlierDM* [7] based in the **R** environment system is available at http://www.r-project.org/package=OutlierDM for conducting outlier detection.

Acknowledgement

This research was supported by the Basic Science Research Program through the National Research Foundation of Korea (NRF) funded by the Ministry of Education, Science and Technology (2010-0007936).

References

1. Cho H, Lee JW, Kim Y-J et al (2008) OutlierD: an R package for outlier detection using quantile regression on mass spectrometry data. Bioinformatics 24:882–884
2. Su X, Tsai C-L (2011) Outlier detection. WIREs Data Mining Knowl Discov 1:261–268
3. Barnett V, Lewis T (1994) Outliers in statistical data, 3rd edn. Wiley, New York
4. Aggarwal CC (2013) Outlier analysis. Springer, New York
5. Zimek A, Schubert E, Kriegel H-P (2012) A survey on unsupervised outlier detection in high-dimensional numerical data. Stat Anal Data Min 5:363–387
6. Grubbs FE (1950) Sample criteria for testing outlying observations. Ann Math Statist 21:27–58
7. Eo S-H, Pak D, Choi J, Cho H (2012) Outlier detection using projection quantile regression for mass spectrometry data with low replication. BMC Res Notes 5:246
8. Tukey JW (1976) Exploratory data analysis. Addison-Wesley, Boston, MA
9. Dudoit S, Yang YH, Callow MJ, Speed TP (2002) Statistical methods for identifying differentially expressed genes in replicated cDNA microarray experiments. Stat Sin 12:111–139
10. Koenker R (2005) Quantile regression. Cambridge University Press, Cambridge
11. Jolliffe IT (2005) Principal component analysis, 2nd edn. Springer, New York
12. Min H-K, Hyung S-W, Shin J-W et al (2007) Ultrahigh-pressure dual online solid phase extraction/capillary reverse-phase liquid chromatography/tandem mass spectrometry (DO-SPE/cRPLC/MS/MS): a versatile separation platform for high-throughput and highly sensitive proteomic analyses. Electrophoresis 28:1012–1021

Part II

Group Comparisons

Chapter 6

Visualization and Differential Analysis of Protein Expression Data Using R

Tomé S. Silva and Nadège Richard

Abstract

Data analysis is essential to derive meaningful conclusions from proteomic data. This chapter describes ways of performing common data visualization and differential analysis tasks on gel-based proteomic datasets using a freely available statistical software package (R). A workflow followed is illustrated using a synthetic dataset as example.

Key words Proteomics, R, Two-dimensional gel electrophoresis, Data visualization, Differential analysis, Feature selection, Multidimensional scaling, Independent component analysis, Heatmap, Hypothesis testing

1 Introduction

Correct inference from proteomic data depends not only on good experimental design and practice, but also on an adequate analysis of the results. There is extensive discussion on both theoretical and practical issues underlying the inference of biological conclusions from protein expression data, both in this book and elsewhere (e.g., *see* refs. 1, 2 and references therein), and so this chapter focuses on illustrating a practical workflow for analysis of proteomic data using a standard and freely available tool.

The open-source statistical computing environment R is increasingly being used as the analysis tool of choice by the users of high-throughput profiling technologies (colloquially dubbed "omics"). This chapter guides the reader through a practical analysis workflow on R, using a synthetic 2DE dataset as example.

A thorough discussion of the R language and its full range of capabilities is outside the scope of this chapter. Nevertheless, given the nontrivial learning curve of any programming language, we advise the reader to consult some of the many books [3, 4] and online tutorials on this language.

Klaus Jung (ed.), *Statistical Analysis in Proteomics*, Methods in Molecular Biology, vol. 1362,
DOI 10.1007/978-1-4939-3106-4_6, © Springer Science+Business Media New York 2016

2 Materials

2.1 Dataset

2DE gel scans are usually analyzed using specialized image analysis software (*see* **Note 1**). Many software packages can be used for this purpose, each of them with a specific type of approach [5–7], but the end result is usually a table containing measurements for a common set of features (usually "protein spots") across all samples (i.e., gels). Though this chapter and the provided examples assume that the user has a 2DE dataset with "protein spots" as features, its methods also generally apply to other types of multivariate "large p, small n" datasets, proteomic or otherwise (*see* **Note 2**).

A common data format, and the one assumed in this chapter, is such that each row represents a sample (in this case, a 2DE gel) and each column represents a measured variable (in this case, a protein spot), as a plain text file (*see* **Note 3**). Furthermore, besides the dataset table, it is useful to have sample metadata and feature metadata tables, with additional information that can be used in the analysis. In the case of sample metadata, it is common to have information on the technical context of the 2DE gel (e.g., IEF and SDS-PAGE batch), along with the relevant information on the corresponding biological sample (experimental factors and other co-measured variables). In the case of feature metadata, any available information on the protein spots (e.g., *XY* position in the gel image, mean spot area, estimated background, protein identity) should be included.

The example files used throughout the chapter can be obtained at http://dx.doi.org/10.6084/m9.figshare.1284620 or at any of the two other mirrors (http://tomesilva.com/2DExample/ and http://tomessilva.github.io/2DExample/).

2.2 Analysis Software

To follow the analysis workflow described in this chapter, you need to have a specific set of software packages installed on your computer:

1. Install the latest version of R for your platform, either from the CRAN website (http://cran.r-project.org/bin/) or from your operating system's repositories.
2. Install the latest version of RStudio for your platform (http://www.rstudio.com/products/rstudio/download/).
3. Run RStudio and install the required R packages for this example (*see* **Note 4**), by running this code:

```
install.packages(c("caret","klaR",  "e1071",
"effsize",
"pROC","pcaPP","mixOmics","fastICA","pheatm
ap","statmod"))
source("http://bioconductor.org/biocLite.R")
biocLite()
biocLite(c("vsn", "qvalue", "limma"))
```

3 Methods

3.1 Importing Protein Expression Data

1. To import a dataset, either use the "Import Dataset" button (if using RStudio) or run one of the following commands:

```
# if using period as decimal separator and TAB as field separator input.data <- read.delim(file.choose(),dec=".",sep="\t")
# if using comma as decimal separator and TAB as field separator input.data <- read.delim(file.choose(),dec=",",sep="\t")
```

For illustrative purposes, we will use an example dataset, by running

```
input.data <-
  read.delim("http://files.figshare.com/1860747/example_dataset.txt",dec=".")
```

2. You should check whether the dataset was correctly imported:

```
View(input.data)
head(input.data) # alternative way
```

3. It is also necessary to import any eventual metadata available on the samples (e.g., experimental factors). For this simple example, metadata consists simply of a binary variable denoting class membership of the samples:

```
input.metadata <-
  read.delim("http://files.figshare.com/1860748/example_sample_metadata.txt", dec=".")
head(input.metadata) # check if it was correctly imported
treatment <- input.metadata$treatment
```

3.2 Normalizing and Transforming Protein Expression Data

Normalization refers to scaling of protein abundance estimates so that comparisons across gels are possible. This step is required due to technical factors (e.g., different protein load per gel, different staining intensity per gel, different dye efficiencies) that make direct comparisons inadequate (*see* **Note 5**). Many methods have been proposed to normalize 2DE protein abundance data, from simple ones (total gel volume, total spot volume, quantile normalization) to more specialized ones (VSN method, LOESS normalization). As part of the normalization process, it is common to transform the data (using, e.g., log, arsinh, power transforms), for variance stabilization purposes. Extensive discussion on the different normalization and transformation approaches is available elsewhere [8].

1. Check the distribution of abundance values for each gel and any dependence of the standard deviation with the mean:

```
boxplot(t(input.data))
require(vsn)
meanSdPlot(t(input.data))
```

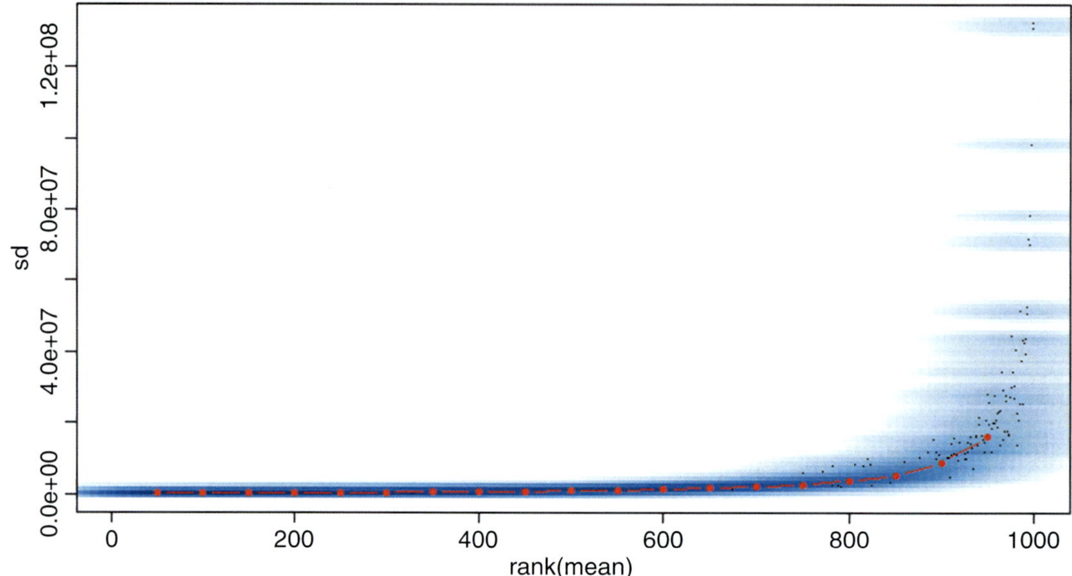

Fig. 1 Scatterplot of each spot's standard deviation against the rank of its mean value, before dataset normalization. The *red line* indicates a running median

2. If you get a trend similar to Fig.1 (i.e., standard deviation increases dramatically for high-abundance spots), then you probably have raw data. A quick way of normalizing and transforming the data (using the VSN method) is

```
require(vsn)
norm.data <- t(justvsn(t(as.matrix(input.
data))))
```

3. You can confirm that the norm.data variable contains your normalized dataset, using the same approach as in **step 1**:

```
boxplot(t(norm.data))
require(vsn)
meanSdPlot(t(norm.data))
```

As can be seen in Fig.2, the mean and the standard deviation of each spot have become decorrelated, making the assumption of heteroscedasticity (required by many regression approaches and hypothesis tests) more plausible.

3.3 Visualizing Protein Expression Data

Data visualization should be seen as an essential step in a proteome analysis workflow, as it allows the researcher to assess data quality and to detect any unexpected heterogeneities that can compromise the validity of downstream statistical tests.

As with other type of multivariate datasets, the classical tools for data visualization (e.g., scatterplots) are inadequate for protein expression data unless some sort of dimensionality reduction step is applied.

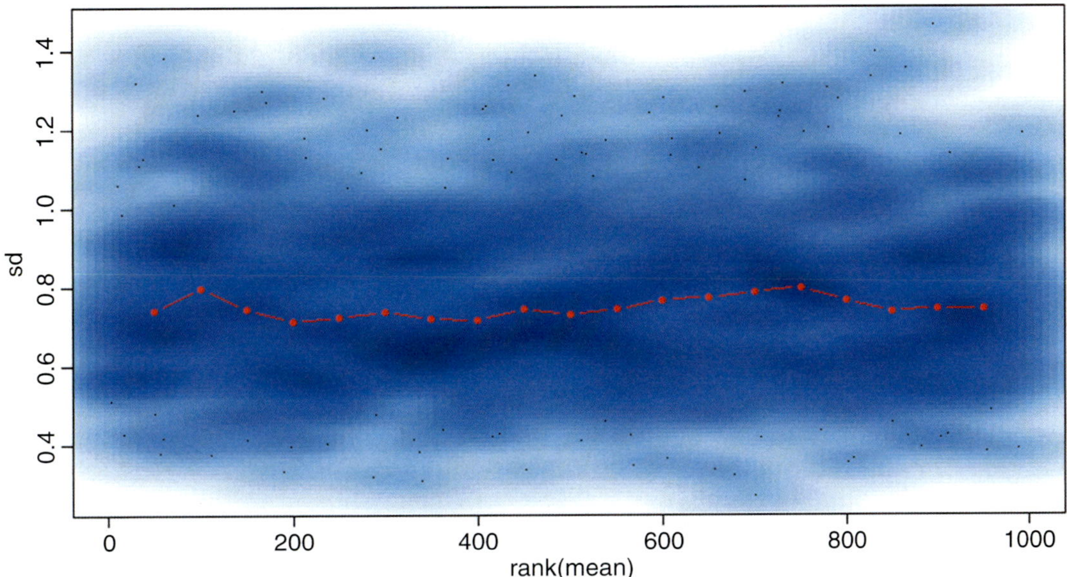

Fig. 2 Scatterplot of each spot's standard deviation against the rank of its mean value, after dataset normalization. The *red line* indicates a running median

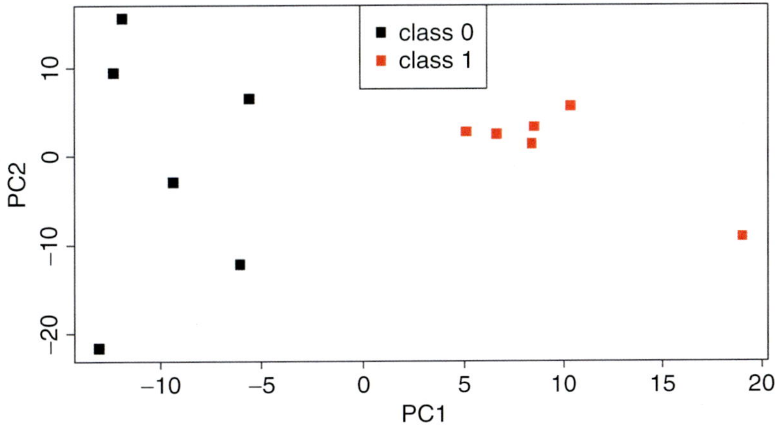

Fig. 3 PCA scores plot, showing the PC1 and PC2 values for each sample (i.e., gel). The color of each *square* (*black* or *red*) denotes class membership

While the most popular approach here consists in performing a principal component analysis (PCA), many other approaches can be undertaken using R.A typical analysis workflow could include steps such as the following:

1. Perform PCA with scaled variables to confirm whether the experimental factor is the main source of variability in terms of protein expression (*see* Fig. 3):

```
results.pca <- prcomp(norm.data,scale=TRUE,
retx=TRUE)
```

```
plot(results.pca) # scree plot
cumsum((results.pca$sdev)^2) / sum(results.
pca$sdev^2)
biplot(results.pca) # biplot of scores and
loadings (not very clear)
plot(results.pca$x[,1],results.
pca$x[,2],col=1+treatment,
  xlab="PC1",ylab="PC2",pch=15)        legend
("top",c("class 0","class 1"),col=1:2,pch=15)
```

2. As an alternative, plot samples according to their similarity in terms of a scale-independent measure (such as Kendall's tau correlation), using multidimensional scaling (MDS):

```
require(pcaPP)
inter.gel.distances <- as.dist(1- cor.
fk(t(norm.data))^2)
results.mds <- cmdscale(inter.gel.distances)
plot(results.mds,col=1+treatment,
  xlab="Dimension 1",ylab="Dimension 2",pch=15)
legend("top",c("class 0","class 1"),col=1:2,
pch=15)
```

3. You can also perform independent component analysis (ICA) to confirm that the direction of maximum non-Gaussianity is orthogonal to the hyperplane that separates samples from the two groups (i.e., that most observed non-Gaussian variation can be attributed to the experimental factor):

```
require(fastICA)
results.ica <- fastICA(norm.data,2)
plot(results.ica$S,col=1+treatment,xlab="IC
1",ylab="IC2",pch=15)
legend("top",c("class 0","class 1"),col=1:
2,pch=15)
```

4. Finally, the use of a heatmap, where both variables and samples are ordinated using hierarchical clustering and expression levels are represented with colors, is also a popular way of visualizing proteomic datasets:

```
require(pheatmap)
pheatmap(t(norm.data),scale="row",show_
rownames=FALSE)
```

3.4 Differential Analysis of Protein Expression Data

When the visualization step supports the hypothesis that a significant fraction of observed variation in protein expression is due to the experimental factor(s), the experimenter is usually interested in assessing which specific proteins are being affected (*see* **Note 6**). This step of "differential analysis" is usually centered on the use of univariate hypothesis testing approaches (e.g., Student's *t*-test, ANOVA, Wilcoxon-Mann-Whitney *U*-test, Kruskal-Wallis test) with some form of multiple comparison correction (in order to control the false discovery rate).

Other forms of feature selection, often based on the use of classifiers, are also available within R.One should have into account, though, that a set of "useful features" (from a classification point of view) does not necessarily contain all "relevant features" (from a biological point of view), which is why the classical approach of (corrected) univariate hypothesis testing is the most widely used for the purposes of "differential analysis."

1. A first step should be univariate hypothesis testing with multiple comparison correction, in this case using Smyth's moderated *t*-test [9] with Storey's *q*-value method [10] to control the FDR, a combination which has been shown to be adequate for the differential analysis of 2DE datasets [11]. It is also important to know (and report) estimates of the effect size (e.g., fold change or, in this case, Cohen's d):

```
require(limma)
require(qvalue)
require(effsize)
design.mtrx<-cbind(rep(1,length(treatment)),
treatment)
data.transposed <- t(norm.data)
fit <- lmFit(data.transposed,design=design.mtrx,
method="robust",maxit=1024)
fit <- eBayes(fit,robust=TRUE)
qval  <-  (qvalue(fit$p.value[,2],pi0.method=
"bootstrap"))$qvalues
fx.size <- apply(norm.data,2,
  function(d,f) cohen.d(d,f)$estimate,f=factor
  (treatment))
sig.spots <- names(qval[qval < 0.1]) # FDR < 10%
n.sig.spots <- length(sig.spots) # number of
significant spots
spot.class <- as.numeric(colnames(norm.data)
%in% sig.spots)
```

2. One of the assumptions of the *t*-test is that errors are normally distributed. It is therefore advisable to check if this assumption is plausible by looking at the distribution of the residuals. A few options are to either studentize the per-spot residuals, pool them all, and compare the resulting distribution with a normal distribution via a Q-Q plot (Fig.4) or look at the skewness and kurtosis distributions of the per-spot residuals (Fig.5):

```
require(e1071)
# pooled studentized residuals approach
fit.residuals <- residuals(fit,data.transposed)
fit.residuals.student  <-  as.vector(scale(fit.
residuals))
qqnorm(fit.residuals.student)
abline(0,1)
# skewness/kurtosis approach
```

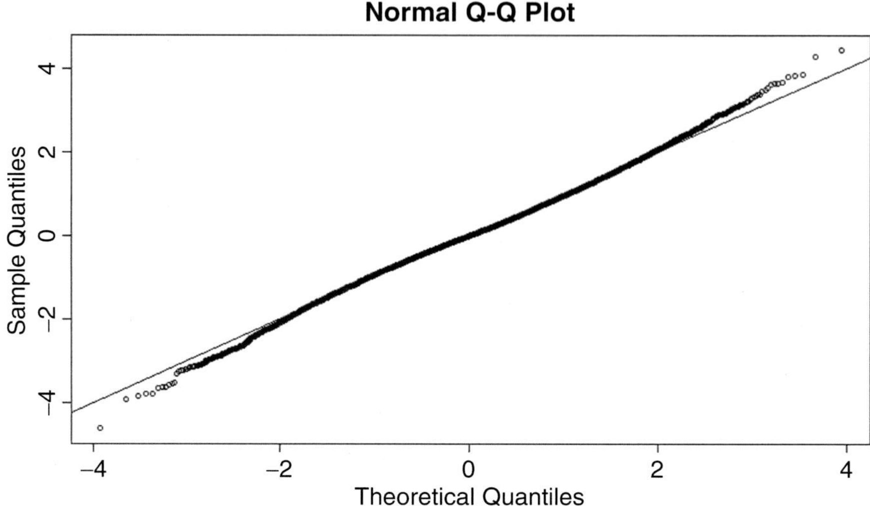

Fig. 4 Quantile-quantile plot (or Q–Q plot) that compares the quantiles of the distribution obtained by pooling all residuals (after studentizing them) against the theoretical quantiles of a standard normal distribution. The *black line* shows where the two agree ($y=x$)

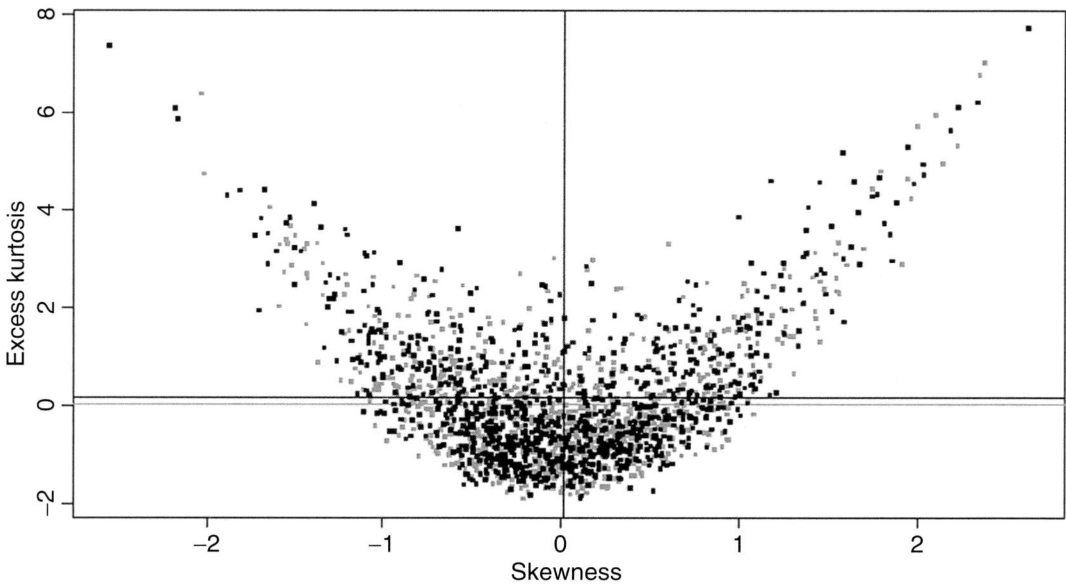

Fig. 5 Scatterplot of excess kurtosis versus skewness, for the residue distributions of each protein spot model. *Black points* show actually observed residue distributions (with the *black vertical* and *horizontal lines* showing the mean value). *Grey points* represent draws of standard normal variates with the same number of degrees of freedom as the empirical residue distributions (with the *grey vertical* and *horizontal lines* showing the mean value)

```
set.seed(1) # make it reproducible
skew2 <- function(x) skewness(x,type=2)
kurt2 <- function(x) kurtosis(x,type=2)
normal.variates <-
  matrix(0,ncol=ncol(fit.residuals),nrow=
  nrow(fit.residuals))
for (i in 1:nrow(fit.residuals)) normal.
variates[i,] <-
  rnorm(ncol(fit.residuals),0,1)
normal.skew <- apply(normal.variates,1,skew2)
normal.kurt <- apply(normal.variates,1,kurt2)
residual.skew <- apply(fit.residuals,1,skew2)
residual.kurt <- apply(fit.residuals,1,kurt2)
plot(NA,xlab="Skewness",ylab="Excess
kurtosis",
  xlim=c(min(c(normal.skew,residual.
  skew)),max(c(normal.skew,residual.skew))),
  ylim=c(min(c(normal.kurt,residual.
  kurt)),max(c(normal.kurt,residual.kurt))))
points(normal.skew,normal.kurt,pch=".",cex=5,
col="grey")
abline(v=mean(normal.skew),col="grey")
abline(h=mean(normal.kurt),col="grey")
points(residual.skew,residual.
kurt,pch=".",cex=5)
abline(v=mean(residual.skew))
abline(h=mean(residual.kurt))
```

Both plots suggest that the distribution of the residuals is approximately normal, which supports the application of statistical tests that assume errors to be normally distributed (*see* **Note 7**).

3. The results of univariate hypothesis testing can be visualized using a volcano plot (Fig.6) and written to a text file (*see* **Note 8**):

```
spot.colours <-
  as.numeric(colnames(norm.data) %in% names
  (qval[qval < 0.05])) + 1 + spot.class
plot(fx.size,-log(qval)/log(10),col=spot.
colours,
  xlab="Effect size (Cohen's d)",ylab=
  "Significance (-log10(q-value))", pch=15,
  cex=0.7)
legend("bottomleft",c("FDR > 10%","FDR <
10%","FDR < 5%"),col=1:3,pch=15) univariate.
results <-
  data.frame(spot.name=colnames
  (norm.data),p.value=fit$p.value[,2],
  q.value=qval,effect.size=fx.size,
  significant=spot.class)
```

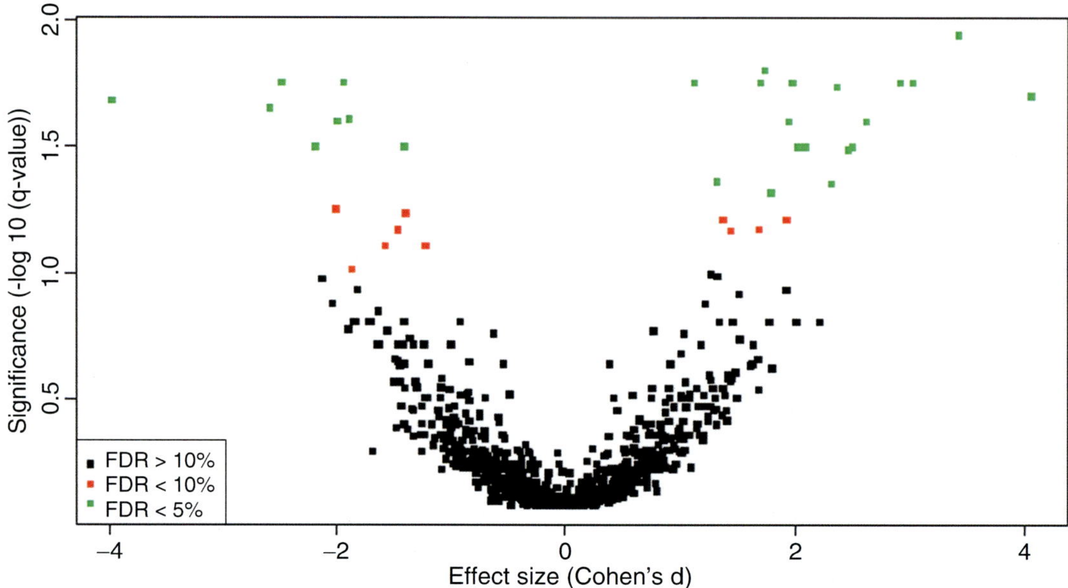

Fig. 6 Volcano plot, where effect significance (−log10(*q*-value)) is plotted against an estimate of effect size (Cohen's d), for each spot. *Green spots* would be considered significant by taking "FDR < 5%" as selection criterion, while both *green* and *red spots* would be considered significant by taking "FDR < 10%" as selection criterion

```
write.table(univariate.results,"univariate_
results.txt",
  row.names=FALSE,sep="\t")
```

4. An alternative method for feature selection is, for example, the application of a classifier with regularization, such as sparse partial least squares discriminant analysis (sPLS-DA) [12] (*see* **Note 9**):

```
require(mixOmics)
results.splsda   <-   mixOmics::splsda(norm.
data, as.factor(treatment),
  ncomp=2, keepX=c(n.sig.spots,n.sig.spots))
tmp.load <- ((results.splsda$loadings)$X)[,1]
sig.spots.splsda <- names((tmp.load)[abs(tmp.
load) > (.Machine$double.eps)])
```

5. Another possible alternative is to use recursive feature elimination (RFE) with some classification method (in this case, with naive Bayes classifiers) [13] (*see* **Note 9**):

```
require(caret)
require(klaR)
require(e1071)
rfeCtrl  <-  rfeControl(functions  =  nbFuncs,
method = "LOOCV")
results.rfe<-rfe(norm.data,factor(treatment),
  sizes=n.sig.spots,rfeControl=rfeCtrl)
```

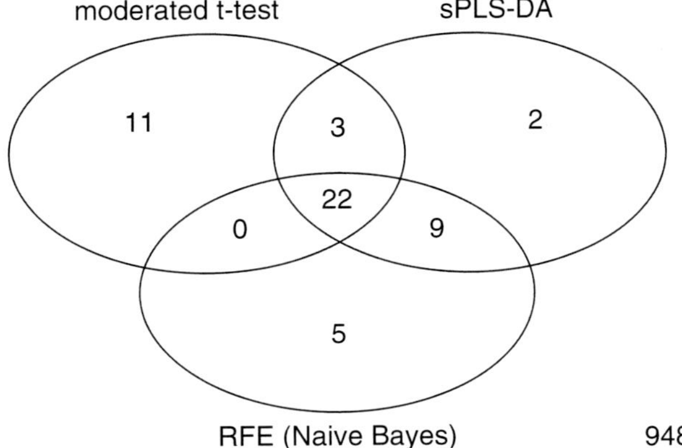

Fig. 7 Venn diagram displaying the number of spots selected by each feature selection approach, along with the level of overlap between the approaches

```
sig.spots.nbrfe <- rownames(varImp(results.
rfe))[(1:n.sig.spots)]
```

6. It can be useful to compare the level of overlap between the different feature selection strategies using a Venn diagram:

```
require(limma)
all.spots <- colnames(norm.data)
tmp.counts <- matrix(0, nrow=length(all.
spots), ncol=3)
for (i in 1:length(all.spots)) {
  tmp.counts[i,1] <- all.spots[i] %in% sig.
  spots
  tmp.counts[i,2] <- all.spots[i] %in% sig.
  spots.splsda
  tmp.counts[i,3] <- all.spots[i] %in% sig.
  spots.nbrfe
}
colnames(tmp.counts) <- c("moderated t-test",
"sPLS-DA","RFE (Naive Bayes)")
vennDiagram(vennCounts(tmp.counts))
```

The results (shown in Fig.7) demonstrate that, though each strategy selects a distinct set of spots, there is a high degree of overlap between these sets.

7. A final summary of the results, showing the 22 spots of the consensus set, can be generated as a heatmap (Fig.8):

```
consensus.set <-
  intersect(sig.spots,intersect(sig.spots.
  splsda,sig.spots.nbrfe))
require(pheatmap)
pheatmap(t(norm.data[,consensus.set]),
scale="row")
```

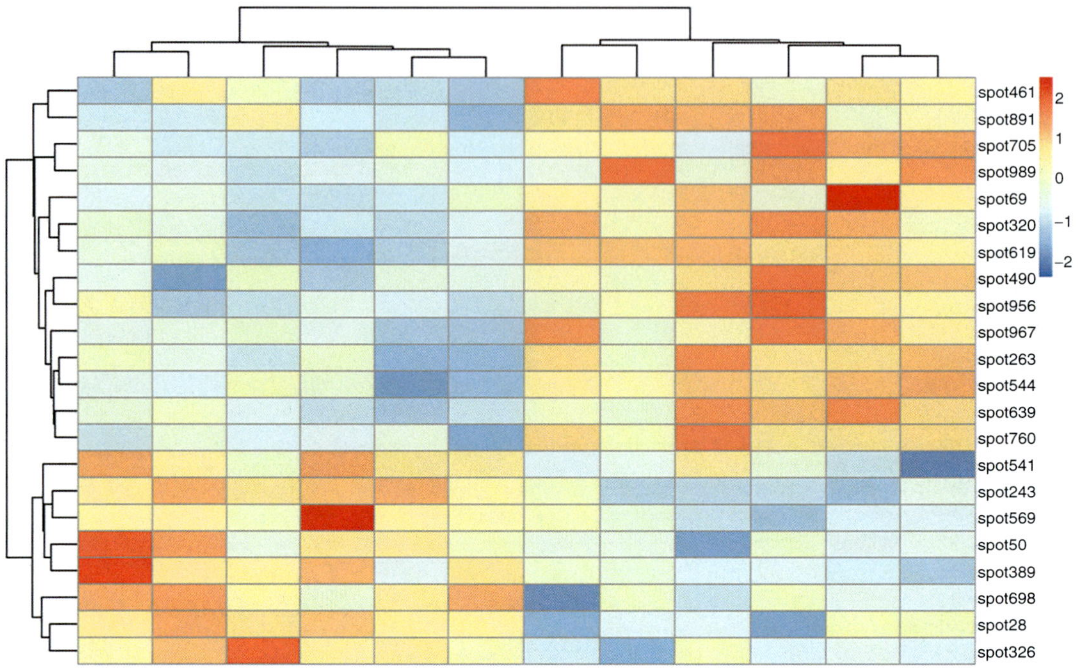

Fig. 8 Heatmap displaying the expression of the 22 most relevant proteins using color-coding (*red* for "high expression" and *blue* for "low expression"). Samples (i.e., gels) and variables (i.e., spots) were ordinated using agglomerative hierarchical clustering

4 Conclusion

In this chapter, we demonstrated how to analyze a 2DE dataset from a simple one-factor experiment using R.Though real proteomic experiments often have more complicated experimental designs and richer metadata (i.e., more measured covariates for each sample/experimental unit), R can usually accommodate any type of analysis with relative ease, given its flexibility and the wide range of statistical tools available as R packages.

5 Notes

1. Some of these software packages can generate datasets with missing values. Though we assume the analysis of datasets without missing values, R has functions to address such situations, namely through either single-imputation (implemented in packages such as impute, yaImpute, pcaMethods, missForest, Amelia, VIM, rrcovNA, missMDA, and softImpute) or multiple-imputation approaches (implemented in packages mi and mice).

2. While the provided example is a standard 2DE dataset, the described analysis approach can also be applied to multiplex 2DE (i.e., DIGE) datasets, even though the type of data is slightly different (abundance ratios rather than raw abundance values).

3. If the dataset does not have the required structure, R has specific functions that can help the user express his or her dataset in an appropriate form for analysis (e.g., function t(), to exchange rows with columns; functions melt() and acast(), from package reshape2, to convert datasets between formats).

4. If problems arise during this step due to operating system permission restrictions, the use of a local library is advised. Alternatively, this type of problem can often be bypassed on Windows by running R as administrator during package installation.

5. Most 2DE analysis software packages already perform some form of normalization. If you know that your analysis software performs adequate data normalization, you can skip this step.

6. On the other hand, when the visualization step suggests that most of the variation is not due to the experimental factors, or otherwise shows some level of heterogeneity (e.g., each of the two experimental groups is seemingly separated in two sub-clusters), it is essential to determine this source of variation. Possible sources of variation can include both biological (e.g., male vs. female, subject weight, subject age) and technical factors (e.g., different IEF/SDS-PAGE runs, different technicians, different dyes in the case of DIGE), which is why it is important to have as much metadata as possible about each "statistical sample" (i.e., each single 2DE gel or each single DIGE channel) and the underlying biological sample ("experimental unit"). In order to accurately assess the effects of the experimental factors, it is important to have into account the concurrent effect of non-experimental factors.

7. If a significant deviation from normality is observed, it suggests that there was either a problem with the data pre-processing (normalization and transformation) or that the classical linear models are not adequate to analyze the dataset. In the latter case, using more flexible models (e.g., generalized linear and generalized additive) or rank-based approaches (Wilcoxon-Mann-Whitney *U*-test, Kruskal-Wallis test) may be advisable. In either case, one should go back to the visualization step in order to determine a viable analysis strategy.

8. To determine the default location where R will write to, run the getwd() command.

9. This code has been adapted to select only the X most "relevant" spots, where X is the number of spots considered differentially expressed by the univariate hypothesis testing approach.

Acknowledgement

Nadège Richard was supported by a postdoctoral grant (SFRH/BDP/65578/2009) from the Portuguese Foundation for Science and Technology (FCT).

References

1. Dowsey AW, Morris JS, Gutstein HB etal (2010) Informatics and statistics for analyzing 2-D gel electrophoresis images. In: Hubbard JS, Jones AR (eds) Proteome bioinformatics. Humana Press, NewYork, pp239–255
2. Silva TS, Richard N, Dias JP etal (2014) Data visualization and feature selection methods in gel-based proteomics. Curr Protein Pept Sci 15:4–22
3. Hothorn T, Everitt B (2009) A handbook of statistical analyses using R, 2nd edn. Chapman & Hall/CRC, London
4. Crawley MJ (2012) The R book. Wiley, Chichester, England
5. Dowsey AW, English JA, Lisacek F etal (2010) Image analysis tools and emerging algorithms for expression proteomics. Proteomics 10: 4226–4257
6. Rye M, Fargestad EM (2012) Preprocessing of electrophoretic images in 2-DE analysis. Chemometr Intell Lab Syst 117:70–79
7. Wheelock M, Buckpitt AR (2005) Software-induced variance in two-dimensional gel electrophoresis image analysis. Electrophoresis 26:4508–4520
8. Chich J-F, David O, Villers F etal (2007) Statistics for proteomics: experimental design and 2-DE differential analysis. J Chromatogr B 849:261–272
9. Smyth GK (2004) Linear models and empirical Bayes methods for assessing differential expression in microarray experiments. Stat Appl Genet Mol Biol 3:1–25
10. Storey JD, Tibshirani R (2003) Statistical significance for genomewide studies. Proc Natl Acad Sci U S A 100:9440–9445
11. Artigaud S, Gauthier O, Pichereau V (2013) Identifying differentially expressed proteins in two-dimensional electrophoresis experiments: Inputs from transcriptomics statistical tools. Bioinformatics 29:2729–2734
12. Cao K-AL, Boitard S, Besse P (2011) Sparse PLS discriminant analysis: biologically relevant feature selection and graphical displays for multiclass problems. BMC Bioinformatics 12:253
13. Guyon I, Elisseeff A (2003) An introduction to variable and feature selection. J Mach Learn Res 3:1157–1182

Chapter 7

False Discovery Rate Estimation in Proteomics

Suruchi Aggarwal and Amit Kumar Yadav

Abstract

With the advancement in proteomics separation techniques and improvements in mass analyzers, the data generated in a mass-spectrometry based proteomics experiment is rising exponentially. Such voluminous datasets necessitate automated computational tools for high-throughput data analysis and appropriate statistical control. The data is searched using one or more of the several popular database search algorithms. The matches assigned by these tools can have false positives and statistical validation of these false matches is necessary before making any biological interpretations. Without such procedures, the biological inferences do not hold true and may be outright misleading. There is a considerable overlap between true and false positives. To control the false positives amongst a set of accepted matches, there is a need for some statistical estimate that can reflect the amount of false positives present in the data processed. False discovery rate (FDR) is the metric for global confidence assessment of a large-scale proteomics dataset. This chapter covers the basics of FDR, its application in proteomics, and methods to estimate FDR.

Key words False discovery rate, Posterior error probability, Target-decoy, Peptide spectrum matches, Statistical validation, Shotgun proteomics

1 Introduction

In any large-scale high-throughput study, including genomics and proteomics, a large number of statistical hypothesis are tested, usually independently, for significance [1]. Each hypothesis tested (a gene, a transcript, a peptide, etc.) yields a *p*-value, *e*-value, or a score that depicts the quantitative measure of that hypothesis being correct. In proteomics, shotgun proteomics data is searched using a database search algorithm that provides such confidence metrics after searching spectra against peptides in a given FASTA database.

While such metrics can reflect the "*goodness of fit*" of an experimental spectrum to the assigned peptide and related chances of error in its identification, it cannot reflect the associated error in the whole dataset. For example, selecting peptide spectral matches (PSMs) or hits with p-value ≤ 0.05 means that each PSM has a 5 % or less chance of incorrectly being assigned as a significant match. This, however, does not mean that all PSMs passing this threshold

Klaus Jung (ed.), *Statistical Analysis in Proteomics*, Methods in Molecular Biology, vol. 1362, DOI 10.1007/978-1-4939-3106-4_7, will have a collective error of 5 %. Any PSM, even with a low estimated error of 5 %, could turn out to be wrong. It could occur by (highly unfortunate and rare) chance that all PSMs (each with 5 % estimated error) turn out to be wrong.

One cannot estimate the percentage of false hits in the accepted PSMs by using *p*-value as a metric because this is a *single spectrum-specific* significance measure. To assess global false hits or error rate, one needs to understand population-level false estimation metrics. This is a classic case of what is called as the *multiple testing problem*; that is, when multiple independent statistical hypothesis tests are conducted, single hypothesis significance measures (like *p*-value) are neither sufficient nor amenable to extrapolation to calculate population error rate. By random chance alone, there will be many hits which may turn out to be false [2] in a collection of hits, each with $p\text{-value} \leq 0.05$ with a final error rate of more than 5 % globally.

1.1 False Discovery Rate

Adjusting for multiple comparisons can be achieved by applying Bonferroni correction which readjusts significance threshold ($\alpha = 0.05$) to control the false positives. For n spectra, the population-level significance threshold becomes $0.05/n$, to adjust for the error rate of 5 % globally. This method is very stringent and false positives are extremely low when n is large. But this occurs at the cost of false negatives. To avoid false positives, this method excludes many true positives with good scores and *p*-values. False discovery rate (FDR) is a measure of the incorrect PSMs among all accepted PSMs [2–4]. Proposed by Benjamini and Hochberg [5] as an alternate to the Bonferroni correction, it is defined as the rate of false positives in accepted hits. FDR is a less stringent metric for global confidence assessment. In the context of proteomics, it is a global estimate of the false positives present in the results obtained by a database search algorithm. There are many different strategies to estimate FDR like the *nonparametric* simple target-decoy (TD) database searches [4, 6] and *parametric* or *semi-parametric* mixture modeling approaches used in the Trans-proteomics pipeline (TPP) [7–10].

For estimating the FDR, a null model is required for which a decoy database search is carried out in proteomics. In the TD search strategy, the database search is carried out on the true (target) as well as null (decoy) database. A decoy database is constructed by shuffling, randomizing or by simply reversing the target database. It is the simplest approach to calculate FDR and requires no distributional assumptions, i.e., *nonparametric* in nature. The basic assumption made for TD approach is that the number of false PSMs in decoy search will be equal to the number of false PSMs in target search above a given threshold score. The database search for this approach can be performed together (concatenated) or separately with the decoy database. A *concatenated* search can be

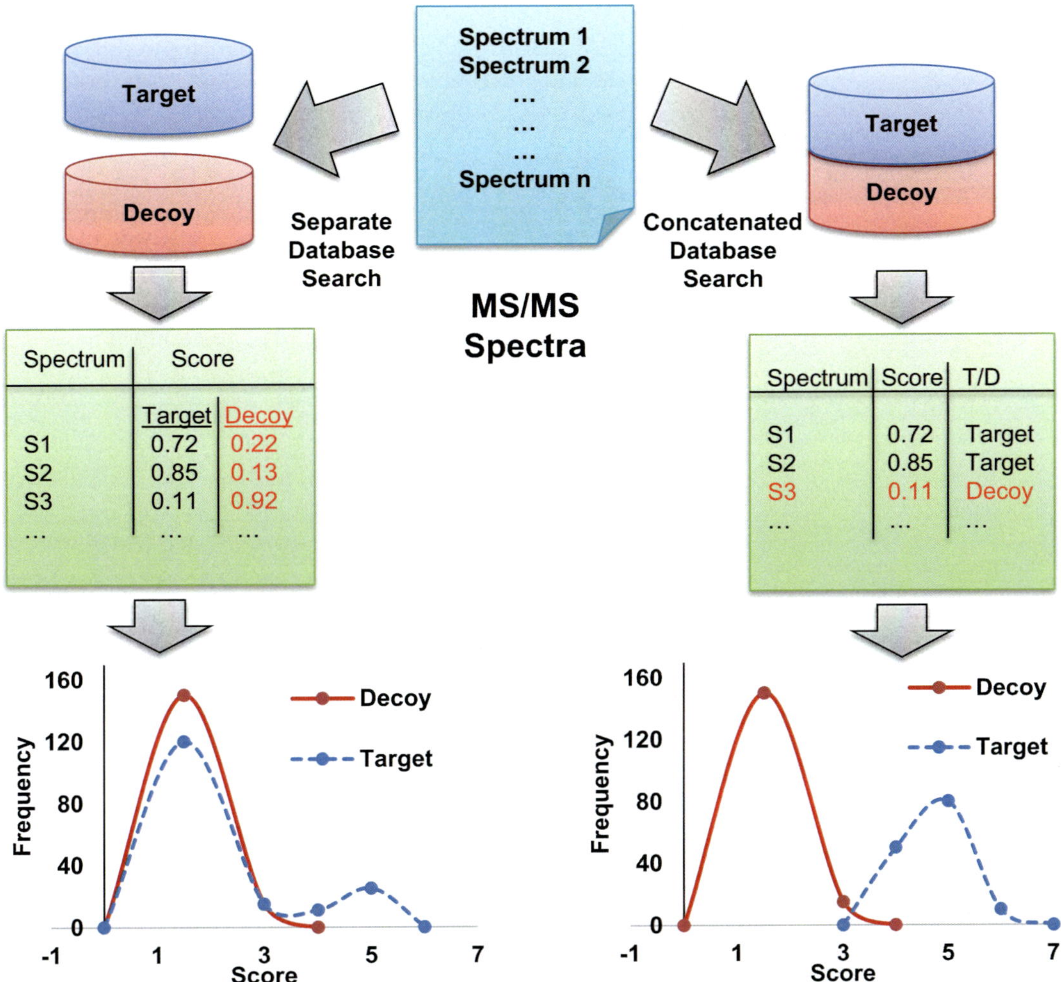

Fig. 1 There are two database search strategies—separate or combined database search. In separate search target and decoy databases are searched separately and FDR is estimated using Kall's method (*see* text). Each spectrum has one target and one decoy best score. In combined/concatenated approach, one unified target-decoy database is searched in which both TD peptides compete with each other. Each spectrum has one best score, either from target or decoy but not both. This also changes the score distributions

conducted by combining the target and decoy databases together (both target and decoy) as proposed by Elias and Gygi [6]. In a *separate* database search [4], as the name implies, both target and decoy databases are searched separately and scored using a search algorithm. The number of false positives divided by the total hits allows for easy calculation of FDR. Overview of this process is shown in Fig. 1.

1.2 q-Value

While FDR is a global measure of population error rate, this communicates nothing about confidence of individual PSMs. A PSM-specific metric is needed which conveys the confidence measure of a particular PSM after FDR correction has been applied. Thus, *q-value* was introduced by Storey and Tibshirani [1], which is defined as the minimum FDR cutoff at which a particular PSM can be accepted. It is the property of a single PSM rather than a set of PSMs. The *q*-value of a PSM provides a direct measure of significance for a particular PSM with respect to the complete dataset and the risk accrued to the total accepted matches if that hit is deemed significant. For example, if a PSM with *q*-value 0.07 seems biologically important, we will need to lower the FDR threshold to 7 % in the dataset to accept this PSM as significant. A sorted list of hits by *q*-value becomes a monotonous function of search score/*p-value* and is thus easily interpretable [2]. A dataset can be revisited to select a biologically important hit without the need to recalculate the FDR.

1.3 Posterior Error Probability

Posterior error probability (PEP) is the probability of a PSM to be incorrect. Borrowing from the example given by Kall et al. [11], a PEP of 1 % would signify that there is 99 % chance for the PSM to be correct. It is also referred to as local FDR, as unlike FDR it measures the error rate associated with a single PSM. From Fig. 2, it can be seen that FDR represents the ratio of area under incorrect (decoy) region for any given score threshold x to the area under the total region for the same threshold. PEP is ratio of the infinitesimally small areas (virtually the height) of incorrect to total hits

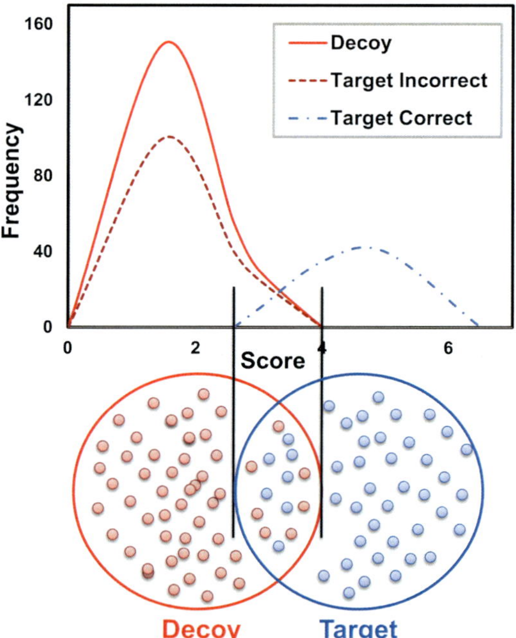

Fig. 2 There is considerable overlap between the correct and incorrect hits shown here as score frequency distribution and the corresponding set-based visualization. FDR is applied to control the proportion of false hits from getting accepted

in a local score region of x. Both are complementary to each other but have different meanings. While the q-value conveys the risk (error introduced) in the whole dataset if we accept the PSM at hand, the PEP on the other hand informs us whether the PSM is likely to be correct or not.

FDR can be calculated from PEP by integrating (summing up) all the PEPs. PEPs can be accurately calculated by using machine learning to learn the model parameters from labeled (correct and incorrect) training data. For any given score x, the PEP can be predicted from the model parameters. This strategy is used in PeptideProphet [8] and ProteinProphet [9].

2 Materials

Any FDR strategy would require a large-scale shotgun proteomics experiment data. This data should be searched using a database search algorithm. A normal desktop computer with enough memory according to dataset size should be useful. Perl programming language and ProteoStats [12] should be configured on the computer.

3 Methods

To calculate FDR, the spectra are searched against a target and a decoy database to obtain the top-ranked PSMs. This can be searched either separately or in a combined database search, each with different assumptions on target decoy competition and false-positive estimation. This section briefly explains how to use ProteoStats library for FDR estimation. ProteoStats requires the data to be searched using separate TD approach as it can perform the TD competition after the search as suggested by Fitzgibbon et al. [13]. More details can be found in ProteoStats documentation, supplementary material of [12], and blog post on ProteoStats [14]. Based on how TD matches are defined (in terms of TD competition) and how false positives are defined, there are five methods for FDR calculation. Mainly separate and concatenated FDR are two main methods based on the mode of search. Several variants of these formulations have been proposed which may provide better results [15, 16]. Though the community majorly uses Kall's or Elias and Gygi's formulae due to ease of calculation, there is no consensus on which formula is better or appropriate. Currently, the proteomics community agrees on any method as long as it is clearly defined. We also propose that users can try all methods in ProteoStats for a specific workflow for standardizing the protocol according to search engine, decoy strategy, data quality, etc. In our hands, we have found FDR with percentage of incorrect target (PIT) correction and refined methods to work better than others.

3.1 FDR Calculation Using ProteoStats

TD searches are completed separately and results in the form of target and decoy top hits provided as input to ProteoStats. When the searches are conducted separately, all different FDR methods can be applied *a posteriori*, but if a concatenated search is used, only concatenated FDR method can be applied as the correspondence between TD top hits is lost. ProteoStats removes the peptides identical in decoy and target considering isoleucine and leucine as identical. The resulting TD sets are sorted separately on the basis of scores/*e*-values/*p*-values from best to worst and depending on the search strategy chosen the FDR, *q*-value, and receiver operating curve (ROC) are calculated [12]. There are other supplementary modules for chart generation and comparing results (*see* **Note 1**).

For the calculation of FDR, there are different methods/formulae available:

1. Separate/simple FDR (FDR_S)
   This method by Kall et al. [4] assumes that the number of decoys passing the threshold (D) represents the number of false positives in the target PSMs (T) above the same threshold (*see* **Note 2**):

$$\mathrm{FDR}_{\mathrm{S}} = \frac{D}{T} \tag{1}$$

2. Concatenated FDR (FDR_C)
   This method by Elias and Gygi [17] assumes that for any number of decoys (D) passing a given threshold, there are equal number of false hits in target PSMs (T) above that threshold. Adding up the false hits in decoy and target, the number of false positives is therefore double of the decoy count above threshold. In this search, TD competition results in either target or a decoy best hit for any given spectrum, the reference population in which FDR is calculated is not the same as other methods:

$$\mathrm{FDR}_{\mathrm{C}} = \frac{2 \times D}{(T + D)} \tag{2}$$

3. FDR with PIT correction (FDR_{PIT})
   Kall's formula for simple FDR did not consider the incorrect target PSMs during simple FDR calculation which tilts the balance of random matches in decoy's favor due to higher decoy population. To correct for this effect, it was suggested to calculate the PIT, which is used as a factor to accurately determine the FDR. Note that the name PIT is a misnomer as it is a fraction and not a percentage. It is similar to the notion of fraction of true negatives, Π_0, as defined in genomics [1]:

$$\mathrm{FDR}_{\mathrm{PIT}} = \mathrm{PIT} \times \frac{D}{T} \tag{3}$$

4. Refined separate FDR (FDR_{RS})
   In this formulation, Navarro et al. [15] propose that FDR should be calculated in the correct reference population (only targets). They argued that the estimated false positives should not be doubled blindly by observing decoy hits above threshold directly. The decoy PSMs above threshold should not be considered false positives if they do not score more than the corresponding target PSM. This causes inflated false-positive estimation and leads to overestimation of FDR. The hits could be above threshold only in target (target only, TO) or only in decoy (decoy only, DO). When both are above threshold, either target could be better (target better, TB) or the decoy (decoy better, DB). The FDR in the correct reference population is calculated by estimating the correct false positives and dividing by corrected total population [15]:

$$\mathrm{FDR}_{\mathrm{RS}} = \frac{(2 \times \mathrm{DB} + \mathrm{DO})}{(\mathrm{TB} + \mathrm{TO} + \mathrm{DB})} \tag{4}$$

5. Refined concatenated method (FDR_{RC})
   In this formulation, Cerqueira et al. [16] argued that since decoy hits are by definition false, they can be disregarded in FDR estimation and thus the FDR_C formula is changed to yield the following formula:

$$\mathrm{FDR}_{\mathrm{RC}} = \frac{D}{(T - D)} \tag{5}$$

3.2 Peptide to Proteins

FDR calculated at PSM level is not the same as protein-level FDR. Although the goal of a shotgun proteomics experiment is to assess protein-level significance, the hypotheses tested from shotgun proteomics data are spectra. Due to variable abundances of different proteins and the non-random distribution of peptides across these proteins, one-to-one correspondence between peptide and protein FDRs does not exist. Due to this incongruity, calculating protein FDRs is complicated.

Since the decoy database is made out of the target database with equal number of proteins and same protein length distribution, we can simply use the same algorithm (*see* **Note 3**) to estimate the number of false positives owing to the identification of the decoy proteins. This step requires a robust protein score which can help in estimation of protein FDR. The FDR for protein estimation is calculated as the ratio of the expected number of false-positive protein identifications (those that have a hit to the decoy database proteins) to that of the total number of protein identifications mapping to the target database at any threshold protein score. For protein FDR, MAYU software can be used which performs protein identification-level FDR on the basis of peptide identifications [18].

4 Notes

1. Calculating FDR using ProteoStats [12] software is accurate and reliable. It is written in Perl and it can be integrated in any kind of pipeline easily. In terms of file format inputs, it provides great flexibility as it can read proteomics output file format from different database searches like Mascot [19], MassWiz [20], OMSSA [21], X!Tandem [22], MyriMatch [23], and Comet [24]. These results may be processed using a rescorer like FlexiFDR [25] for MassWiz, Percolator [26–28] for Mascot, and OScore for Sequest [29] to improve identification results. Tab-delimited files can also be used for FDR to support other algorithms. It provides with CSV/Excel-based output files which can easily be interpreted and used for further analysis in R. It also contains plotting functionalities for visual analysis of the results obtained. Comparison of results and Venn charts can also be generated.
2. A general algorithm is described for FDR estimation using Kall's method. The pictorial representation is shown in Fig. 3. Please note that this is a generic algorithm for simple FDR calculation. This code is provided within ProteoStats for all FDR formulae, so it need not be manually calculated.
   (a) Sort target results on score/*p*-value/*e*-value from best to worst hit.
   (b) Sort decoy results on score/*p*-value/*e*-value from best to worst hit.

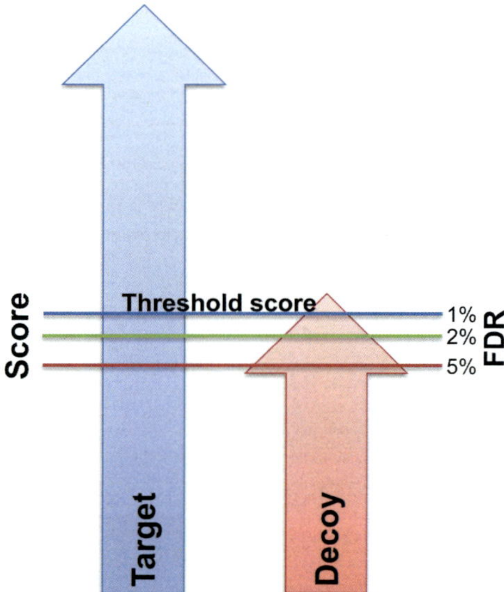

Fig. 3 Target and decoy scores are sorted and iteratively different thresholds can then be applied until a desirable level of FDR is achieved. FDR is the ratio of decoy/target hits at any particular threshold

   (c) For every target score as threshold, count D and T, and the number of decoys and targets above threshold.
   (d) Calculate FDR using Eq. 1. This algorithm explains simple FDR using Kall's method. Other methods differ in definition of false-positive counts, so should be accordingly calculated.
   (e) This FDR is also the q-value for the PSM serving as score threshold, and other PSMs with same score.
   (f) A tabulated list of number of targets and corresponding q-value can be used to create an ROC plot.

3. After FDR is calculated, the PSMs and peptides are used for protein inference. Similar TD approaches can be used for estimating protein-level FDR estimates although the correspondence between peptide and protein FDR is not same due to non-random distribution of peptides across proteins due to varying abundances. MAYU [18] is a tool for protein FDR calculation though it does not infer proteins. IDPicker [30] is another tool which integrates the process but can only perform FDR for concatenated search. ProteoStats will be updated in near future to support protein-level FDR calculation.

Acknowledgement

S.A. is supported by SRF grant and A.K.Y. is supported by Innovative Young Biotechnologist Award (IYBA) grant and DDRC-SFC grant from Department of Biotechnology (DBT), India. Authors acknowledge Manu Kandpal for proofreading the manuscript.

References

1. Storey JD, Tibshirani R (2003) Statistical significance for genomewide studies. Proc Natl Acad Sci U S A 100:9440–9445
2. Choi H, Nesvizhskii AI (2008) False discovery rates and related statistical concepts in mass spectrometry-based proteomics. J Proteome Res 7:47–50
3. Nesvizhskii AI (2010) A survey of computational methods and error rate estimation procedures for peptide and protein identification in shotgun proteomics. Proteomics 73:2092–2123
4. Kall L, Storey JD, MacCoss MJ et al (2008) Assigning significance to peptides identified by tandem mass spectrometry using decoy databases. J Proteome Res 7:29–34
5. Benjamini Y, Hochberg Y (1995) Controlling the false discovery rate: a practical and powerful approach to multiple testing. J R Stat Soc B 57:289–300
6. Elias JE, Gygi SP (2007) Target-decoy search strategy for increased confidence in large-scale protein identifications by mass spectrometry. Nat Methods 4:207–214
7. Choi H, Ghosh D, Nesvizhskii AI (2008) Statistical validation of peptide identifications in large-scale proteomics using the target-decoy database search strategy and flexible mixture modeling. J Proteome Res 7:286–292
8. Keller A, Nesvizhskii AI, Kolker E et al (2002) Empirical statistical model to estimate the accuracy of peptide identifications made by MS/MS and database search. Anal Chem 74:5383–5392
9. Nesvizhskii AI, Keller A, Kolker E et al (2003) A statistical model for identifying proteins by tandem mass spectrometry. Anal Chem 75:4646–4658

10. Tabb DL (2008) What's driving false discovery rates? J Proteome Res 7:45–46
11. Kall L, Storey JD, MacCoss MJ et al (2008) Posterior error probabilities and false discovery rates: two sides of the same coin. J Proteome Res 7:40–44
12. Yadav AK, Kadimi PK, Kumar D et al (2013) ProteoStats—a library for estimating false discovery rates in proteomics pipelines. Bioinformatics 29:2799–2800
13. Fitzgibbon M, Li Q, McIntosh M (2008) Modes of inference for evaluating the confidence of peptide identifications. J Proteome Res 7:35–39
14. Yadav AK, Perez-Riverol Y (2014) ProteoStats: computing false discovery rates in proteomics. BioCode's notes, computational proteomics & bioinformatics. http://computationalproteomic.blogspot.com/2014/08/proteostats-computing-false-discovery.html
15. Navarro P, Vazquez J (2009) A refined method to calculate false discovery rates for peptide identification using decoy databases. J Proteome Res 8:1792–1796
16. Cerqueira FR, Graber A, Schwikowski B et al (2010) MUDE: a new approach for optimizing sensitivity in the target-decoy search strategy for large-scale peptide/protein identification. J Proteome Res 9:2265–2277
17. Elias JE, Gygi SP (2010) Target-decoy search strategy for mass spectrometry-based proteomics. Methods Mol Biol 604:55–71
18. Reiter L, Claassen M, Schrimpf SP et al (2009) Protein identification false discovery rates for very large proteomics data sets generated by tandem mass spectrometry. Mol Cell Proteomics 8:2405–2417
19. Perkins DN, Pappin DJ, Creasy DM et al (1999) Probability-based protein identification by searching sequence databases using mass spectrometry data. Electrophoresis 20: 3551–3567
20. Yadav AK, Kumar D, Dash D (2011) MassWiz: a novel scoring algorithm with target-decoy based analysis pipeline for tandem mass spectrometry. J Proteome Res 10:2154–2160
21. Geer LY, Markey SP, Kowalak JA et al (2004) Open mass spectrometry search algorithm. J Proteome Res 3:958–964
22. Craig R, Beavis RC (2004) TANDEM: matching proteins with tandem mass spectra. Bioinformatics 20:1466–1467
23. Tabb DL, Fernando CG, Chambers MC (2007) MyriMatch: highly accurate tandem mass spectral peptide identification by multivariate hypergeometric analysis. J Proteome Res 6:654–661
24. Eng JK, Jahan TA, Hoopmann MR (2013) Comet: an open-source MS/MS sequence database search tool. Proteomics 13:22–24
25. Yadav AK, Kumar D, Dash D (2012) Learning from decoys to improve the sensitivity and specificity of proteomics database search results. PLoS One 7, e50651
26. Brosch M, Yu L, Hubbard T et al (2009) Accurate and sensitive peptide identification with Mascot Percolator. J Proteome Res 8:3176–3181
27. Spivak M, Weston J, Bottou L et al (2009) Improvements to the percolator algorithm for peptide identification from shotgun proteomics data sets. J Proteome Res 8:3737–3745
28. Wright JC, Collins MO, Yu L et al (2012) Enhanced peptide identification by electron transfer dissociation using an improved mascot percolator. Mol Cell Proteomics 11:478–491
29. Shao C, Sun W, Li F et al (2009) Oscore: a combined score to reduce false negative rates for peptide identification in tandem mass spectrometry analysis. J Mass Spectrom 44:25–31
30. Ma ZQ, Dasari S, Chambers MC et al (2009) IDPicker 2.0: improved protein assembly with high discrimination peptide identification filtering. J Proteome Res 8:3872–3881

Chapter 8

A Nonparametric Bayesian Model for Nested Clustering

Juhee Lee, Peter Müller, Yitan Zhu, and Yuan Ji

Abstract

We propose a nonparametric Bayesian model for clustering where clusters of experimental units are determined by a shared pattern of clustering another set of experimental units. The proposed model is motivated by the analysis of protein activation data, where we cluster proteins such that all proteins in one cluster give rise to the same clustering of patients. That is, we define clusters of proteins by the way that patients group with respect to the corresponding protein activations. This is in contrast to (almost) all currently available models that use shared parameters in the sampling model to define clusters. This includes in particular model based clustering, Dirichlet process mixtures, product partition models, and more. We show results for two typical biostatistical inference problems that give rise to clustering.

Key words Dirichlet process, Pólya urn, Protein expression, Reverse phase protein array, Random partitions

1 Introduction

1.1 The Problem

We review a nonparametric Bayesian approach to clustering of experimental units that was introduced in [1, 2]. The main feature that distinguishes the proposed method from the previous literature is that clusters of experimental units $C = \{1, \ldots, n_C\}$ are defined by giving rise to a common (nested) clustering of a second set of experimental units $R = \{1, \ldots, n_R\}$. In one motivating application $c = 1, \ldots, n_C$ indexes proteins and $r = 1, \ldots, n_R$ indexes patients. That is, we define clusters of proteins as sets of proteins with respect to which patients can be grouped into the same clusters. The data are protein activation y_{rc} for protein c in patient r. If we arrange the data in a matrix $Y = [y_{rc}]$ with columns c corresponding to proteins and rows r corresponding to patients, then the desired clustering can be described as clustering of columns and a nested clustering of rows for each column cluster. That is, column clusters are *defined* by sharing the same arrangement of (nested) row clusters. Figure 1 illustrates these features with a stylized data matrix, arranged according to the cluster membership of proteins

Klaus Jung (ed.), *Statistical Analysis in Proteomics*, Methods in Molecular Biology, vol. 1362,
DOI 10.1007/978-1-4939-3106-4_8,

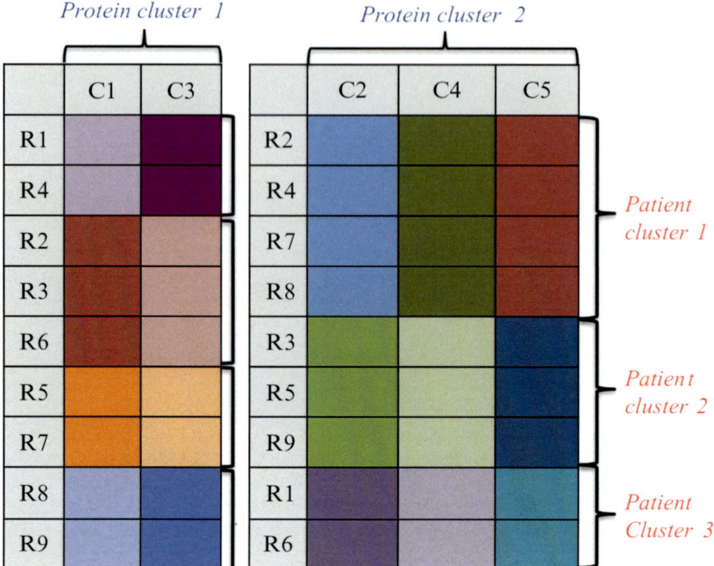

Fig. 1 Illustration of a nested clustering with $n_R = 9$ patients (*rows*) and $n_C = 5$ proteins (*columns*). Different *colors* represent different protein activation levels

and patients. Colors represent protein activation levels. Protein clusters 1 and 2 consist of proteins 1 and 3, and proteins 2, 4, and 5, respectively. The two protein clusters define their own patient partitions. For example, protein cluster 2 groups patients into four patient clusters. Within each of those four patient clusters, the colors are the same across the patients but vary across proteins, because protein clusters are defined by matching partitions of patients across all proteins rather than by matching activation levels. This is in contrast to previously proposed probability models for random partitions, which are (almost) all based on common parameters in the sampling model. Some of these approaches are reviewed below. However, in many applications investigators have in mind some common biologic process when interpreting clusters. For example, in the application with protein activation investigators want to find groups of proteins that are related to some common underlying process. We argue that defining clusters of proteins by a common pattern of grouping patients is closer to this desired biologic interpretation than the use of a common activation level shared by all proteins in a cluster. For example, a group of proteins might be part of the same molecular pathway and thus group patients in similar ways. But the activation of some proteins in the pathway might be down-regulated, while others are up-regulated. This would make it difficult to discover such grouping on the basis of, for example, a common location parameter in a sampling model for protein activation.

In the following discussion, for simplicity we will refer to experimental units as "proteins" and "patients," keeping in mind that the models are general, and assuming that the goal is to define a partition of proteins into subsets that give rise to a shared (nested) partition of patients. That is, we will refer to the columns as proteins and to the rows as patients. Using the particular example we will without any loss of generality greatly simplify the description. A partition or clustering of proteins is a family $\mathscr{C} = \{C_1, \ldots, C_{K_C}\}$ of subsets $C_k \subset C$ with $C_{k_1} \cap C_{k_2} = \varnothing$ and $\bigcup_{k=1}^{K_C} C_k = \{1, \ldots, n_C\}$. When we want to highlight that $\mathscr{C}$ is a partition of n_C units we write $\mathscr{C}_{n_C}$. Nested within each subset C_k we consider a partition $\mathscr{R}_k = \{R_{k\ell}; \ell = 1, \ldots, L_k\}$ of subsets $R_{kl} \subset \{1, \ldots, n_R\}$, again with $R_{k\ell_1} \cap R_{k\ell_2} = \varnothing$ and $\bigcup_{l=1}^{L_k} R_{kl} = \{1, \ldots, n_R\}$. It is often convenient to use an alternative notation for the random partitions by introducing cluster membership indicators. We will use $\gamma_c = k$ if $c \in C_k$ and $\rho_{kr} = l$ if $r \in R_{kl}$. We will use $n_k^\star = | C_k |$ and $n_{k\ell}^\star = | R_{k\ell} |$ to denote the size of the k-th column cluster and the l-th row cluster (nested within C_k). Finally, we will usually use a convention of indexing clusters by appearance. That is, $\gamma_1 = 1$ and $\gamma_{c+1} \le \max\{\gamma_1, \ldots, \gamma_c\} + 1$.

In summary $\mathscr{C}, \{C_k \text{ for } k = 1, \ldots, n_C\}, \gamma_c, n_k^\star$ characterizes the column partitions, and $\mathscr{R}_k, \{R_{k\ell} \text{ with } \ell = 1, \ldots, n_{Rk}\}, \rho_{kr}, n_{k\ell}^\star$ characterize the nested row partitions.

1.2 Random Partition Models

Similar problems of nested clustering arise frequently in applications. Several approaches have been proposed in the recent literature. Naturally, models for nested clustering build on an underlying model for marginal clusters. We therefore briefly review some of the common models that are used for Bayesian inference on random partitions.

A common way to define priors $p(\mathscr{C})$ on partitions is through an indirect definition by sampling from a discrete random probability measure. Let $\delta_x(\cdot)$ denote a point mass at x. Consider a discrete distribution $G(\theta) = \sum_h w_h \delta_{m_h}(\theta)$ with probability masses w_h in locations m_h. A random sample $\theta_c \mid G \overset{\text{iid}}{\sim} G$, $c = 1, \ldots, n_C$, implicitly defines a partition $\mathscr{C}$ by the following construction. Sampling from a discrete distribution implies a positive probability for ties, i.e., $\theta_{c_1} = \theta_{c_2}$ for some $c_1 \neq c_2$. Let $\theta_k^\star$, $k = 1, \ldots, K_C$, denote the $K_C \le n_C$ unique values of θ_c's and define $C_k = \{c : \theta_c = \theta_k^\star\}$. Then $\{C_1, \ldots, C_K\}$ defines a random partition of $\{1, \ldots, n_C\}$. In other words, i.i.d. sampling from G implies a random partition $p(\mathscr{C} \mid G)$. If G is a random probability measure, then $p(\mathscr{C})$ can be defined by marginalizing w.r.t. G.

The reader might now wonder whether the use of a random probability measure G is not an unnecessarily complicated way to define $p(\mathscr{C})$. There are some good reasons for this construction. First, the indirect construction ensures that the random partition for n experimental units arises from marginalizing a random partition for $n + 1$ units. That is, by ignoring the cluster membership of the $(n + 1)$-st unit. Let $\mathscr{C}_n$ denote a partition of n units. Then $p(\mathscr{C}_n) = \sum_{\gamma_{n+1}=0}^{n_{C+1}} p(\mathscr{C}_{n+1})$ (here n_C is the number of clusters for n units). Such coherence across sample sizes is desirable. It would be embarrassing if inference on a partition $\mathscr{C}$ for n units were to depend upon whether or not we will ever consider an $(n + 1)$-st experimental unit. Another important reason is the following. Any exchangeable sequence of random partitions $p(\mathscr{C}_n)$ can be represented as arising from random sampling under a discrete random probability measure G. Here $p(\mathscr{C}_n)$ is called exchangeable if it is invariant under permutations of the indices $\{1,\ldots,n_C\}$. The result is known as Kingman's representation theorem [3, 4]. See, for example, [5], for a recent review.

1.2.1 DP Random Partition

The perhaps most popular random partition model $p(\mathscr{C})$ is the random partition that is induced by random sampling from a Dirichlet process (DP) random probability measure [6]. See, for example, [7] for a recent review of the DP and related models. The generative model for the implied random partition is as follows. First we generate $G \mid M, G_0 \sim \text{DP}(MG_0)$ from a DP prior. Next generate an i.i.d. sample $\theta_c \mid G \overset{\text{iid}}{\sim} G$, $c = 1,\ldots,n$. Now record the configuration of ties to define clusters S_j. The implied prior $p(\mathscr{C})$ is known as the Pólya urn. It is characterized by the following sequential rule. Let $n_k^{c\star} = |\{c' \le c : s_{c'} = k\}|$ denote the size of the k-th cluster up to and including the c-th unit. Similarly, let K_C^c denote the number of clusters among the first c units. Conditional on the cluster membership of the first n units, $(\gamma_1,\ldots,\gamma_n)$, the distribution of the cluster membership for the $(n + 1)$-th unit is

$$p(\gamma_{c+1} = k \mid \gamma_1,\ldots,\gamma_c) = \begin{cases} n_k^{c\star} / (c+M) & \text{for } k \le K_C^c \\ M / (c+M) & \text{for } k = K_C^c + 1, \end{cases}$$

for $c = 2,\ldots,n_C$, and $\gamma_1 = 1$. Multiplying over $c = 2,\ldots,n_C$ we find

$$p(\mathscr{C}) = \frac{M^{K_C - 1} \prod_{k=1}^{K_C} (n_k^\star - 1)!}{(M+1)\cdot\ldots\cdot(M+n-1)}. \tag{1}$$

There is an important subtlety about the representation of the partition. Recall that clusters are indexed by order of appearance. If, however, the partition is recorded without any restrictions on the labels to identify the clusters, then an additional factor $1/K_C!$ needs to be added to (1). This is because, for example, $\gamma = (1,1,1,2,3,2)$ and $\gamma' = (2,2,2,3,1,3)$ record the same partition. The additional factor accounts for the $K_C!$ different ways of coding the partition with cluster membership indicators. Usually, an additional convention is added, requiring labeling by order of appearance. In the last example, this would single out partition γ only. Given γ, $\theta_k^\star$, $k=1,\ldots,K_C$ is a random sample of size K_C from G_0 and let $\theta_c=\theta_k^\star$ for any c with $\gamma_c=k$. Some other random partition models can be used (*see* **Note 1**). Also, *see* **Note 2** for some comments on a choice of G_0.

1.2.2 Model Based Clustering

A similarly popular prior on random partitions is model based clustering [8, 9]. The basic idea is simple. Consider sampling from a mixture model $y_c \overset{iid}{\sim} \sum_{h=1}^{H} w_h f(y_c \mid \theta_h^\star)$, $c=1,\ldots,H$, for example, a location mixture of normal distributions with $f(\cdot \mid \theta_h^\star) = \mathrm{N}(\cdot \mid \theta_h^\star, \sigma^2)$ for known σ^2. Next we replace the mixture model with an equivalent hierarchical model by introducing latent variables $\gamma_c \in \{1,\ldots,H\}$ with

$$\begin{aligned} y_c \mid \gamma_c = h, \theta^\star &\overset{\text{indep}}{\sim} f(y_c \mid \theta_h^\star), \\ p(\gamma_c = h) &= w_h. \end{aligned} \tag{2}$$

If we interpret $\gamma = (\gamma_1,\ldots,\gamma_n)$ as cluster membership indicators, then (2) implicitly defines a prior $p(\mathscr{C})$. A minor detail is the labeling of clusters. If order of appearance labeling is desired, then the labels γ_c might have to be re-arranged. Also, it is possible that only $K \le H$ distinct labels are generated (this is why we used H instead of K for the number of terms in the mixture model). There is a connection between model based clustering and the DP random partitions [10]. Consider model based clustering with a Dirichlet prior $(w_1,\ldots,w_H) \sim \mathrm{Dir}(\delta_H,\ldots,\delta_H)$ for the weights in (2) and $\theta_h^\star \mid G_0 \overset{\text{iid}}{\sim} G_0$. For fixed n consider now a limit with $H \to \infty$ and $\delta_H \to 0$, such that $H\delta \to \alpha > 0$. The limiting model on $(y_1,\ldots,y_n)$ is identical to a DP mixture model $y_c \mid G \overset{\text{iid}}{\sim} \int f(y_c \mid \theta)\, G(d\theta)$, with $G \mid \alpha, G_0 \sim \mathrm{DP}(\alpha G_0)$.

Another large class of random partition models are the product partition models (PPM) [11, 12]. See also [13] for a review. Like the DP prior and like model based clustering the PPM model introduces cluster-specific parameters in the sampling model that are usually shared across all units in the cluster.

1.3 Nested Random Partition Models

Rodríguez et al. [14] define the nested DP, which can be used to define nested random partitions. We start with a partition of the columns. The prior $p(\mathscr{C})$ is exactly the Pólya urn prior of (1). Next, for each column cluster C_k the model defines a (discrete) random probability measure G_k. Using G_k we then define a random partition of the rows. The critical detail is that each column has its own column-specific partition of the rows, $\mathscr{R}_c$. The prior on $\mathscr{R}_c$ is defined by G_k, that is $p(\mathscr{R}_c \mid G_k)$. All columns $c \in C_k$ share the same *prior* $p(\mathscr{R}_c \mid G_k)$. This is meaningful for the motivating application in [14], but would not be useful for our two motivating applications.

Wade et al. [15] define the enriched DP, which can be used to define a prior on column clusters and nested row-clusters, with the same limitation (for our application) as the nested DP. The partitions $\mathscr{R}_c$ are column-specific and only share a common prior across all $c \in C_k$ in a column cluster.

There are many other approaches to inference on nested partitions that have been proposed in the recent literature. See, for example, [1] for more discussion. A common feature of these approaches is that they define a random partition indirectly, through the specification of a (discrete) random probability measure G. A random partition is implied by sampling some parameter $\theta_i \sim G$. The discrete nature of G implies a positive probability for ties. The arrangement of ties then defines a partition. In contrast, Lee et al. [1] define a Bayesian approach to local clustering that defines a prior on the partitions only, completely unrelated to the prior assumptions for cell-specific parameters. The prior on the nested partition is defined directly.

Rodríguez and Ghosh [16] define a model that comes closer to the desired inference. Their model includes $\rho_c = \rho_{c'}$ for any $\{c, c'\} \subset C_k$. The critical limitation for the desired application in our motivating examples is that the model requires common parameters $\theta_{rc} = \theta_{rc'}$. Recall the initial discussion. For a biologically meaningful interpretation of protein clusters we want proteins to be reported in the same cluster when they give rise to the same partition of patients. This could include a situation where protein c is activated in the same set of patients in which protein c' is deactivated.

2 Motivating Applications

In the following discussion we review an approach proposed in [1, 2]. The proposed model is motivated by the following two case studies.

2.1 Protein Activation

Lee et al. [1] consider the following problem. Figure 2 shows levels of protein activation for $n_C = 55$ proteins in samples from $n_R = 256$ breast cancer patients. The data come from a reverse phase protein array (RPPA) experiment that allows to record protein activation for a large number of samples and proteins simultaneously. The idea is to identify sets of proteins that give rise to a clinically meaningful clustering of patients. The eventual goal is to exploit these subgroups for the development of optimal adaptive treatment allocations. This goal is in the far future. The immediate goal is to identify subsets of proteins that are characterized by a common partition of patients. Different biologic processes give rise to different clusterings of the patient population. If subsets of proteins characterize different biologic processes, this is formalized as different partitions of patients with respect to each protein subset.

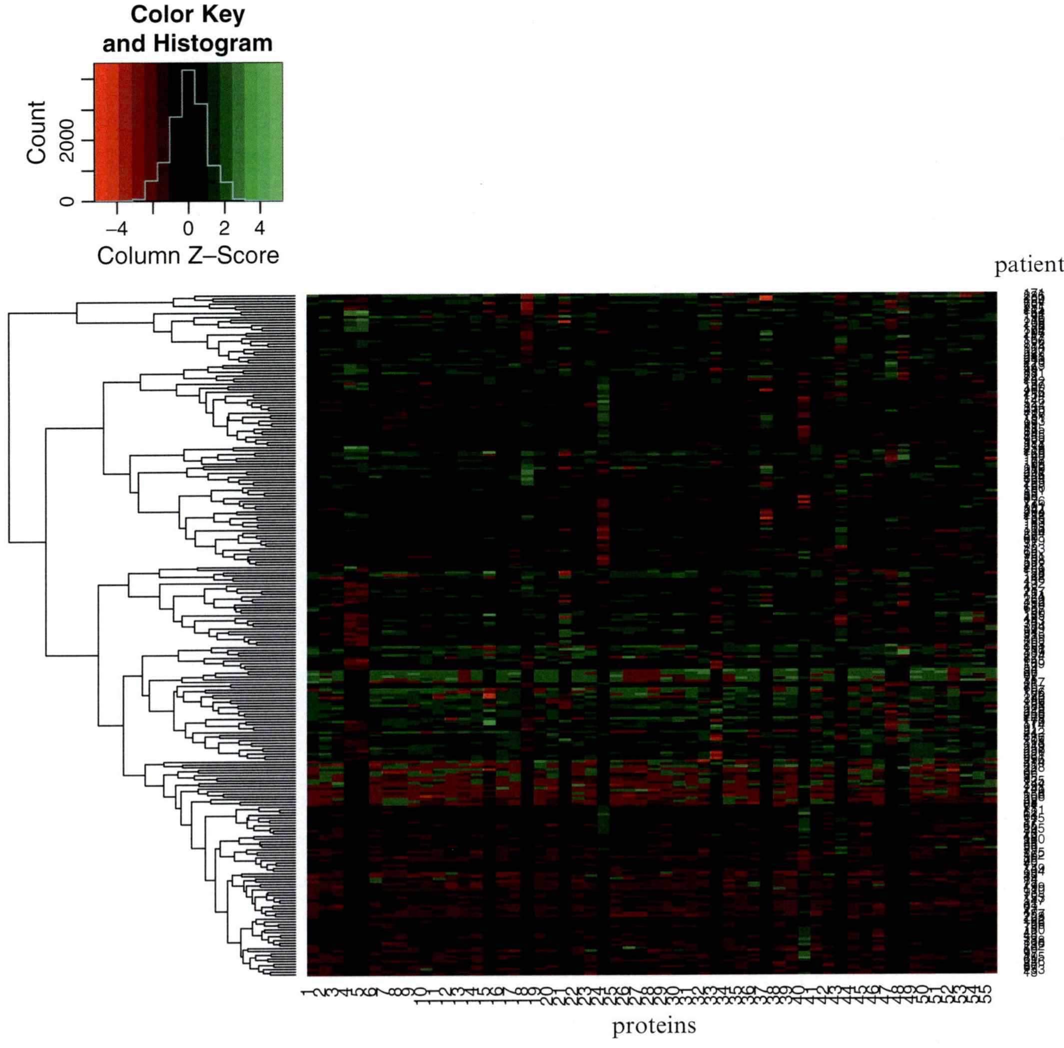

Fig. 2 RPPA Data. Protein activation for $n_R = 256$ patients (*rows*) and $n_C = 55$ proteins (*columns*). The dendrograms are estimated using hierarchical clustering

2.2 Clustering Histone Modifications

Xu et al. [2] consider bi-clustering of data for histone modification (HM) counts. Histones are small proteins around which DNA strands are wrapped around, in a way that has been characterized as "beads on a string." These histone proteins undergo modifications which play an important role in transcription [17, 18]. The detail process by which these histone modifications (HM) help to control gene transcription is known as histone code [19], details of which remain largely unknown. In particular, not much is known about how combinations of different HMs work together.

To learn about some of these details, Xu et al. [2] analyze ChIP-Seq data on HMs. ChIP-Seq is an experimental platform that allows, among many other uses, to report counts of HMs by genomic position. We analyze ChIP-Seq data from CD4+ T lymphocytes [20, 21], in which 39 types of HMs are reported. We consider a small subset of the ChIP-Seq data covering randomly selected 50 genomic locations in promoter regions and 50 genomic locations in insulator regions. That is, the data is a 100 × 39 matrix $[y_{rc}]$ with rows corresponding to $n_R = 100$ genomic loci and columns corresponding to the $n_C = 39$ HMs. The goal of this analysis is to find clusters of HMs that might share a similar function in the histone code. We formalize the desired aim as finding clusters of HMs that give rise to a common grouping of genomic locations. That is, we consider a nested partition of loci within a partition of HMs. The data are shown in Fig. 4.

3 Model

Let γ_c denote cluster membership indicators to index a partition of proteins into $\{C_k; k = 0, \ldots, K_C\}$. For each subset C_k we define a protein-set specific partition of patients with respect to that protein set. Let ρ_{kr} denote cluster membership indicators for the partition with respect to the k-th protein set. We seek to identify a partition of proteins and a nested partition of patients, nested within each protein set. Figure 3 indicates protein clusters by vertical white lines, and the nested patient clusters by horizontal lines. There are two more important details. We allow for some proteins to be unrelated to any biologic process of interest, i.e., they do not give rise to a nested partition of patients. We collect these proteins in a special cluster $\gamma_c = 0$ and define nested patient partitions only for clusters $k = 1, \ldots, K_C$. Similarly, we allow for the fact that some patients might not meaningfully fall into clusters with respect to a particular protein set. We combine these patients in the special patient cluster $\rho_{kr} = 0$.

Lee et al. [1] define a prior model $p(\gamma)$ for protein clusters, and conditional on γ a prior $p(\rho \mid \gamma) = \prod_{k=1}^{K_C} p(\rho_k)$ on random patient clusters with respect to each of the protein sets. Recall that $n_k^\star$ denotes the size of the k-th protein cluster. We assume

$$p(\gamma) = \pi_0^{n_0^\star}(1-\pi_0)^{n_C - n_0^\star} \frac{M^{K_C - 1} \prod_{K_C}^{k=1} (n_k^\star - 1)!}{(M+1)\cdot \ldots \cdot (M + n_C - n_0^\star - 1)} \tag{3}$$

and similarly we define a prior for the partition ρ_k of patients with respect to the k-th protein set,

$$p(\rho_k \mid \gamma) = \pi_1^{n_{k0}^\star}(1-\pi_1)^{n_R - n_{k0}^\star} \frac{\alpha^{L_k - 1} \prod_{\ell=1}^{L_k} (n_{k\ell}^\star - 1)!}{(\alpha+1)\cdot \ldots \cdot (\alpha + n_R - n_{k0}^\star - 1)}. \tag{4}$$

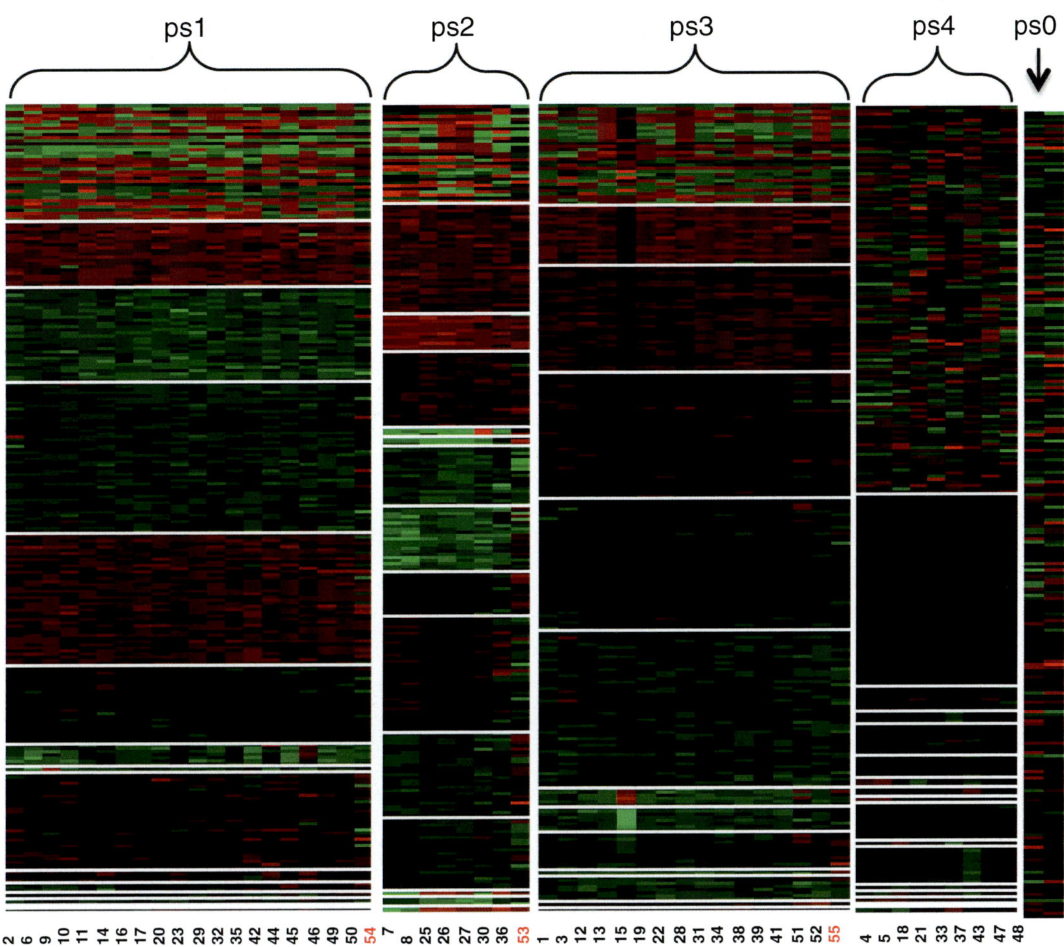

Fig. 3 RPPA Data. The image shows protein activation by patient (*rows*) and proteins (*columns*). The *vertical white lines* separate subsets W_k of proteins that give rise to different partitions of patients. Partitions of patients are indicated by *horizontal white lines*

The prior $p(\gamma)$ can be characterized as a zero-inflated Pólya urn. Each protein is in cluster C_0 with probability π_0. The proteins that are not in C_0 form clusters $C_1,\ldots,C_{K_C}$ using the Pólya urn defined in (1), with total mass parameter M. Similarly, $p(\rho_k \mid \gamma)$ is another zero-inflated Pólya urn, with probability π_1 for the zero cluster and a Pólya urn with total mass parameter α for the remaining patients.

The prior model for the nested clusters on columns and rows is completed to a full inference model by defining a sampling model for observed data y_{rc}. Any sampling model $p(y_{rc} \mid \gamma, \rho)$ can be used. Details depend on the problem. Typically, the sampling models will include additional hyperparameters, cluster-specific and/or unit-specific parameters. Below we briefly summarize the sampling models that were used with the two motivating applications.

4 Results

4.1 RPPA Data

We implement inference on the desired nested clustering of patients with respect to different protein sets using the prior model (3) and (4), together with a hierarchical normal sampling model.

Specifically, we use a normal sampling model with protein and patient cluster-specific location parameters θ_{rc}. We assume that all patients r in the same cluster R_{kl} share the same parameters $\theta_{rc} = \theta^{\star}_{\ell c}$ for all proteins in a protein set, $c \in C_k$. Importantly, we only assume shared parameters across patients in a patient cluster R_{kl}, but allow for different parameters across the proteins $c \in R_k$. This is important, as it allows to cluster together proteins that are consistently down- and up-regulated together, as desired. In summary, we assume

$$y_{rc} \mid \theta^{\star}_{\ell c} \overset{\text{ind}}{\sim} \mathsf{N}(\theta^{\star}_{\ell c}, \sigma^2_{\ell}), \text{ for all } r, c \text{ with } \gamma_c = k \geq 1 \text{ and } \rho_{kr} = \ell \geq 1. \tag{5}$$

For idle clusters, that is $\gamma_c = 0$ and/or $\rho_{kr} = 0$ we use separate models. We use $y_{rc} \overset{\text{ind}}{\sim} \mathsf{N}(m, \sigma^2_2)$ for idle patients, that is, for $\gamma_c = k \geq 1$ and $\rho_{kr} = 0$. We use $y_{rc} \overset{\text{ind}}{\sim} \mathrm{N}(m, \sigma^2_3)$ for idle proteins, that is, $\gamma_c = 0$. The model is completed with conjugate priors on $\theta^{\star}_{\ell c}$, m and σ^2_{ℓ}, $\ell = 1, 2, 3$.

The posterior distribution $p(\gamma \mid y)$ and $p(\rho_k, k = 1, \ldots, K_C \mid \gamma, y)$ is a probability model on a partition of patients nested within protein sets. Protein sets are formed to give rise to meaningful patient clusters. Figure 3 shows a summary of the posterior distribution on the random partitions. The displayed partition is found by the algorithm proposed in [22]. The code developed for this example can be available upon request (*see* **Note 3**).

4.2 Histone Modifications

We implement inference under the proposed nested clustering model for clustering the $n_C = 39$ HMs based on nested partitions of the $n_R = 100$ genomic loci. Here we denote with $\mathscr{C} = \{C_1, \ldots, C_{K_C}\}$ the partition of HMs and with $\mathscr{R}_k = \{R_{k1}, \ldots, R_{kL_k}\}$ the partition of loci with respect to the HMs in the k-th cluster. Equivalently $\mathscr{C}$ is be represented by γ and $\mathscr{R}_k$ is represented by ρ_k.

We define a sampling model for the observed HM counts y_{rc}. Let $\mathsf{Poi}(\theta)$ denote a Poisson distribution with mean θ. We assume a Poisson sampling model for the count data, $y_{rc} \sim \mathsf{Poi}(\theta_{rc})$. The prior probability model for θ_{rc} makes use of the clustering by assuming a common value for all loci in the same cluster, $r \in R_{kl}$. That is, $\theta_{rc} = \theta^{\star}_{\ell c}$ for all $r \in R_{kl}$ and $c \in C_k$. Importantly, HMs in the same HM cluster C_k share the same partition R_k of rates only, not the same rate.

We complete the model with a hierarchical prior $\theta^{\star}_{\ell c} \sim \text{Gamma}(\kappa_{0c}, \lambda_{0c})$. Here $\text{Gamma}(a, b)$ denotes a gamma

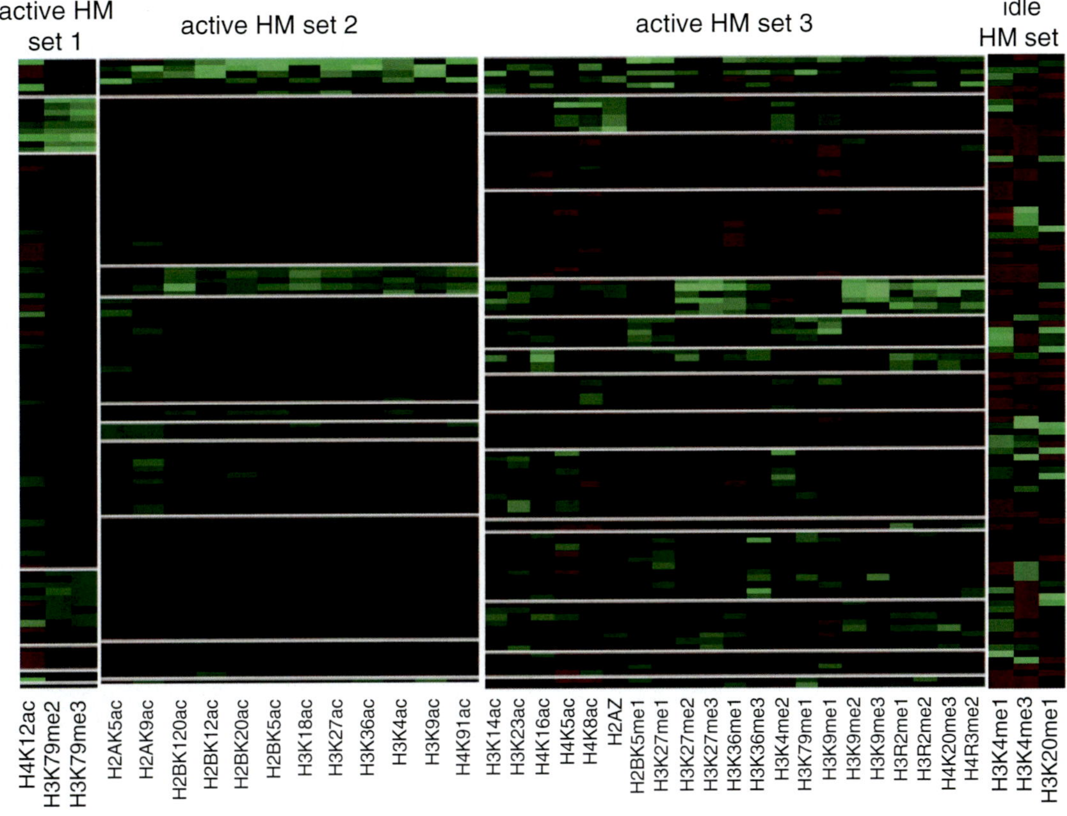

Fig. 4 HM Data. Heatmaps for $K_C = 3$ active HM sets. *White horizontal lines* indicate division of location clusters R_{kl}. The *rightmost column* block of HMs is the set of idle HMs

distribution with mean *a/b*. For the idle genomic locations in the active HM sets, i.e., locations with $\rho_{kr} = 0$ for $k \geq 1$, we assume $\theta_{rc} \sim \text{Gamma}(\kappa_{1c}, \lambda_{1c})$. For idle HMs, that is, $\gamma_c = 0$, we assume $\theta_{rc} \sim \text{Gamma}(\kappa_{2c}, \lambda_{2c})$, i.i.d.

Figure 4 shows the posterior estimated partition $\mathscr{C}$ of HMs, together with the nested partition of loci that gives rise to $\mathscr{C}$. The code developed for this example can be available upon request (*see* **Note 3**).

5 Conclusions

We have reviewed an approach to model-based inference on nested clustering. We illustrated the proposed method in two case studies with protein activation data from RPPA experiments and histone modification counts from ChIP-Seq experiments. The substantially different nature of the data types and the required sampling models in these two case studies highlights the principled nature of the proposed model construction. We first construct a probability model for the underlying partitions, still without exploiting any details of the sampling model. Then only, conditional on the assumed partitions, we define the sampling model, as needed for the application. We believe that it is good practice in statistical modeling to separate the prior on the random partition and the construction of the sampling model.

6 Notes

1. The Dirichlet process used in the proposed method for random clustering can be replaced with other nonparametric Bayesian priors such as the two-parameter Poisson-Dirichlet process for generalization.
2. The sampling distribution and G_0 (the distribution of the DP base measure) should be carefully chosen for analysis. Joint calibration of G_0, M, and α may be needed.
3. The code developed for the examples presented in the chapter is available upon request (contact Juhee Lee at juheelee@soe.ucsc.edu).

Acknowledgements

Yuan Ji and Peter Müller's research is partially supported by NIH R01 CA132897.

References

1. Lee J, Müller P, Zhu Y et al (2013) A nonparametric Bayesian model for local clustering with application to proteomics. J Am Stat Assoc 108:775–788
2. Xu Y, Lee J, Yuan Y et al (2013) Nonparametric Bayesian bi-clustering for ChIP-Seq count data. Bayesian Anal 8:1–22
3. Kingman JFC (1978) The representation of partition structures. J Lond Math Soc 18:374–380
4. Kingman JFC (1982) The coalescent. Stoch Process Appl 13:235–248
5. Lee J, Quintana F, Müller P et al (2013) Defining predictive probability functions for species sampling models. Stat Sci 28:209–222
6. Ferguson TS (1973) A Bayesian analysis of some nonparametric problems. Ann Stat 1:209–230
7. Ghoshal S (2010) The Dirichlet process, related priors and posterior asymptotics. In: Hjort NL, Holmes C, Müller P, Walker SG (eds) Bayesian nonparametrics. Cambridge University Press, Cambridge, pp 22–23
8. Dasgupta A, Raftery AE (1998) Detecting features in spatial point process with clutter via model-base clustering. J Am Stat Assoc 93:294–302
9. Fraley C, Raftery AE (2002) Model-based clustering, discriminant analysis, and density estimation. J Am Stat Assoc 97:611–631
10. Green PJ, Richardson S (2001) Modelling heterogeneity with and without the Dirichlet process. Scand J Stat 28:355–375
11. Hartigan JA (1990) Partition models. Commun Stat Theory Methods 19:2745–2756
12. Barry D, Hartigan JA (1992) Product partition models for change point problems. Ann Stat 20:260–279

13. Quintana FA, Iglesias PL (2003) Bayesian clustering and product partition models. J R Stat Soc Ser B 65:557–574
14. Rodríguez A, Dunson DB, Gelfand AE (2008) The nested Dirichlet process, with discussion. J Am Stat Assoc 103:1131–1144
15. Wade S, Mongelluzzo S, Petrone S (2011) An enriched conjugate prior for Bayesian nonparametric inference. Bayesian Anal 6:359–385
16. Rodríguez A, Ghosh K (2012) Modeling relational data using nested infinite relational models. Technical Report, Department of Applied Mathematics and Statistics, University of California, Santa Cruz
17. Bernstein BE, Humphrey EL, Erlich RL et al (2002) Methylation of histone H3 Lys 4 in coding regions of active genes. Proc Natl Acad Sci USA 99:8695–8700
18. Roh TY, Cuddapah S, Zhao K (2005) Active chromatin domains are defined by acetylation islands revealed by genome-wide mapping. Genes Dev 19:542–552
19. Strahl BD, Allis CD (2000) The language of covalent histone modifications. Nature 403:41–45
20. Barski A, Cuddapah S, Cui K et al (2007) High-resolution profiling of histone methyla- tions in the human genome. Cell 129:823–837
21. Wang Z, Zang C, Rosenfeld JA et al (2008) Combinatorial patterns of histone acetylations and methylations in the human genome. Nat Genet 40:897–903
22. Dahl DB (2006) Model-based clustering for expression data via a Dirichlet process mixture model. In: Vannucci M, Do K-A, Müller P (eds) Bayesian inference for gene expression and proteomics. Cambridge University Press, Cambridge

Chapter 9

Set-Based Test Procedures for the Functional Analysis of Protein Lists from Differential Analysis

Jochen Kruppa and Klaus Jung

Abstract

The analysis of most high-throughput proteomics experiments involves the selection of differentially expressed proteins or peptides between two different sets of samples, e.g., from two experimental groups. As a result, a large list of selected features is reported, typically sorted by a measure for the expression fold change and a p-value from a statistical test. The biological interpretation of such a list is usually difficult since the features can typically be assigned to a large variety of biological classes. To facilitate the biological interpretation, set-based procedures focus on the analysis of feature subsets that all belong to the same biological class (e.g., same cellular component, biological process, molecular function, or pathway). Set-based procedures can roughly be divided into "enrichment methods" and "global test procedures," where the first involve all features of an experiment and the second only those features of a particular set. In this chapter we detail the working principle of these kind of statistical methods and describe how features can be classified into molecular subsets. We illustrate the use of the methods on a data example from a proteomics Parkinson study.

Key words Enrichment test, Differentially expressed proteins, Gene ontology, Global test, Pathway database

1 Introduction

Many experiments in proteomics focus on the comparison of two experimental groups by searching for differentially expressed proteins or peptides. The term "differentially expressed" means that the identified features are significantly up- or downregulated in the one group compared to the other. The motivations for this kind of experiment can be diverse. Differentially expressed proteins can for example serve as biomarkers or as "molecular signatures" for diagnosis, prognosis, and prediction [1–4]. In other medical examples, differentially expressed proteins are supposed to serve as targets for new drugs [5, 6].

Typically, a list of differentially expressed proteins can comprise hundreds of features which can play various roles in the

Klaus Jung (ed.), *Statistical Analysis in Proteomics*, Methods in Molecular Biology, vol. 1362,
DOI 10.1007/978-1-4939-3106-4_9, © Springer Science+Business Media New York 2016

molecular processes. These features can for example be related to different cellular components, biological processes, and molecular functions [7], or they can be related to different molecular pathways [8]. This diversity often makes the interpretation of a list of differentially expressed features difficult. One type of statistical and computational tool that can help to analyze such lists with respect to the molecular role of the included features is set-based procedures.

Set-based procedures can roughly be divided into two types. One type of set-based procedure is enrichment tests [9]. These tests study whether features with a certain molecular characteristic are more frequent among the set of differentially expressed features compared to their occurrence among the non-differential ones. Since this type of procedure includes all features of an experiment these tests are also called "competitive." In contrast, "self-contained" tests only utilize the expression levels of those features that belong to a specified molecular characteristic, e.g., a biological process or a molecular pathway. Self-contained procedures usually are also called global tests [10–12]. One advantage of global test procedures is also that they can detect a global effect in the case that none of the features in a set is individually significant but if many features of the set show a rather weak effect.

While set-based procedures are very frequently applied and well known in the area of genomics, they are not regularly used in proteomics. In this chapter, we review available methods for set-based analysis in proteomics and illustrate their use in a data example. In Subheading 2 we describe how a proteomic feature set can be defined by database information and detail next the basic principles of enrichment and global test procedures, respectively. In Subheading 3 we apply the described methods to a data example from a proteomics study on samples from patients with Parkinson's disease. We close the chapter with a short discussion and notes.

2 Methods

2.1 Functional Annotations from Public Databases

The usage of set-based procedures requires at first a definition of the sets. There are many publicly available databases for different kinds of biological background. In the following, four popular databases are presented: the UniProt database as a representative for a protein database, the GO and KEGG database with gene-pathway information, and finally the Ensemble database with a broad overview on gene information.

The UniProt database [13] (http://www.uniprot.org/) consists of different tools and knowledgebase like the UniProt Knowledgebase (UniProtKB) as a central hub for the functional information on proteins, the UniParc database as a comprehensive and non-redundant source of publicly available protein sequences, and the UniProt Reference Cluster (UniRef) with access to clustered sets of sequences from the UniProtKB and selected UniParc records. The information is clustered and can be sorted by organisms from human to animal like mouse to popular plant organisms like *Arabidopsis thaliana* or rice. Different types of information on the protein records are available like amino acid sequence, protein name or description, taxonomic data, citation information, and even connection to GO categories or gene entries. Therefore as much as possible information on the annotation is added to each record. There are mainly two types of records to be distinguished: reviewed (Swiss-Prot) and therefore manually annotated records with information extracted from literature and evaluated analysis (548,208 records over all organisms) or unreviewed (TrEMBL) and therefore computationally analyzed records without manually checks (46,714,516 records over all organisms). As an example 47,827 reviewed records (1,991,882 unreviewed records) are available for human (http://www.uniprot.org/uniprot/?query=Human). Finally, the Uniprot database allows an easily applicable access to the data export by a fast parameter selection and a high variety of download file possibilities: Excel, XML, FASTA, plain text, FTP server, and more. Unresolved questions can be solved by many tutorials and video support.

The Gene Ontology project [7] (http://geneontology.org/) database consists of hierarchical gene ontologies with the effort to describe gene products across different databases. Three structured and controlled ontologies have been developed: associated biological processes, cellular components, and molecular functions. Different organisms are available including the world's major repositories for human (including 43,410 annotated gene products), plant, animal, and microbial genomes. It has to be mentioned that the GO terms are all nested and therefore smaller GO categories are included in bigger ones and so on. Therefore it might be interesting to use GO slim (http://geneontology.org/page/go-slim-and-subset-guide) as a cutdown version of the GO ontologies without the specific fine-grained terms. In general, there are two types of GO slims: maintained and achieved. Both can be achieved from the home page for different species and science topics. In addition, there is the possibility to generate own GO slim databases. GO terms are also available in the UniprotKB database for each protein. Finally, for a sophisticated use of the GO database Blake (2013) announced ten quick tips for using the gene ontology [14].

The Kyoto Encyclopedia of Genes and Genomes (KEGG) database [8] (http://www.genome.jp/kegg/) includes information on different pathways and is divided into four categories: systems information, genomic information, chemical information, and health information. The mapping can be done by an internal KEGG mapper. The number of available pathways is very large and the generated pathways complex and detailed. According to the design principle KEGG is a "computer representation" of the biological system. In contrast to the GO categories, the KEGG database is not as hierarchical. Still, due to the nature of the pathways there are some overlaps, but the KEGG pathways tend to be larger than the GO categories. The FTP access is available by subscription; still the website delivers a free access to the overall data.

The Ensemble database [15] (http://www.ensembl.org/) produces genome databases for vertebrates and other eukaryotic species. The information gained among others includes the Ensemble Gene database, the Ensemble Variation database, the Ensemble Regulation database, and the Vega database. A broad spectrum of information can be achieved given a wide set of filters and attributes. Among others DNA sequence information can be downloaded and the ENCODE data (Encyclopedia of DNA Element) is implemented and can be achieved. One strength of the database is the close connection to the statistic software R (http://www.r-project.org/) by the R-package "biomaRt" [16] with easy access and data handling. Among much other information the GO categories connected to specific genes or locations can be achieved by the Ensemble database.

2.2 Enrichment Tests

When comparing two experimental groups with regard to differentially regulated proteins, one statistical test is typically performed per protein reflecting the significance of differential expression. The raw *p*-values from this multiple testing procedure then are adjusted to control the so-called false discovery rate (FDR), i.e., the proportion of false-positive detections among all positive test decision. By setting a threshold for the FDR one can divide all proteins of the experiment into those which are differentially expressed and those that are not.

Since the differentially expressed features usually are related to different biological annotations, the interpretation of the resulting list is often difficult. Therefore, the researchers might be interested whether certain biological annotations appear relatively more often among the differentially expressed proteins compared to their appearance among the non-differential features. This question can be studied by set-based enrichment tests. The set definition depends strongly on the biological background of the experiment (*see* previous section). In the context of genomics a variety of enrichment tests have been proposed. In the following we want to detail some of them.

One of the first presented enrichment tests was based on Fisher's exact test [17]. This test can be used to compare the proportion of differentially regulated proteins related to the functional annotation of interest to the proportion on non-differentially regulated proteins related to this annotation. Consider for example an experiment including 1000 proteins in total, where 200 of them were detected as differentially expressed between two experimental groups. Further let 25 (12.5 %) of the differential features be related to apoptosis compared to 40 (5 %) proteins of the 800 non-differential features. Fisher's exact test would yield a *p*-value of<0.01, meaning that apoptosis-related proteins are relatively more frequent in the set of differentially regulated features. Note that the resulting *p*-value can change after adjusting again for multiple testing (since multiple annotations can be tested simultaneously) (*see* **Note 1**).

Another class of enrichment tests compares the distribution of *p*-values between the set of features related to a certain annotation and the set of features that are not related to the annotation of interest. In most cases the nonparametric Mann–Whitney *U* test or the Kolmogorov-Smirnov goodness-of-fit test are used to compare these distributions [18]. It was proposed to perform these tests with one-sided hypothesis to be able to conclude that *p*-values for the proteins related to the specified annotation are significantly higher than the *p*-values for the protein not related to that annotation.

While enrichment tests are widely used in the area of genomics, only very few methods have been proposed for enrichment tests in proteomics [19, 20].

2.3 Global Test Procedures

While enrichment tests are mainly based on the order of features derived by a differential analysis (e.g., order of *p*-values or fold changes)—and thus are depended on the procedure for ordering—global test procedures are directly based on their expression levels. Global tests on a defined set of proteins can detect a global group effect for example when many proteins of the set are not significant individually but the effect is present for the set as a whole.

Statistically, global test procedures can be performed in different ways. One way is to perform a test to compare two multivariate means, where in particular the number of features is larger than the group sizes. In the context of gene expression data Mansmann and Meister [11] proposed an analysis of covariance (ANCOVA) model for comparing the groups' means as one null hypothesis and for testing a (group × gene)-interaction as a second null hypothesis. To overcome the problem of the high dimensionality they propose to perform permutation tests. Their ANCOVA model also allows for adjustment of covariate effects.

Another way of performing a global test is to regard the group levels (coded, e.g., by 0 for control and 1 for treatment) as the dependent variable of a logistic regression model and the expression data as independent variables. One can than test the hypothesis that the group level is independent from the expression data [12].

Both the ANCOVA approach and the logistic regression approach are implemented in R-packages, "GlobalAncova" and "globaltest," and are available from the website http://www.bioconductor.org/.

Currently only very few approaches have been proposed for global test procedures specifically for protein expression data. One approach is based on Hotelling's T statistic and focuses on protein expression data from mass spectrometry [21]. Another global approach considers protein expression data from 2-dimensional gel electrophoresis with missing values [12]. The latter one is available in the R-package "RepeatedHighDim" from the website http://cran.r-project.org/.

3 Analysis of Data Example

3.1 Parkinson Data and Functional Annotation

We illustrate the use of set-based procedures on a data example from a Parkinson study on $n1 = 72$ patients and $n2 = 72$ healthy controls [22]. Expression data was generated using protein microarrays. In total, 9483 proteins were observed. Data were normalized the quantile method and as quality control hierarchical cluster analysis with single linkage was performed to detect potential outliers.

Functional annotation in the form of GO terms [7] was retrieved from the ensemble database using the R-package "biomaRt." In total, annotation was available for only 4839 proteins. Since some proteins are related to multiple GO terms, 9083 GO terms were found in total. Most of these GO terms were only related to one protein. In the following, only GO terms related to more than one gene were considered in the set-based analyses. These were 5231 terms. The distribution of set sizes is illustrated in Fig. 1, where set sizes > 100 were excluded. In fact, the third quartile of the distribution was given at a size of 8, while the maximum size observed for the 5231 terms was 4481. Over 4000 sets had a size of ≤10.

3.2 Enrichment Analysis

Before doing enrichment analysis proteins need to be divided into those which are significantly different between the samples from patients and controls. Therefore, a differential analysis was performed using the linear models implemented in the R-package "limma." Besides a statistical test result, the log fold change was reported for each protein. Using the test results and the fold changes, different criteria can be applied to divide proteins into a group of selected and a group of non-selected proteins. Here, we use two different criteria: (a) a threshold for the p-values, and (b) a threshold for the fold changes. While the p-value criterion reflects statistical significance, the fold change criterium reflects biological relevance. It is also possible to combine the two criteria to a third one; that is, a protein is selected if it is statistically significant according to a defined significance level and biologically relevant according to a specified fold change threshold. In the Parkinson data 316 proteins showed a p-value < 0.05. According to hard FDR

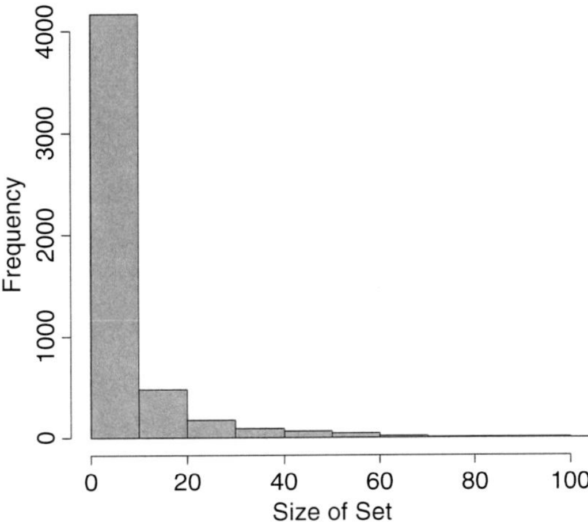

Fig. 1 Distribution of set sizes in the Parkinson data example. Most protein sets were rather small, with the third quartile of the distribution being 8. The *x*-axis was cut at 100; also there were some few sets with a size larger than 100 proteins (*see* main text)

criteria; that is, after *p*-value adjustment for multiple testing, none of the proteins would have been selected. Nevertheless, in order to be able to perform exemplarily the enrichment analysis, we use here the unadjusted *p*-values for selection, yielding 316 selected proteins. Since the observed expression changes are rather small in the Parkinson data, we use exemplarily also a small threshold for the log fold change of $\log 2(1.1) = 0.14$. With this threshold, 1439 proteins were selected. The selection process is also illustrated in the "volcano plot" in Fig. 2.

When using the unadjusted *p*-values as selection criterion (i.e., with 316 selected and 9167 non-selected proteins), 98 GO terms are enriched according to the unadjusted *p*-values of Fisher's exact test, and 348 GO terms according to Mann–Whitney *U* tests. After FDR adjustment, none of these test results remains significant. The top five findings are presented in Tables 1 and 2, respectively.

When using the fold change criterion (i.e., with 1439 selected and 8044 non-selected proteins) 177 GO terms are enriched according to the unadjusted *p*-values of Fisher's exact test (the top five are listed in Table 3), and 348 GO terms according to Mann–Whitney *U* tests. After FDR adjustment, five proteins remain for the analysis with Fisher's exact test and none with the Mann–Whitney *U* test. The Mann–Whitney *U* test yields of course the same result for both selection criteria since it is independent from a selection criterion.

As can be seen by the above results, enrichment analysis with Fisher's test and the Mann–Whitney *U* test can yield very different results. In fact, when we plot the *p*-values of both methods against each other (Fig. 3), only a moderate correlation becomes obvious.

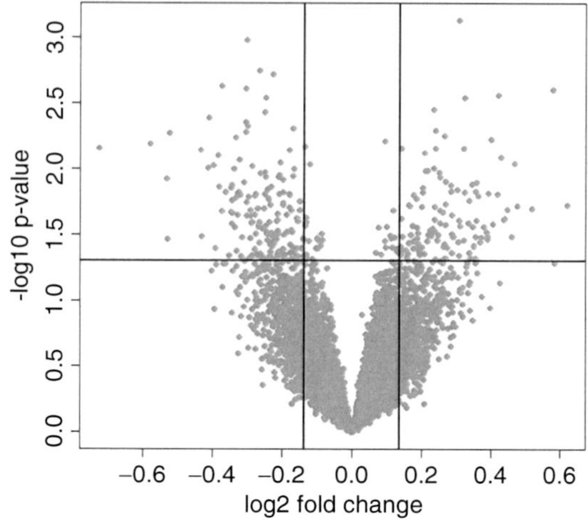

Fig. 2 Volcano plot showing −log10 *p*-values versus the log2 fold changes from the differential comparison of diseased samples versus control samples in the Parkinson data. The *horizontal line* marks the *p*-value threshold of 0.05. The *vertical lines* mark the fold change cutoff of log2(1.1) = 0.14. Proteins on the *top left* and the *top right* area can be considered as statistically significant and biologically relevant. Set-based enrichment analysis can for example be performed on this selection

3.3 Global Test Analysis

As global test procedure, we used here the globaltest approach implemented in the R-package "globaltest." For performing a global test procedure, no prior selection criterion is necessary, because the procedure only focuses on the expression levels of proteins related to each GO term. According to the unadjusted *p-values* of the globaltest approach, 153 GO terms were statistically significant. However, after FDR adjustment none of these terms remained significant. The GO terms with the five smallest unadjusted *p*-values are listed in Table 4.

4 Discussion

Although set-based analysis can bring new insights into a differential analysis, the interpretation of the selected GO terms or pathway can still be problematic. In most cases, a deep biological background is necessary to interpret selected sets in a reasonable biological context. There is a high risk for misinterpreting the role of selected GO terms to give them too much weight. We have also seen that the results can strongly differ depending on the chosen selection procedure in the differential analysis but also on the type of enrichment or global test procedure. Furthermore, the definition of the feature sets based on public databases has a strong influence on the results (*see* **Note 2**). However, with necessary biological prior knowledge, set-based analysis can be helpful to get a better

Table 1
Top five GO terms, selected by enrichment analysis with Fisher's exact test. Proteins that were studied for enrichment of GO terms were selected by statistical tests

GO ID	GO description	p	Adjusted p	Annotated/significant (absolute)	Annotated/non significant (absolute)	Annotated/ significant (relative)	Annotated/ nonsignificant (relative)
GO:0001071	Nucleic acid-binding transcription factor activity	0.0006	0.6386	17/316	193/9167	5.4	2.1
GO:0017124	SH3 domain binding	0.0010	0.6386	5/316	19/9167	1.6	0.2
GO:0015101	Organic cation transmembrane transporter activity	0.0011	0.6386	2/316	0/9167	0.6	0.0
GO:0015695	Organic cation transport	0.0011	0.6386	2/316	0/9167	0.6	0.0
GO:0019534	Toxin transporter activity	0.0011	0.6386	2/316	0/9167	0.6	0.0

Table 2
Top 5 GO terms, selected by enrichment analysis with Fisher's exact test. Proteins that were studied for enrichment of GO terms were selected by a threshold for the log fold change

GO ID	GO description	p	Adjusted p	Annotated/significant (absolute)	Annotated/nonsignificant (absolute)	Annotated/significant (relative)	Annotated/nonsignificant (relative)
GO:0005634	Nucleus	3.3314×10^{-8}	0.0002	355/1439	1478/8044	24.7	18.4
GO:0005654	Nucleoplasm	8.5256×10^{-8}	0.0002	182/1439	657/8044	12.6	8.2
GO:0000278	Mitotic cell cycle	6.9564×10^{-6}	0.0121	45/1439	110/8044	3.1	1.4
GO:0005829	Cytosol	1.8389×10^{-5}	0.0240	202/1439	822/8044	14	10.2
GO:0007049	Cell cycle	2.5193×10^{-5}	0.0264	88/1439	295/8044	6.1	3.7

Table 3
Top 5 GO terms, selected by enrichment analysis using the Mann–Whitney *U* test

GO ID	GO description	*p*	Adjusted p	Annotated/ significant (absolute)	Annotated/ nonsignificant (absolute)	Annotated/ significant (relative)	Annotated/ nonsignificant (relative)
GO:0031252	Cell leading edge	0.0006	0.7414	2/316	5/9167	0.6	0.1
GO:0034446	Substrate adhesion-dependent cell spreading	0.0012	0.7414	0/316	11/9167	0.0	0.1
GO:0030512	Negative regulation of transforming growth factor beta receptor signaling pathway	0.0013	0.7414	3/316	16/9167	0.9	0.2
GO:0034366	Spherical high-density lipoprotein particle	0.0019	0.7414	1/316	3/9167	0.3	0.0
GO:0034384	High-density lipoprotein particle clearance	0.0019	0.7414	1/316	3/9167	0.3	0.0

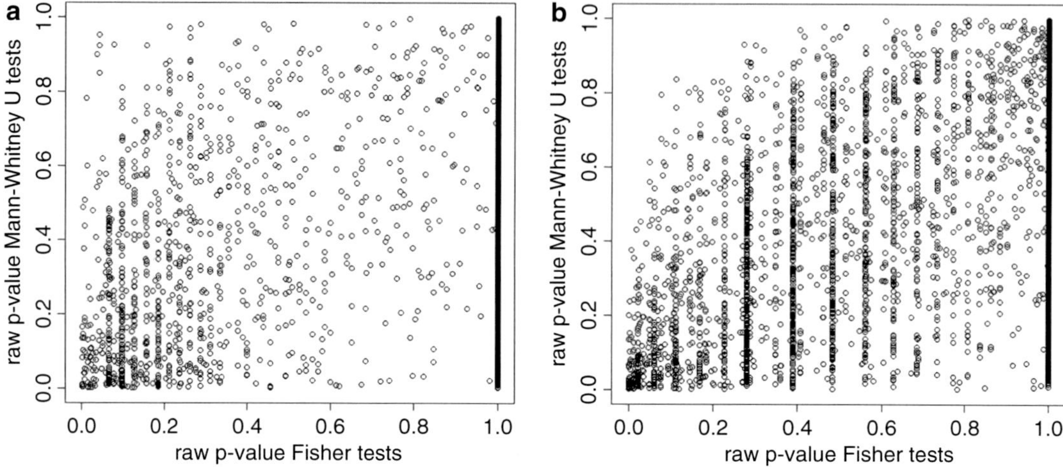

Fig. 3 Comparison of *p*-values from the enrichment analysis with Fisher's test and the Mann–Whitney *U* test. In general, there is only a moderate correlation between the two approaches. *Subplot A* shows the results when using the statistical test result from differential analysis as selection criterion; *subplot B* shows the results when using the fold change as selection criterion

Table 4
Top 5 GO terms, selected a global test procedure

GO term	GO description	Raw *p*-value	Adjusted *p*-value
GO:0002042	Cell migration involved in sprouting angiogenesis	0.0014	0.9989
GO:0043129	Surfactant homeostasis	0.0042	0.9989
GO:0000982	RNA polymerase II core promoter proximal region sequence-specific DNA-binding transcription factor activity	0.0062	0.9989
GO:0061154	Endothelial tube morphogenesis	0.0068	0.9989
GO:0061314	Notch signaling involved in heart development	0.0069	0.9989

understanding of differentially expressed proteins or peptides. In addition, the further range of applications where global test procedures can be used involve the analysis of peptides as a protein set (*see* **Note 3**) or the integration of expression data from different areas (*see* **Note 4**).

5 Notes

1. Like in a standard differential analysis, multiple test corrections need to be done when performing multiple set-based tests simultaneously. Typically, a large number of molecular classes

are selected from a set-based analysis, including most likely many false-positive detections. Therefore, the p-values from enrichment or global testing have to be adjusted using the classical methods for controlling the FDR [23].

2. It is important to mention that set-based analyses are very sensitive to the definition of the set itself or—in the case of enrichment analyses—to the definition of the feature set outside the set of interest. It is therefore recommended to always report the date when the pathway or GO information was retrieved from a specified database.
3. The global approaches are also an adequate tool to analyze the expression levels of peptides that can be assigned the same protein. Thus, these peptides define an own set that can be analyzed by a global test. This may be applicable to data from 2-D gel electrophoresis, where several spots can be related to the same protein.
4. Set-based procedures can also be used for integrating different types of molecular high-throughput data [24], e.g., from genomics and proteomics. If gene and protein expression data was measured simultaneously on the same biological samples one can for example proceed as follows. First, the differentially expressed genes are selected by multiple hypothesis testing. Next, all proteomic features that belong to the same gene that has been selected are defined as a set. This set is tested by a set-based procedure within the analysis of the proteomics data. The p-value for the gene and the p-value for the protein set can then be combined, e.g., by Fisher's combination methods which is typically used in meta-analysis to combine results of different clinical studies.

Acknowledgements

The authors would like to thank Maike Ahrens and Martin Eisenacher (Medizinisches Proteom-Center, Ruhr-University Bochum, Germany) for providing their data from the Parkinson study.

References

1. Soares H, Chen Y, Sabbagh M et al (2009) Identifying early markers of Alzheimer's disease using quantitative multiplex proteomic immunoassay panels. Ann N Y Acad Sci 1180: 56–67
2. Pan S, Chen R, Brand RE et al (2012) Multiplex targeted proteomic assay for biomarker detection in plasma: a pancreatic cancer biomarker case study. J Proteome Res 11:1937–1948
3. Baas T, Baskin CR, Diamond DL et al (2006) Integrated molecular signature of disease: analysis of influenza virus-infected macaques through functional genomics and proteomics. J Virol 80:10813–10828

4. Paweletz CP, Trock B, Pennanen M et al (2001) Proteomic patterns of nipple aspirate fluids obtained by SELDI-TOF: potential for new biomarkers to aid in the diagnosis of breast cancer. Dis Markers 17:301–307
5. O'Connell K, Prencipe M, O'Neill A et al (2012) The use of LC-MS to identify differentially expressed proteins in docetaxel-resistant prostate cancer cell lines. Proteomics 12:2115–2126
6. Kjellin H, Johannsson H, Höög A et al (2014) Differentially expressed proteins in malignant and benign adrenocortical tumors. PLoS One 9, e87951
7. The Gene Ontology Consortium (2000) Gene ontology: tool for the unification of biology. Nat Genet 25:25–29
8. Kanehisa M, Goto S (2000) KEGG: Kyoto encyclopedia of genes and genomes. Nucleic Acids Res 28:27–30
9. Subramanian A, Tamayo P, Mootha VK et al (2005) Gene set enrichment analysis: a knowledge-based approach for interpreting genome-wide expression profiles. Proc Natl Acad Sci U S A 102:15545–15550
10. Goeman JJ, van de Geer SA, de Kort F et al (2004) A global test for groups of genes: testing association with a clinical outcome. Bioinformatics 20:93–99
11. Mansmann U, Meister R (2005) Testing differential gene expression in functional groups. Goeman's global test versus an ANCOVA approach. Methods Inf Med 44:449–453
12. Jung K, Dihazi H, Bibi A et al (2014) Adaption of the global test idea to proteomics data with missing values. Bioinformatics 30:1424–1430
13. The UniProt Consortium (2015) UniProt: a hub for protein information. Nucleic Acids Res 43:D204–D212
14. Blake JA (2013) Ten quick tips for using the gene ontology. PLoS Comput Biol 9, e1003343
15. Cunningham F, Amode MR, Barrell D et al (2015) Ensemble 2015. Nucleic Acids Res 43:D662–D669
16. Durinck S, Moreau Y, Kasprzyk A (2005) BioMart and bioconductor: a powerful link between biological databases and microarray data analysis. Bioinformatics 21:3439–3440
17. Beißbarth T, Speed T (2004) Gostat: find differentially overrepresented Gene Ontologies within a group of genes. Bioinformatics 20:1464–1465
18. Naeem H, Zimmer R, Tavakkolkhah P et al (2012) Rigorous assessment of gene set enrichment tests. Bioinformatics 28:1480–1486
19. Isserlin R, Merico D, Alikhani-Koupaei R et al (2010) Pathway analysis of dilated cardiomyopathy using global proteomic profiling and enrichment maps. Proteomics 10:1316–1327
20. Al-Shahrour F, Carbonell J, Minguez P (2008) Babelomics: advanced functional profiling of transcriptomics, proteomics and genomics experiments. Nucleic Acids Res 36:W341–W346
21. Chen LS, Paul D, Prentice RL et al (2011) A regularized Hotelling's T^2 test for pathway analysis in proteomic studies. J Am Stat Assoc 106:1345–1360
22. Ahrens M, Turewicz M, Casjens S et al (2013) Detection of patient subgroups with differential expression in omics data: a comprehensive comparison of univariate measures. PLoS One 8, e79380
23. Benjamini Y, Hochberg Y (1995) Controlling the false discovery rate: a practical and powerful approach to multiple testing. J R Stat Soc Series B 75:289–300
24. Artmann S, Jung K, Bleckmann A et al (2012) Detection of simultaneous group effects in microRNA expression and related target gene sets. PLoS One 7, e38365

Part III

Classification Methods

Chapter 10

Classification of Samples with Order-Restricted Discriminant Rules

David Conde, Miguel A. Fernández, Bonifacio Salvador, and Cristina Rueda

Abstract

In recent years, mass spectrometry techniques have helped proteomics to become a powerful tool for the early diagnosis of cancer, as they help to discover protein profiles specific to each pathological state. One of the questions where proteomics is giving useful practical results is that of classifying patients into one of the possible severity levels of an illness, based on some features measured on the patient. This classification is usually made using one of the many discrimination procedures available in statistical literature. We present in this chapter recently developed restricted discriminant rules that use additional information in terms of orderings on the means, and we illustrate how to apply them to mass spectrometry data using R package dawai. Specifically, we use proteomic prostate cancer data, and we describe all steps needed, including data preprocessing and feature extraction, to build a discriminant rule that classifies samples in one of several disease stages, thus helping diagnosis. The restricted discriminant rules are compared with some standard classifiers that do not take into account the additional information, showing better performance in terms of error rates.

Key words Mass spectrometry, Preprocessing, Feature extraction, Mean spectrum, Supervised classification, Order restrictions, Restricted discriminant rules, R dawai package

1 Introduction

Proteomics has become a powerful tool for the early diagnosis of cancer, allowing to characterize proteins and therefore to identify diagnostic biomarkers from tissues and body fluids [1]. Recent advances in mass spectrometry (MS) techniques have made it possible to discover protein profiles specific to each pathologic state from high-dimensional MS data [2]. In association with other approaches, MS has become the central technique used by most proteomic biomarker discovery platforms [3]. In cancer, one important proteomic issue is the identification of a panel of biomarkers suitable for discriminating different pathological states, which helps to improve the early detection of cancer [4].

Klaus Jung (ed.), *Statistical Analysis in Proteomics*, Methods in Molecular Biology, vol. 1362, DOI 10.1007/978-1-4939-3106-4_10, © Springer Science+Business Media New York 2016

Once the biomarkers are identified, the classification of the patients is done using discrimination techniques. The general discrimination problem deals with the prediction of the group a patient belongs to, based on some features measured on the patient. In supervised classification, the discriminant rule is built using a training sample, that is, a set of patients for which both the features and the group membership are known. When the underlying distribution of the data is known, the optimal classification rule is the so-called Bayes rule. When it is assumed that the measurements from each population are normally distributed, linear discriminant analysis (LDA) and quadratic discriminant analysis (QDA) are two of the most commonly used in applications discriminant analysis methods. LDA and QDA are obtained replacing in the Bayes rule the unknown parameters, that is, the means vectors and the covariance matrix, by their usual estimators, assuming for LDA that the covariance matrices of the populations are identical. The relaxation of this assumption leads to the QDA rule, where no equality of covariance matrices is assumed.

These two classical discriminant rules were followed by a great deal of classification algorithms such as nearest neighbors [5], classification trees [6], neural networks [7], support vector machines (SVM) [8], or random forests (RF) [9]. All of them build the corresponding classification rules solely from the information in the training sample.

In this chapter we describe and show the usefulness of recently developed discriminant rules [10–13] that use not only the training sample but also other information called additional information. For instance, in a certain application it can be known that certain features of the patients take, on average, higher values in some groups than in others. Let us suppose that we want to classify patients in one of the following groups: G_1—healthy, G_2—early-stage disease, and G_3—advanced-stage disease. We know from previous studies that the mean of variable V_1 increases with the severity of the disease and that the mean of variable V_2 decreases. This additional information can be expressed in terms of restrictions on the model parameters: if $\mu j,i$ represents the mean of variable V_i in group Gj, $i=1, 2, j=1, 2, 3$, then the additional information can be expressed as $\mu_{1,1} \leq \mu_{2,1} \leq \mu_{3,1}$, $\mu_{1,2} \geq \mu_{2,2} \geq \mu_{3,2}$.

Restricted linear and quadratic discriminant rules [10–13] are obtained plugging into the respective Bayes rules the estimators of the unknown parameters, defined from the training sample and the restrictions on the means, via an iterative procedure [10, 11, 13] to ensure that the estimators fulfill the restrictions. In applications as cancer diagnosis, patients are intended to be classified into one of various diagnosis groups based on gene expression or proteomic data. Often, some of the predictors are known to take higher values in some groups with respect to the others. Taking into account this underlying order could potentially result in significantly lower misclassification rates with respect to the standard classification methods that do not take it into account. In this chapter, we use proteomic prostate cancer data from patients with the following diagnosis groups: normal prostate, benign prostate conditions, prostate cancer and prostate-specific antigen (PSA) between 4 and 10, and prostate cancer and PSA levels above 10.

The layout of the chapter is as follows. In Subheading 2 we start describing briefly the data set and the software with which we show the use of the restricted discriminant rules. In Subheading 3 we detail the standard methods used to perform the preprocessing and feature extraction, in order to obtain the data matrix (patients in rows, features in columns) needed for classification. Also in this section, a brief summary of the restricted discriminant rules is presented. The R library dawai [13] is used in Subheading 4 with the purpose of showing how the restricted discriminant rules can be used in practice when additional information is present. These rules, applied on the mentioned data set, show a significantly better performance than other usual discrimination rules not considering the additional information such as LDA, RF or SVM. In Subheading 5, short step-by-step instructions are provided and some important issues concerning the analysis are highlighted.

2 Materials

MS techniques allow the identification of the amount and type of proteins present in a sample by measuring the mass-to-charge ratio (m/z) and abundance of gas-phase ions. A mass spectrum is a plot of the ion signal (intensity) versus m/z. These m/z ratios can be used to calculate the molecular weights of protein or peptide. Two broadly used MS techniques for proteome screening are matrix-assisted laser desorption and ionization (MALDI) and surface-enhanced laser desorption and ionization (SELDI) with time-of-flight (TOF) tubes.

In this chapter we apply our restricted discriminant rules [11–13] to the proteomic prostate cancer data of *JNCI Data 7-30-02.zip* [14], which are publicly available at http://home.ccr.cancer.gov/ncifdaproteomics/ppatterns.asp. The data consist of 322 serum spectra measuring peak amplitudes at 15,154 m/z values in the range 0–20,000 Da. Serum samples provided by patients have the following frequencies: normal prostate (63), benign prostate conditions (190), prostate cancer and PSA levels between 4 and 10 [26], and prostate cancer and PSA levels above 10 (43). Samples were applied to a C16 hydrophobic interaction protein chip (Ciphergen Biosystems, Freemont, CA) and analyzed as described in Petricoin et al. [14]. Data were generated using the SELDI-TOF MS techniques and are provided with baseline subtracted. In Fig. 1, the mean spectrum, computed averaging over all raw spectra, is shown.

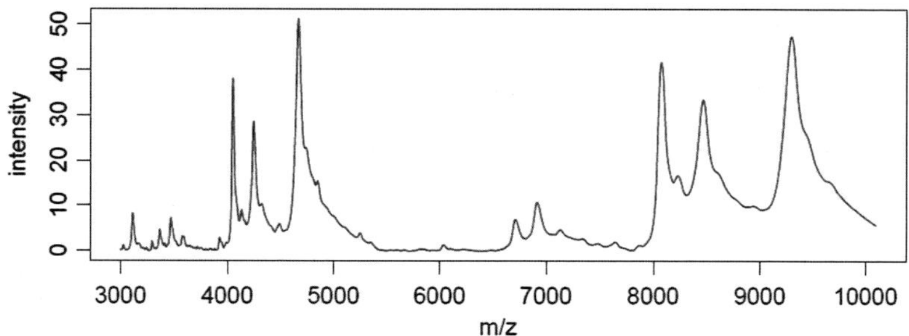

Fig. 1 Mean spectrum

We will illustrate how, when additional information on the means of the populations (normal prostate, benign prostate conditions and prostate cancer) is known, our restricted discrimination rules can be applied to MS data using the R package dawai [13], which can be downloaded from http://cran.r-project.org/web/packages/dawai/.

3 Methods

3.1 Preprocessing

Data from MS measurements usually contain a substantial amount of noise and show large inter-measurement variation [15]. A preprocessing stage is needed, including, as a first task, noise reduction.

3.1.1 Baseline Reduction

Commercial mass spectrometers implement basic noise reduction methods. Some studies explore methods for reducing noise, especially baseline noise [16, 17]. Most of the studies employing a baseline reduction method estimate the baseline noise and subtract the estimated baseline from the original mass spectrum.

The data we will use to illustrate our methods are already provided with baseline subtracted [14], so we do not need to care about baseline reduction.

3.1.2 Normalization

A peak in mass spectra indicates the relative abundance of a protein, but mass spectra cannot be directly compared with each other as MS spectra of similar samples are not always quantified within the same amplitude range. A second task in the preprocessing stage is normalization, needed to compare the real intensities, converting all the spectra to the same intensity ranges.

Many different approaches have been proposed and used to handle this issue. Normalization of mass spectra typically involves subtracting an offset and dividing by a scaling factor. Such offset and scaling parameters can be defined and applied globally or locally using a window. One of the most frequently used approaches is normalization with respect to the total ion current (TIC), i.e., dividing each intensity by the sum of all the intensities in a mass spectrum [18–20]. This is equivalent to normalization with respect to the mean of the intensities in the spectrum [17]. One alternative to TIC normalization is scaling by the sum of the squares of the intensities so the spectrum forms a unit vector [21]. Other studies perform normalization with respect to the largest peak [16] or linear scaling with the largest and the smallest peak intensities [14, 22], known as min range.

Meuleman et al. [23] compare eight such normalization procedures for both global or local normalization, according to two objectives: inter-spectra variance minimization and classification performance maximization. They state that in general it is better to use a local than a global normalization method. As for the distinct methods, they show that global mean SD (subtracting the global mean and dividing by the standard deviation) is the best method attending the classification maximization objective and one of the best regarding the variance minimization objective. Here, we follow their advice and normalize our data with this method.

3.1.3 Smoothing

An ion peak may be spread across many data points, so each m/z data point should not be regarded as the record of a distinct peptide. To reduce this noise, we smooth the normalized spectra using a Gaussian kernel with a full width at half maximum (FWHM) of 11 m/z values [17, 24, 25], with the convention FWHM $\approx 2.355\sigma$. In this way, for each m/z value, the normalized intensity x is replaced by a weighted average of the form $\Sigma_t\, t\, N(t;x;\sigma)$ where the summation is over all the 11 m/z values around x, 5 each side, and $N(t; x, \sigma)$ is a Gaussian kernel with mean x and variance σ^2.

Other common approaches are smoothing filters [26], wavelet transform (WT) [27], deconvolution filter [28], and moving average filter [29].

3.2 Feature Extraction

Feature extraction is the process of selecting small sets of relevant features. Because of the high measurement variation in MS data, peaks are the most suitable biomarkers [24]. A peak is defined as an m/z value with higher intensity than the nearby values around it and than the average intensity at those nearby values.

Peak detection deals with identifying peaks in a mass spectrum, which is not simple due to variability between samples in intensity and location. Peak alignment is the process of matching peaks that represent the same protein species in distinct spectra.

Morris et al. [30] propose a method for performing feature extraction that uses the average spectrum for peak detection. The mean spectrum is computed averaging over all raw spectra. After the mean spectrum is normalized and smoothed, peaks are detected. Then the peaks are quantified in the individual spectra. We perform peak detection in this way, using the mean spectrum, and obtain the following 14 m/z peaks: 3116, 3468, 4052, 4133, 4245, 4483, 4662, 4847, 6709, 6912, 8073, 8228, 8462, and 9297.

Once the peaks are identified, they usually do not correspond to local maxima in the individual processed spectra. Wagner et al. [16] consider the local maximum within 30 measurements points of the peak mass from the processed spectra as the features. Petricoin et al. [22] also consider the maximum peak height, but other metrics as average or median peak height [19] can be considered too. We proceed as in [16], looking for the maximum heights within 30 measurements points of the mentioned 14 identified peaks for each of the processed spectra. These 14 values and the disease stage for each patient serum sample are rendered to a text file.

For short step-by-step instructions about preprocessing and feature extraction (*see* **Note 1**).

3.3 Restricted Classification Rules

Let us consider k disease stages $G_1,\ldots,Gk$, also called groups or populations throughout this chapter, so that each patient belongs to one and only one of them. Let $\pi_1,\ldots,\pi k$, be the a priori probabilities of the groups, with $\pi_1+\ldots+\pi k=1$. Let us suppose that a p-dimensional vector X of features is measured for each patient, and that X is normally distributed (*see* **Note 2**) with mean $\boldsymbol{\mu}_j$ for group Gj, $j=1,\ldots,k$, and common covariance matrix $\boldsymbol{\Sigma}$. Then, if $\boldsymbol{U}$ is the vector of features measured for a patient whose disease stage is unknown, the optimal classification rule is the Bayes rule:

Classify $\boldsymbol{U}$ in Gj if and only if

$$\log\pi_j+(\boldsymbol{U}-\mu_j)^T\Sigma^{-1}(\boldsymbol{U}-\mu_j)\leq\log\pi_l+(\boldsymbol{U}-\mu_l)^T\Sigma^{-1}(\boldsymbol{U}-\mu_l),$$

$l=1,\ldots,k$.

Parameters μj, $j=1,\ldots,k$, and $\boldsymbol{\Sigma}$ are usually unknown, but they can be estimated from a training sample (a set of patients for which the features values and the group they belong to are known) by the sample vector means $\bar{\boldsymbol{X}}_j$, $j=1,\ldots,k$, and the pooled sample covariance matrix $\boldsymbol{S}$:

$$\bar{\boldsymbol{X}}_j=\frac{1}{n_j}\sum_{l=1}^{n_j}\bar{\boldsymbol{X}}_{j_l},\ j=1,\ldots,k,\ \ \boldsymbol{S}=\frac{1}{n-k}\sum_{j=1}^{k}\sum_{l=1}^{n_j}(\bar{\boldsymbol{X}}_{j_l}-\bar{\boldsymbol{X}}_j)(\bar{\boldsymbol{X}}_{j_l}-\bar{\boldsymbol{X}}_j)^T,$$

where nj is the sample size of group Gj, $j=1,\ldots,k$, and $n=n_1+\ldots+nk$.

The linear discriminant rule (LDA) or Fisher's rule is obtained plugging estimators $\bar{\boldsymbol{X}}_j$, $j=1,\ldots,k$, and $\mathbf{S}$ into the Bayes rule.

In applications, it is usual that some additional information is available, often through order restrictions on the populations means. Let us suppose that we want to classify patients in one of the following groups: G_1—healthy, G_2—early-stage disease, and G_3—advanced-stage disease ($k=3$), and that two variables V_1 and V_2 are measured for each patient ($p=2$).

If we know that the patients from G_1 (the control group) take, in mean, lower values than those coming from any of the other groups for all variables, in the usual statistical terminology we can say that there is a "tree order" among the means of the variables: $\mu_{1,1} \le \mu_{j,1}$, $j=2, 3$, $\mu_{1,2} \le \mu_{j,2}$, $j=2, 3$.

Another common situation appears when it is known that there is an increase in the means of the variables. We can say now that there is a "simple order" among the groups means: $\mu_{1,1} \le \mu_{2,1} \le \mu_{3,1}$, $\mu_{1,2} \le \mu_{2,2} \le \mu_{3,2}$.

Let us denote as C the subset of the parameter space where the restrictions on the means are fulfilled.

The family of restricted linear classification rules [11] that we apply here to the proteomics data considers estimators for μ_j, $j=1,\ldots,k$, that take into account the additional information known about the parameters. When the sample means do not verify the restrictions, an iterative procedure starting from vector $\bar{X} = (\bar{X}_1^T, \ldots, \bar{X}_k^T)^T$ is used to obtain an estimator of the means that verifies the additional information contained in set C.

A first approach would be to consider the value in C closest to $\bar{X}$, that is, the projection of $\bar{X}$ onto C. We call this value $\hat{\mu}^0$ (*see* Fig. 2). However, it is known that that estimator lacks good statistical properties as it is not admissible [31]. For this reason, we consider an estimator that is inside the set C (parameter $\gamma \in [0,1]$ will control how much inside C is the estimator considered). However, as it can be seen in Fig. 2, we need an iterative procedure to ensure that the final estimator is inside the set C, as it might happen that trying to put the estimator inside C takes it to the other side of C. In each iteration, if v is the vector obtained in the previous iteration and w is the projection of v over C, the new vector is defined as $w - \gamma(v - w)$. The procedure ends when the new vector verifies the restrictions, i.e., when $w - \gamma(v - w)$ belongs to C. We denote as $\hat{\mu}^\gamma$ the limit of the procedure. Figure 2 illustrates this process for a set C and an initial estimator $\bar{X}$ not belonging to C for three different values of γ (0, 0.5, 1). The results in [10, 11]

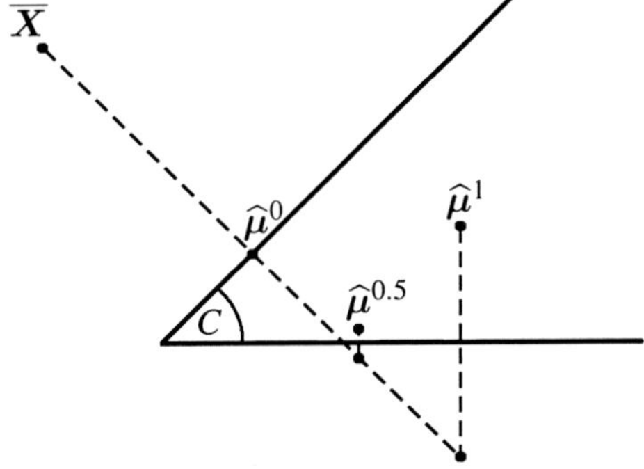

Fig. 2 Set C and estimators $\hat{\mu}^0$, $\hat{\mu}^{0.5}$ and $\hat{\mu}^1$

ensure the convergence of this scheme and the good properties of these estimators.

If $\hat{\mu}^{\gamma} = (\hat{\mu}_1^{\gamma T}, \ldots, \hat{\mu}_k^{\gamma T})^T$, the restricted linear classification rules are:

Classify **U** in *Gj* if and only if

$\log \pi_j + (U - \hat{\mu}_j^{\gamma})^T S^{-1} (U - \hat{\mu}_j^{\gamma}) \le \log \pi_l + (U - \hat{\mu}_l^{\gamma})^T S^{-1} (U - \hat{\mu}_l^{\gamma})$, $l = 1, \ldots, k$.

When the covariance matrices are not assumed to be equal, an analogous family of restricted quadratic classification rules is proposed in ref. 13.

4 Results

Now we go back to our data set and detail how to use the restricted discriminant rules in a case like this. As it was seen in previous section, after normalization and feature extraction, the maximum heights within 30 measurements points of 14 identified peaks for each patient serum sample were rendered to a text file. It consists of a matrix with 322 rows and 15 columns, the 14 features and the group label, i.e., the group each patient belongs to: 1—normal prostate, 2—benign 8 prostate conditions, 3—prostate cancer and PSA between 4 and 10, and 4—prostate cancer and PSA levels above 10.

In this section we illustrate the restricted linear discriminant rules on the mentioned data set using the R package dawai [13] that we have developed. R is a free software environment for statistical computing that runs on a broad variety of platforms including UNIX, Windows, and MacOS. It is widely used for developing and sharing statistical software, which makes it highly extensible through user-created packages, and can be easily installed without any cost. R package dawai depends on boot [32], ibdreg [33], and mvtnorm [34] R packages, which must be installed before loading dawai.

As we lack the expert's knowledge regarding additional information for our data set, in this example we will consider as restrictions the order restrictions verified by the total sample. Then we split the sample into training and test sample, and use the training sample to build the restricted linear discriminant rules and the test sample to evaluate the accuracy of the rules.

We first load package dawai and read the data file, called *data.txt*.

```
R> library(dawai)
R> data <- read.table("data.txt", header=TRUE)
```

We separate the first variable (data$Class), with the group label for each patient, from the other ones with the 14 peak heights.

```
R> class <- as.factor(data$Class)
R> dataset <- data[, 2:15]
```

Variable class contains the group label for each patient.

```
R> table(class)
 1    2   3   4
63  190  26  43
```

These are the number of patients in each of the four groups. The groups correspond to increasingly advanced levels of the disease. We join the original groups 3 and 4 into one (the ones with prostate cancer), relabeling them, so that the groups have enough elements to split the sample into training and test data sets of reasonable sizes.

```
R> levels(class) <- c(1, 2, 3, 3)
R> table(class)
 1   2   3
63 190  69
```

Let us have a look to the order restrictions verified by the total sample.

```
R> means <- colMeans(dataset[class == 1,])
R> means <- rbind(means, colMeans
                           (dataset[class == 2,]))
R> means <- rbind(means, colMeans
                           (dataset[class == 3,]))
R> rownames(means) <- 1:3
R> round(t(means), 2)
         1    2    3
P3116 0.06 0.60 0.26
P3468 1.64 0.22 0.03
P4052 3.33 3.35 3.65
P4133 0.40 0.47 0.65
P4245 0.88 3.03 2.18
P4483 0.02 0.28 0.09
P4662 3.14 5.76 4.38
P4847 0.54 1.50 0.93
P6709 1.09 0.30 0.11
P6912 2.31 0.59 0.35
P8073 3.35 4.02 3.82
P8228 0.93 1.32 1.44
P8462 1.19 3.58 2.74
P9297 2.50 5.30 3.98
```

Except for variables 2, 9, and 10, i.e., P3468, P6709, and P6912, all means are lower for group 1 than for groups 2 and 3. As for variables 2, 9, and 10, means are higher for group 1 than for groups 2 and 3. Group 1 (normal prostate) can be regarded as the control group in a decreasing tree order among the mean values of variables 2, 9, and 10, and an increasing tree order among the rest of the variables. We change the sign of these three variables so that there is the same tree order on all predictors.

```
R> dataset[, c(2, 9, 10)] <- -dataset[, c(2, 9, 10)]
```

In this way, restrictions in the training sample can now be easily specified by just `restext = "t<1,2,3,4,5,6,7,8,9,10,11,12,13,14"` (*see* **Note 3**). See dawai package [13] help files (`help(rlda)`) for advice.

We split the data set into a randomly selected training set and a test set, fixing a seed in order to get the same results as the reader. We are doing it here in order to have a test set and to show in an easy way that our rules outperform the usual ones.

```
R> set.seed(4100)
R> values <- runif(dim(dataset)[1])
R> trainsubset <- (values < 0.5)
R> testsubset <- (values >= 0.5)
```

Now we can build the restricted linear discriminant rules on the training sample, using $\gamma = 0$, 0.5, 1 (*see* **Note 4**).

```
R> obj <- rlda(dataset, class, subset = trainsubset,
restext =
"t<1,2,3,4,5,6,7,8,9,10,11,12,13,14",
gamma = c(0, 0.5, 1))
```

We have not specified `prior` parameter, so the group proportions of the training set have been used (by default) as the prior probabilities of group membership (*see* **Note 4**).

```
R> obj$prior
   class1     class2     class3
0.1863354 0.5838509 0.2298137
```

Now, let us consider the test set and classify its observations. We know the groups that the observations in the test set belong to, so we can estimate the true error rates as the proportion of observations in the test set wrongly classified by the restricted discrimination rules. The first command below classifies the observations in the test set (*see* **Note 4**). The second command yields the percentages of wrong classification of these observations. See dawai package [13] help files (`help(predict.rlda)`) for advise.

```
R> pred <- predict(obj,newdata = dataset
[testsubset,], grouping = class[testsubset])
R> pred$error
                  gamma=0  gamma=0.5  gamma=1
True error rate (%): 11.80124  11.18012 10.55901
```

These results can also be compared with the error rates for some standard classifiers that do not take into account the additional information considered such as LDA in R MASS package [35], RF in R randomForest package [36], and SVM in e1071 package [37], packages that have to be loaded first. The following commands load these three packages:

```
R> library(MASS)
R> library(randomForest)
R> library(e1071)
```

Now we build the rule and compute the LDA error as we did with the restricted rules:

```
R> lda_out <- lda(dataset, class, subset =
trainsubset)
R> lda_error <- mean(predict(lda_out,newdata =
dataset[testsubset,]) $class != class
[testsubset])*100
R> lda_error
[1] 13.04348
```

Also for RF:

```
R> rf_out <- randomForest(class ~ ., data = dataset,
subset = trainsubset)
R> rf_error <- mean(predict(rf_out, newdata =
dataset[testsubset,]) != class[testsubset])*100
R> rf_error
[1] 14.28571
```

And also for SVM, using always the default parameters:

```
R> svm_out <- svm(class ~ ., data = dataset,
subset = trainsubset)
R> svm_error <- mean(predict(svm_out, newdata =
dataset[testsubset,]) != class[testsubset])*100
R> svm_error
[1] 13.04348
```

We can see that, for $\gamma = 1$, the test error rates for the restricted linear rules are 19.04 % lower than for LDA and SVM, and 29.09 % lower than for RF.

We finish showing the behavior of the rules when variable selection is performed (*see* **Note 5**). We ask ourselves if we can dispense with redundant or irrelevant variables and if we can reduce possible overfitting by performing variable selection. We have searched for the variables maximizing the Mahalanobis [38] distances among G_1, G_2, and G_3 means. For $p = 6$ variables, variables selected are P4052, P4133, P4847, P6709, P6912, and P8462. For $p = 10$ variables, variables selected are P3468, P4052, P4245,

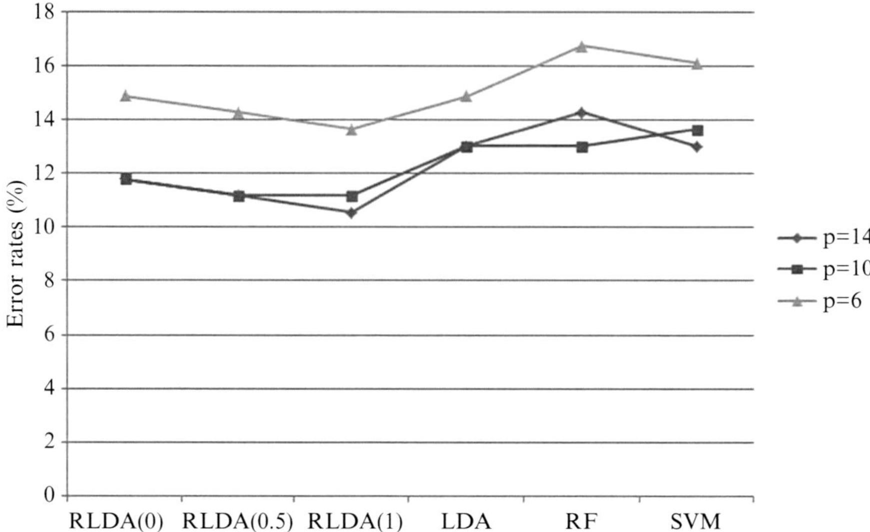

Fig. 3 Error rates for $p = 6, 10, 14$

P4483, P4847, P6709, P6912, P8228, and P8462. For these two reduced sets of variables we perform the same analysis we have just detailed for the full 14 variables set. The corresponding error rates for $p = 6, 10, 14$ are represented in Fig. 3.

We can see that error rates are quite similar for $p = 10$ and $p = 14$, being significantly lower for both sets of variables than for $p = 6$. This means that we can reduce the number of variables from $p = 14$ to $p = 10$ without a significant loss of prediction accuracy, but not from $p = 10$ to $p = 6$. In all cases, the restricted rules perform better than those procedures that do not consider additional information and the lowest test error rates correspond to the restricted linear rules for $\gamma = 1$ (RLDA(1)).

5 Notes

1. Preprocessing and feature extraction.
   - Baseline reduction: Commercial mass spectrometers implement basic noise reduction methods; if not, the baseline noise must be estimated and subtracted from the original mass spectrum.
   - Normalization: The global mean is subtracted from each spectrum and divided by the standard the standard deviation.
   - Smoothing: The normalized spectra are smoothed replacing the normalized intensity x by a weighted average of the form $\Sigma_t t N(t; x;, \sigma)$ where the summation is over FWHM

(full width at half maximum) 11 m/z values around x, 5 each side, and $N(t; x, \sigma)$ is a Gaussian kernel with mean x and variance σ^2 ($\sigma \approx \text{FWHM}/2.355 \approx 4.67$).

- Feature extraction: The mean spectrum is computed averaging over all raw spectra; after the mean spectrum is normalized and smoothed, peaks are detected. Then, the maximum heights within 30 measurements points of the identified peaks for each of the processed spectra are the features measured. These values and the disease stage for each patient serum sample have to be rendered to a text file.

2. Normality

   The restricted discriminant rules have shown good performance under normality assumptions [10–13], so, prior to building the rules, it is important to check if the variables (the extracted features) used for classification are approximately normally distributed. If they are not, data transformation as the logarithm or square root transformation should be considered. Nevertheless, these restricted rules have also good robustness properties against different types of contamination [39].

3. Incorporation of the additional information

   In applications, experts knowledge provides some additional information that often can be written in terms of inequality restrictions on the groups means. Let us suppose that we want to classify patients in one of the following groups: G_1—healthy, G_2—early-stage disease, and G_3—advanced-stage disease ($k = 3$), and that two variables V_1 and V_2 are measured for each patient ($p = 2$). Let $\mu_1 = \begin{pmatrix} \mu_{1,1} \\ \mu_{1,2} \end{pmatrix}$, $\mu_2 = \begin{pmatrix} \mu_{2,1} \\ \mu_{2,2} \end{pmatrix}$ and $\mu_3 = \begin{pmatrix} \mu_{3,1} \\ \mu_{3,2} \end{pmatrix}$ be the corresponding means vectors. The available additional information can be incorporated to the restricted classification rules in terms of inequality restrictions on the means: $\mathbf{A} \cdot \begin{pmatrix} \mu_1 \\ \mu_2 \\ \mu_3 \end{pmatrix} \leq 0$, where $\mathbf{A}$ is a q x kp matrix with q inequality restrictions:

   – If the restrictions are, for example, $\mu_{1,1} \leq \mu_{2,1} \leq \mu_{3,1}$, $\mu_{1,2} \geq \mu_{2,2} \geq \mu_{3,2}$, these four restrictions can be expressed with the restrictions matrix $\mathbf{A} = \begin{pmatrix} 1 & 0 & -1 & 0 & 0 & 0 \\ 0 & 0 & 1 & 0 & -1 & 0 \\ 0 & -1 & 0 & 1 & 0 & 0 \\ 0 & 0 & 0 & -1 & 0 & 1 \end{pmatrix}$, and the additional information can be incorporated to the restricted classification rules by simply resmatrix = A.

- If there is an increasing tree order among the means of variable V_2: $\mu_{1,2} \leq \mu_{2,2}$, $\mu_{1,2} \leq \mu_{3,2}$, it can be expressed with $A = \begin{pmatrix} 0 & 1 & 0 & -1 & 0 & 0 \\ 0 & 1 & 0 & 0 & 0 & -1 \end{pmatrix}$ and it can be incorporated to the rules by `resmatrix = A`, or, easier, by just `restext = "t<2"`.
- If there is a decreasing simple order among the means of variables V_1 and V_2, i.e., $\mu_{1,1} \geq \mu_{2,1} \geq \mu_{3,1}$, $\mu_{1,2} \geq \mu_{2,2} \geq \mu_{3,2}$, then it can be easily incorporated by `restext = "s>1,2"`.

4. Restricted discriminant rules

   In order to use other prior probabilities of group membership than the group proportions of the training set, `prior` parameter has to be specified (otherwise, the class proportions for the training set are used). For example, if we want to use equal a priori probabilities, we need to specify the following: `prior = c(1/3, 1/3, 1/3)`.

   In `rlda` function, if `gamma` parameter is not specified, the restricted discriminant rules are by default built for both $\gamma = 0, 1$.

   In `predict` function, if we know the class for each observation to classify, we can estimate the true error rate from `newdata` by adding `grouping = class[testsubset]`.

   In `predict` function, if `gamma` parameter is not specified, all rules built with `obj$gamma` will be used. If present, `gamma` must be contained in `obj$gamma`.

5. Variable selection

   The objectives of variable selection are to dispense with irrelevant or redundant variables, reduce possible overfitting and reduce the number of variables without a significant loss of prediction accuracy. Once the features have been extracted, they can already be examined to determine whether it is possible to dispense with any a priori irrelevant variables. We may also want to study the behavior of the rules when variable selection is performed, thus wondering after discriminant analysis if we can reduce possible overfitting, or if we can reduce the number of variables without a significant loss of prediction accuracy. One way to do this is, for example, searching for the variables maximizing the Mahalanobis [38] distances among the means of the groups for several values of p variables.

References

1. Toss A, DeMatteis E, Rossi E et al (2013) Ovarian cancer: can proteomics give new insights for therapy and diagnosis? Int J Mol Sci 14:8271–8290
2. Yasui Y, Pepe M, Thompson ML et al (2003) A data-analytic strategy for protein biomarker discovery: profiling of high-dimensional proteomic data for cancer detection. Biostatistics 4:449–463
3. Paul D, Kumar A, Gajbhiye A et al (2013) Mass spectrometry-based proteomics in molecular diagnostics: discovery of cancer biomarkers using tissue culture. BioMed Res Int 2013, Article ID 783131
4. Khadir A, Tiss A (2013) Proteomics approaches towards early detection and diagnosis of cancer. J Carcinog Mutagen S14:002
5. Cover T, Hart P (1967) Nearest neighbor pattern classification. IEEE Trans Inf Theory 13:21–27
6. Buntime W (1992) Learning classification trees. Stat Comput 2:63–72
7. Bishop CM (1995) Neural networks for pattern recognition. Oxford University Press, New York
8. Cortes C, Vapnik V (1995) Support-vector networks. Mach Learn 20:273–297
9. Breiman L (2001) Random forests. Mach Learn 45:5–32
10. Fernandez M, Rueda C, Salvador B (2006) Incorporating additional information to normal linear discriminant rules. J Am Stat Assoc 101:569–577
11. Conde D, Fernandez MA, Rueda C et al (2012) Classification of samples into two or more ordered populations with application to a cancer trial. Stat Med 31:3773–3786
12. Conde D, Salvador B, Rueda C et al (2013) Performance and estimation of the true error rate of classification rules built with additional information. An application to a cancer trial. Stat Appl Genet Mol Biol 12:583–602
13. Conde D, Fernandez MA, Salvador B et al (2014) dawai: Discriminant analysis with additional information. http://cran.r-project.org/package=dawai
14. Petricoin EF, Ornstein DK, Paweletz CP et al (2002) Serum proteomic patterns for detection of prostate cancer. J Natl Cancer Inst 94:1576–1578
15. Semmes OJ, Feng Z, Adam B-L et al (2005) Evaluation of serum protein profiling by surface-enhanced laser desorption/ionization time-of-flight mass spectrometry for the detection of prostate cancer: I. Assessment of platform reproducibility. Clin Chem 51:102–112
16. Wagner M, Naik D, Pothen A (2003) Protocols for disease classification from mass spectrometry data. Proteomics 3:1692–1698
17. Zhu W, Wang X, Ma Y et al (2003) Detection of cancer-specific markers amid massive mass spectral data. Proc Natl Acad Sci U S A 100:14666–14671
18. Baggerly KA, Morris JS, Wang J et al (2003) A comprehensive approach to the analysis of matrix assisted laser desorption/ionization-time of flight proteomics spectra from serum samples. Proteomics 3:1667–1672
19. Bhattacharyya S, Siegel ER, Petersen GM et al (2004) Diagnosis of pancreatic cancer using serum proteomic profiling. Neoplasia 6:674–686
20. Li J, Zhang Z, Rosenzweig J et al (2002) Proteomics and bioinformatics approaches for identification of serum biomarkers to detect breast cancer. Clin Chem 48:1296–1304
21. Alfassi ZB (2004) On the normalization of a mass spectrum for comparison of two spectra. Journal Am Soc Mass Spectrom 15:385–387
22. Petricoin EF, Ardekani AM, Hitt BA et al (2002) Use of proteomic patters in serum to identify ovarian cancer. Lancet 359:572–577
23. Meuleman W, Engwegen JYMN, Gast M-CW et al (2008) Comparison of normalisation methods for surface-enhanced laser desorption and ionisation (SELDI) time-of-flight (TOF) mass spectrometry data. BMC Bioinformatics 9:88
24. Bhanot G, Alexe G, Venkataraghavan B et al (2006) A robust meta-classification strategy for cancer detection from MS data. Proteomics 6:592–604
25. Tibshirani R, Hastie T, Narasimhan B et al (2004) Sample classification from protein mass spectrometry, by 'peak probability contrasts'. Bioinformatics 20:3034–3044
26. Wang MZ, Howard B, Campa MJ et al (2003) Analysis of human serum proteins by liquid phase isoelectric focusing and matrix-assisted laser desorption/ionization-mass spectrometry. Proteomics 3:1661–1666
27. Taskin V, Dogan B, Olmez T (2013) Prostate cancer classification from mass spectrometry data by using wavelet analysis and Kernel Partial Least Squares Algorithm. Int J Biosci Biochem Bioinforma 3:98–102
28. Malyarenko DI, Cooke WE, Adam B-L et al (2005) Enhancement of sensitivity and resolution of surface-enhanced laser desorption/

ionization time-of-flight mass spectrometric records for serum peptides using time-series analysis techniques. Clin Chem 51:65–74

29. Liu Q, Krishnapuram B, Pratapa P et al (2004) Identification of differentially expressed proteins using MALDI-TOF mass spectra. Conf Rec Asilomar Conf Signals Syst Comput 2:1323–1327
30. Morris JS, Coombes KR, Koomen J et al (2005) Feature extraction and quantification for mass spectrometry in biomedical applications using the mean spectrum. Bioinformatics 21:1764–1775
31. van Eeden C (2006) Restricted parameter space estimation problems: admissibility and minimaxity properties. Springer, New York
32. Canty A, Ripley B (2014) boot: bootstrap functions (originally by Angelo Canty for S). http://cran.r-project.org/package=boot
33. Sinnwell JP, Schaid DJ (2013) ibdreg: regression methods for IBD linkage with covariates. http://cran.r-project.org/package=ibdreg
34. Genz A, Bretz F, Miwa T et al (2014) mvtnorm: multivariate normal and t distributions. http://cran.r-project.org/package=mvtnorm
35. Ripley B, Venables B, Bates DM et al (2011) Support functions and datasets for venables and Ripley's MASS. http://cran.r-project.org/package=MASS
36. Breiman L, Cutler A, Liaw A et al (2014) randomForest: Breiman and Cutler's random forests for classification and regression. http://cran.r-project.org/package=randomForest
37. Meyer D, Dimitriadou E, Hornik K et al (2014) e1071: Misc Functions of the Department of Statistics (e1071), TU Wien. http://cran.r-project.org/package=e1071
38. Mahalanobis PC (1936) On the generalised distance in statistics. Proc Natl Inst Sci India 12:49–55
39. Salvador B, Fernandez MA, Martin I et al (2008) Robustness of classification rules that incorporate additional information. Comput Stat Data An 52:2489–2495

Chapter 11

Application of Discriminant Analysis and Cross-Validation on Proteomics Data

Julia Kuligowski, David Pérez-Guaita, and Guillermo Quintás

Abstract

High-throughput proteomic experiments have raised the importance and complexity of bioinformatic analysis to extract useful information from raw data. Discriminant analysis is frequently used to identify differences among test groups of individuals or to describe combinations of discriminant variables. However, even in relatively large studies, the number of detected variables typically largely exceeds the number of samples and the classifiers should be thoroughly validated to assess their performance for new samples. Cross-validation is a widely approach when an external validation set is not available. In this chapter, different approaches for cross-validation are presented including relevant aspects that should be taken into account to avoid overly optimistic results and the assessment of the statistical significance of cross-validated figures of merit.

Key words Proteomics, Cross-validation, Double cross-validation, Discriminant analysis, Partial least squares-discriminant analysis

1 Introduction

In recent years, proteomics has become one of the most widely used research tools in high-throughput biology. Proteomic analysis plays a key role not only in basic, but also in biomedical research in fields such as drug discovery or clinical diagnosis. Availability of large scale "omics" data from genomics, transcriptomics, proteomics, or metabolomics increases the importance and complexity of bioinformatic and statistical analysis to get insight into huge amounts of raw data and extract useful information. Proteomic data is frequently analyzed using discriminant analysis (DA) to assess differences among groups of individuals and identify for which combination of variables they are most distinct. For example, in a typical clinical proteomics study, the "case" group includes subjects diagnosed with a disease while a second group includes subjects classified as "healthy" or "control." In this type of study, the objective is frequently the identification and interpretation of

Klaus Jung (ed.), *Statistical Analysis in Proteomics*, Methods in Molecular Biology, vol. 1362,
DOI 10.1007/978-1-4939-3106-4_11,

proteomic disease biomarkers to get further insight into biological processes related to the disease. A second objective could be to pinpoint biomarkers to enable the construction of accurate classifiers. One of the most relevant challenges for the analysis of high-throughput proteomic data is their high dimensionality. Even in relatively large studies with a few hundred biological samples, the number of detected variables in most of the cases largely exceeds the number of samples. Besides, the majority of the detected proteomic variables are irrelevant for outcome prediction and their elimination improves the classifier performance. In addition, variables may be correlated, being multivariate statistical analysis generally required to extract information contained in the dataset.

In proteomic studies, overfitting is a potential pitfall where the classifier models random variation in the data. Because of that, DA models should be subjected to thorough statistical validation to estimate the generalization accuracy of the classifier and to ensure that the model will work for new samples. This validation evaluates the performance of the classifier and yields an accurate estimation of the prediction error and the probability of a chance result. Statistical validation can be carried out by external and internal cross validation. External validation uses test samples not included in the calibration set of samples used to build the classifier and it is considered the "gold standard." But, when the sample size is scarce, if we use sufficient samples to develop a reliable classifier we can find that we might have not enough samples for testing its performance and vice versa. In this situation, cross-validation is used for testing as a suboptimal approximation to external validation that, in spite of its limitations, is one of the most practical methods during model development to provide a point estimate of the performance of a classifier [1].

In this chapter we explain basic procedures used to assess the generalization accuracy of discriminant classifiers using cross-validation. In Subheading 2, partial least squares-DA (PLS-DA) is outlined. Subheading 3 presents cross-validation and double cross-validation including relevant aspects that should be taken into account for the selection of the cross-validation strategy, as well as permutation testing for the assessment of the statistical significance of cross-validated figures of merit.

2 Partial Least Squares-Discriminant Analysis

PLS-DA is currently one of the most popular classification methods in multivariate analysis. This methodology is the discriminant version of PLS regression extensively used in chemometrics. It aims to model the linear relationship between the matrix $\mathbf{X}$ ($N \times J$), where N and J are the number of samples and predictive variables, respectively, and the corresponding vector of responses y ($N \times 1$)

[2]. In short, PLS extracts a series of latent variables explaining the variation in $\mathbf{X}$ correlated with y. In a PLS-DA model, the relation between the predictors $\mathbf{X}$ ($N \times J$) and the response $\boldsymbol{y}$ ($N \times 1$) can be described as

$$y = \mathbf{X}\boldsymbol{b}^T + \boldsymbol{e}$$

where b ($1 \times J$) is the vector of regression coefficients, and $\boldsymbol{e}$ ($N \times 1$) is the error vector (i.e., residuals).

For the application of PLS to DA in proteomics, the $\boldsymbol{y}$ vector consists of dummy variables (e.g., +1, −1) indicating the class of each sample (e.g., control vs. disease or treatment vs. placebo). Parameters to be selected during the development of a PLS-DA model include the type of data pretreatment and the model complexity (i.e., the number of latent variables). The number of latent variables determines the complexity of the model, and thus the selection of a high or a low number of latent variables could lead to overfitting or under-fitting data, respectively. The analysis of cross-validated figures of merit of PLS-DA classifiers obtained using different numbers of latent variables can be used to select the optimum value. Hence, cross-validation is not only used for estimating the generalization accuracy of the classifier, but also during method development.

2.1 Figures of Merit

Figures of merit are used to describe the performance of the classifier during method development and validation. Outcomes of a binary classifier may be, e.g., $y>0$ or $y<0$ to differentiate disease and healthy. The prediction error is an estimate of how well the classifier will predict the outcome (i.e., the class in a discriminant analysis) of future, unknown observations drawn from the same population. It can be defined as the probability of incorrect classification using a DA model, or as the expected difference between the theoretical and the predicted responses from the model in a regression model. Besides, a number of measures are available to assess the performance of a classifier. After selecting a threshold value to classify samples (e.g., if $y_{\text{threshold}} = 0$, samples with $y>0$ are classified as "disease"), the proportion of misclassified samples can be used when the classes are of similar size. The number of true positive (TP), false positives (FP), true negatives (TN), and false negatives (FN) can be used to build a more informative confusion or contingency table summarizing combinations of the predicted and actual classes. Using these values, sensitivity, as the ratio TP/(TP + FN), and specificity, as TN/(FP + TN), can be defined. Both, sensitivity and selectivity values can be used to build a receiver operator characteristic (ROC) curve by plotting the sensitivity versus (1-specificity) of the classifier using different threshold values. The area under the ROC curve (AUROC) is a frequently employed figure of merit. Alternatively, the Q^2 or discriminant-Q^2 statistics can also be used as figures of merit of classification models. Q^2 quantifies the closeness of predicted y values to the theoretical y values. It is calculated as one minus the ratio of the prediction error sum of squares (PRESS) to the total sum of squares (TSS). The discriminant-Q^2 computes the PRESSD (PRESS discriminant), a PRESS that does not take into account accurately classified samples [3]. The use of discriminant-Q^2 is preferable as it provides a more reliable diagnosis of the generalization accuracy of the classifier [4].

3 Validation Methods

Validation of a classifier assesses its generalization accuracy. To ensure a non-biased estimation of the classification error, the most rigorous approach of testing the predictive performance of a classifier consists in computing model predictions for an independent set of samples (i.e., the external validation set). This set of observations should not be used during the model development and so, in practice, one of the first steps is to split the initial sample set into a calibration and a validation subset. This split should be carefully selected considering that samples used for both, calibration and validation should be representative of the whole population. The use of an external validation set is illustrated in Fig. 1.

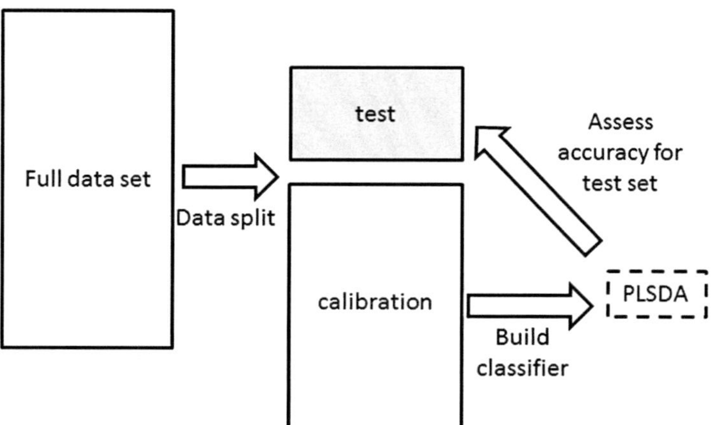

Fig. 1 Scheme of the estimation of PLS-DA model performance using a calibration and an independent test set (i.e., external validation). The initial data set is split into a calibration and a test set. A model is build using the calibration data set. Then, class prediction of samples included in the test set is used to estimate the generalization accuracy of the classifier

3.1 Cross-Validation

As mentioned before, cross-validation is one of the most practical methods to estimate the predictive model performance when an external validation set is not available or simply not even possible. Besides, it is also widely used during the development of the classifier, for example, for the selection of the number of latent variables in PLS-DA model development. When using other methods such as support vector machines (SVM-DA), cross-validation is also employed during model development to select, e.g., the regularization parameter C and kernel parameters such as the γ parameter for a standard radial Gaussian kernel.

During cross-validation the sample set is split usually at random into a number of folds or subsets (k-fold CV). The class of samples included in one subset k is predicted using a model classifier build on samples belonging to the other (k-1) subsets. This step is repeated until each subset has been predicted and then the model performance is estimated using the resulting set of test predictions (*see* Fig. 2). There is not a straightforward way for establishing the fraction of samples that should be used for model development and testing and the optimal value strongly depend on the case under study (e.g., number of samples, sample heterogeneity) (*see* **Note 1**). When the number of k splits is equal to the number of samples, it is called leave-one-out cross-validation. A rule of thumb recommended in machine learning is to use 2/3 of the samples for calibration and 1/3 for testing. The number of k subsets also depends on the size of the calibration set and hence the number of samples. For example, whereas for large data sets a threefold cross-validation (k = 3) is usually appropriate, for small data sets we may have to select leave-one-out cross-validation.

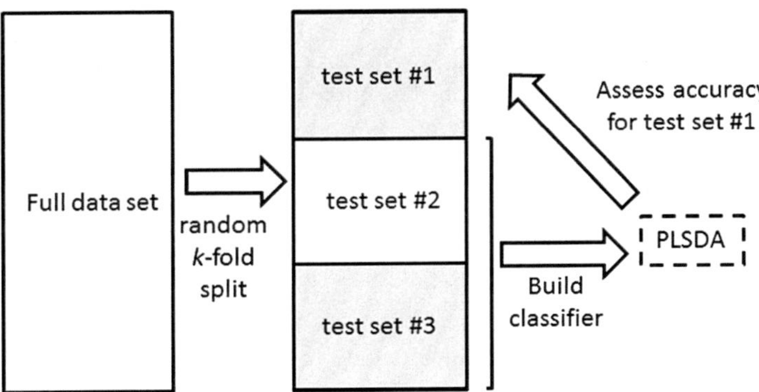

Fig. 2 Scheme of PLS-DA model validation using k-fold cross-validation for k = 3. The full data set is randomly split into k subsets. Then each of these subsets is used to estimate the accuracy of the classifier build from the remaining k-1 subsets. The figure describes the prediction of one of the test sets (test set #1) resulting from a threefold CV. Likewise, model calculation is repeated employing subsets #1 and #2 and #1 and #3 for model calculation, using subsets #3 and #2 for prediction

This division can be performed randomly or using sampling methodologies such as Kennard-Stone, which aims to improve the representativeness of the sample space in both, calibration and test subsets [5]. An important point to keep in mind is that the number of independent samples is not necessarily equal to the number of objects if the study includes sample replicates or repeated measurements (*see* **Note 2**). To avoid overly optimistic results, sample replicates must be kept together during cross-validation in the same subset (calibration or test). On the other hand, it is usually useful to develop a PCA or PLS-DA model including all sample replicates. A scores plot from the obtained model will provide an overview of the variation within replicates of the same sample, and its relation to the overall variance [6].

Cross-validation is straightforward but has several disadvantages. It can lead to overoptimistic results especially when the sample to variable ratio is low [7] and the samples used in calibration or test sets may be too few for being representative of the population, leading to an inaccurate estimate of the generalization accuracy of the classifier. Moreover, the figures of merit may vary depending on the subsets chosen during cross-validation. Monte Carlo cross-validation (MCCV) uses a repeated random selection of the CV subsets to circumvent this drawback. The first step in an MCCV is a random selection of the calibration and test set, without replacement. Then, the classifier is build using the calibration set and the error is calculated for the samples retained in the test set. These two steps are repeated *n* times and an average error is finally calculated from the distribution of *n* figures of merit.

3.2 Double Cross-Validation

Cross-validation provides internal figures of merit because test samples are used to develop the model (e.g., to select the number of latent variables of the model in PLS-DA) [7, 8]. Double cross-validation (2CV) or cross model validation, is a cross-validation strategy that overcomes this potential pitfall providing unbiased external figures of merit [7–10]. In double cross-validation, a subset of objects is set aside as a test set. The remaining set of objects are again split into training and test sets (i.e., internal calibration and test sets) in a *k*-fold cross-validation procedure for the selection of the number of latent variables of the inner classifier used to predict samples included in the test set (*see* Fig. 3). In double cross-validation, samples used for prediction are not used for the building of the classifier (i.e., scaling, selection of the number of LV), which improves the generalizability of the accuracy estimates.

3.3 Assessment of CV Figures of Merit Using a Permutation Test

Some regression and discriminant analysis methods such as PLS-DA or SVM are very potent tools for finding correlation when the number of variables exceeds the number of samples. If method development is not carried out carefully, overfitted models lacking predictive capabilities for future samples may be obtained.

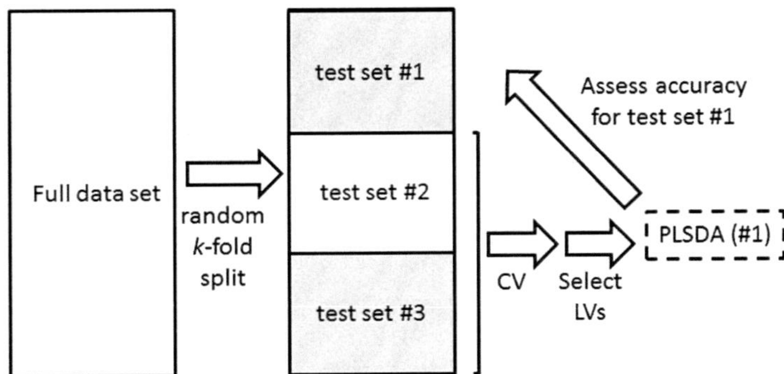

Fig. 3 Scheme of PLS-DA model validation using double cross-validation (for $k=3$). The full data set is randomly split into k subsets. Then each of these subsets is used to estimate the accuracy of the classifier build from the remaining k-1 subsets. The selection of the number of latent variables of each PLS-DA sub-model is based on CV figures of merit calculated using each training set. The figure shows the prediction of one of the test sets (test set #1)

Both, external validation and cross-validation help to identify model overfitting. However, a significant limitation of cross-validation procedures is that they do not assess the statistical significance of the figures of merit estimating the predictive power of the classifier. For example, if the performance of a classifier is assessed by the AUROC, results will range between 0 and 1 and the higher the value, the better the model. However, it is difficult to define a threshold value that corresponds to a "good" classifier.

To obtain an estimate of the statistical significance of the classifier, one may perform a permutation test. This type of hypothesis testing is nonparametric and does not imply assumptions about the distribution of the data. In this test, cross-validated figures of merit calculated using the original class labels (i.e., y vector) are compared to a distribution of the same estimators obtained after a random rearrangement of class labels. In practice, the number of all possible permutations is too large but we can approximate the permutation distribution by using enough random permutations. The statistical significance of the figures of merit, as expressed in a p-value, is then empirically calculated as the fraction of permutation values that are at least as extreme as the statistic obtained from non-permuted data [11]. Results from the test will assess to what extent the classifier is finding chance correlations between the proteomic data and the classes and it is therefore capable of identifying overfitting of the data. Low p-values indicate that the original class label configuration is relevant with respect to the data and assure the significance of the model. A typical plot showing the outcome from a permutation testing for the assessment of CV and calibration is illustrated in Fig. 4. The use of permutation tests in combination with 2CV has been repeatedly shown as a suitable approach to assess the statistical significance of figures of merit [3, 8, 12].

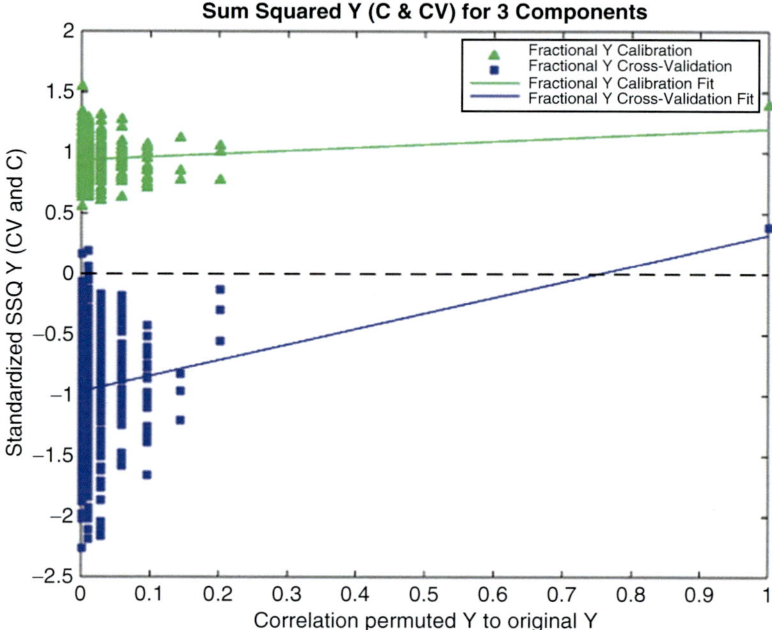

Fig. 4 Typical results from a permutation test to assess the statistical significance of a PLSDA model. The plot displays the fractional *y*-variance captured for self-prediction (calibration) and cross-validation versus the correlation of the permuted *y* to the original *y* vector, and shows the regression lines

3.4 Variable Selection

The identification and elimination of variables irrelevant for classification is usually included in the workflow of data analysis. Variable selection improves the predictive capabilities of multivariate models and facilitates a biological interpretation of the models that might be relevant for further research based on the results (e.g., development of target methods). However, when the training data is used to perform variable selection, figures of merit calculated by cross-validation will lead to specifically overoptimistic and non-generalizable performance estimates. To avoid this, when no external validation set is available, the cross-validation procedure may include the variable selection step [13]. However, this approach is computing intensive and so the use of an external validation set is usually preferable. In any case, the data analysis strategy including the CV method employed should be reported (*see* **Note 3**).

4 Notes

1. Selecting the type of cross validation. There are a number of different cross validation strategies available such as leave-one-out, *k*-fold, venetian blinds, h-block, and random subsets.

Whereas each type of scheme has its advantages and drawbacks, the decision of which is the most effective for the data at hand and the employed classifier is very much data and problem dependent. It is often recommended to take into account some attributes of the data including the underlying distribution of the data (for example, if the samples in the data set are sorted by, e.g., class, collection time, instrument, the distribution may influence CV figures of merit); the total number of samples and the distribution of samples within each class; the presence of replicate measurements (for example, if there are protein profiles of samples collected from the same individual, or instrumental replicates). Double cross validation protects against over-optimistic figures of merit and yields a more realistic estimate of the generalization performance of the classifier.

k-fold and leave-one-out CV. One of the most commonly used CV approaches is k-fold CV in which N/k samples are removed and the classifier is build up on the remaining samples. Then, the classes of the left-out samples are predicted using the classifier and the process is repeated until every sample is predicted once. Finally, the predictive capabilities of the classifier are estimated using, e.g., the number of misclassified samples, AUROC or other figures of merit. In leave-one-out CV, k is the number of samples in the data set.

Random subset selection. The data set is randomly split into k subsets and the same procedure as in k-fold cross-validation is applied. This type of cross-validation offers the advantage of repeating the procedure. The use of iterations is very useful for a reliable model assessment as it helps to reduce and evaluate variance due to instability of the training models and to obtain an interval of the estimate.

Venetian blinds, interleaved, or striped splitting select segments of consecutive samples or interleaved segments as training and test subsets. It is a very straightforward way of stratifying that is useful when samples are randomly distributed in the data set and test and training samples span the same data space as far as possible.

2. Avoid traps! Regardless whether one uses LOO-CV, random subset selection or any other cross-validation approach, two well-known traps must be avoided: the *ill-conditioned* and the *replicate traps* [14]. *Ill-conditioning* is present when the test set of samples is not representative of training samples and leads to overly pessimistic figures of merit of the classifier. The *replicate trap* occurs when replicates of the same sample are included in both training and test sets. In this case cross-validation errors are biased downward. Also, cross-validated errors will be biased downward if an initial selection of the

most discriminant variables is performed using the entire data set. However, unsupervised screening procedures, like removing variables with near-zero variance, have been included prior to cross-validation [15].

3. Reporting results: To report an estimation of the efficiency of a classifier using cross-validated figures of merit, the cross-validation parameters such as type of approach (*k*-fold, LOO, venetian blinds, etc.), the sample size and the number of iterations or data scaling should be specified. In addition, data treatment in, e.g., Matlab or R allows the creation of code integrating the whole data preprocessing and analysis procedure. When reporting the results of an analysis, the availability of the dataset and a detailed description of the code and/or software used improve the reproducibility of the results.

References

1. Esbensen KH, Geladi P (2010) Principles of proper validation: use and abuse of re-sampling for validation. J Chemometr 24:168–187
2. Wold S, Sjöström M, Eriksson L (2001) PLS-regression: a basic tool of chemometrics. Chemometr Intell Lab Syst 58:109–130
3. Westerhuis JA, Velzen EJJ, van Hoefsloot HCJ et al (2008) Discriminant Q2 (DQ2) for improved discrimination in PLSDA models. Metabolomics 4:293–296
4. Szymańska E, Saccenti E, Smilde AK et al (2012) Double-check: validation of diagnostic statistics for PLS-DA models in metabolomics studies. Metabolomics 8:3–16
5. Kennard RW, Stone LA (1969) Computer aided design of experiments. Technometrics 11: 137–148
6. Esbensen KH, Guyot D, Westad F et al (2004) Multivariate data analysis—in practice. An introduction to multivariate data analysis and experimental design, 5th edn. CAMO Process AS, Oslo
7. Rubingh CM, Bijlsma S, Derks EPPA et al (2006) Assessing the performance of statistical validation tools for megavariate metabolomics data. Metabolomics 2:53–61
8. Westerhuis JA, Hoefsloot HCJ, Smit S et al (2008) Assessment of PLSDA cross validation. Metabolomics 4:81–89
9. Filzmoser P, Liebmann B, Varmuza K (2009) Repeated double cross validation. J Chemometr 23:160–171
10. Gidskehaug L, Anderssen E, Alsberg BK (2008) Cross model validation and optimisation of bilinear regression models. Chemometr Intell Lab Syst 93:1–10
11. Knijnenburg TA, Wessels LFA, Reinders MJT et al (2009) Fewer permutations, more accurate p-values. Bioinformatics 25:161–168
12. Wongravee K, Lloyd GR, Hall J et al (2009) Monte-Carlo methods for determining optimal number of significant variables. Application to mouse urinary profiles. Metabolomics 5: 387–406
13. Kuligowski J, Perez-Guaita D, Escobar J et al (2013) Evaluation of the effect of chance correlations on variable selection using Partial Least Squares-Discriminant Analysis. Talanta 116:835–840
14. Bakeev K (ed) (2010) Process analytical technology: spectroscopic tools and implementation strategies for the chemical and pharmaceutical industries, 2nd edn. Wiley, New York
15. Krstajic D, Buturovic LL, Leahy DE et al (2010) Cross validation pitfalls when selection and assessing regression and classification models. J Cheminform 6:10

Chapter 12

Protein Sequence Analysis by Proximities

Frank-Michael Schleif

Abstract

Sequence data are widely used to get a deeper insight into biological systems. From a data analysis perspective they are given as a set of sequences of symbols with varying length. In general they are compared using nonmetric score functions. In this form the data are nonstandard, because they do not provide an immediate metric vector space and their analysis using standard methods is complicated. In this chapter we provide various strategies for how to analyze these type of data in a mathematically accurate way instead of the often seen ad hoc solutions. Our approach is based on the scoring values from protein sequence data although be applicable in a broader sense. We discuss potential recoding concepts of the scores and discuss algorithms to solve clustering, classification and embedding tasks for score data for a protein sequence application.

Key words Protein sequence analysis, Proximity data, Indefinite-kernel, Machine learning

1 Introduction

The area of machine learning and computational intelligence has developed a large set of algorithms to define models for challenging problems like classification, clustering, embedding, regression, and time series prediction [1–3]. Prominent examples are kernel- and prototype-based learning methods [1, 4, 5] which are sometimes summarized under relational learning. These approaches take the natural notion of the pairwise similarity or dissimilarity (in the following simple proximities) between objects to define the models and do not necessarily need the data in an underlying vector space.

In general the models are defined by a combination of proximities from the training set, *see* e.g. [6]. In this way it is also possible to use own user defined metric proximity functions such that very flexible data representation becomes possible. The principle of this encoding procedure is depicted in Fig. 1.

Many of these models permit a clear communication of the model decision process, because the decision function is based on identifiable training points. Thereby the model is inherently or

Klaus Jung (ed.), *Statistical Analysis in Proteomics*, Methods in Molecular Biology, vol. 1362, DOI 10.1007/978-1-4939-3106-4_12,

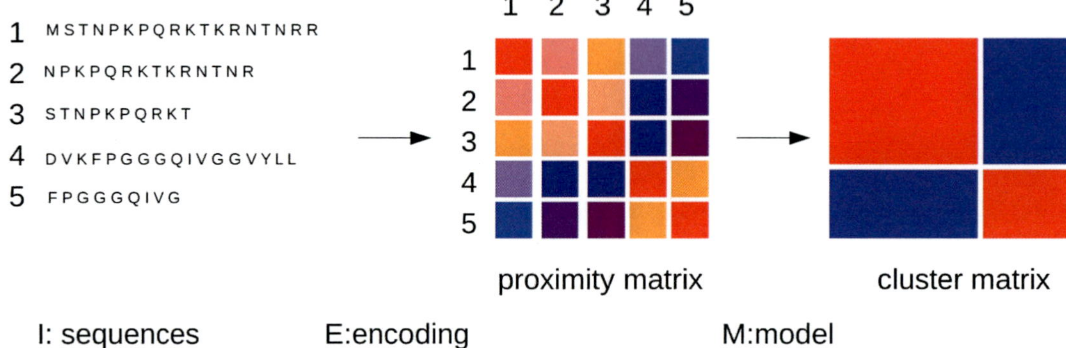

Fig. 1 Schematic view of the encoding and modeling step of protein sequences. First the sequences are compared using a similarity measure like the Smith-Waterman alignment, leading to a proximity matrix. Subsequently the proximity matrix can be used to obtain for example a clustering, classification, or embedding model. For the modeling step the original sequences are not any longer needed but the model can be solely defined from the proximities

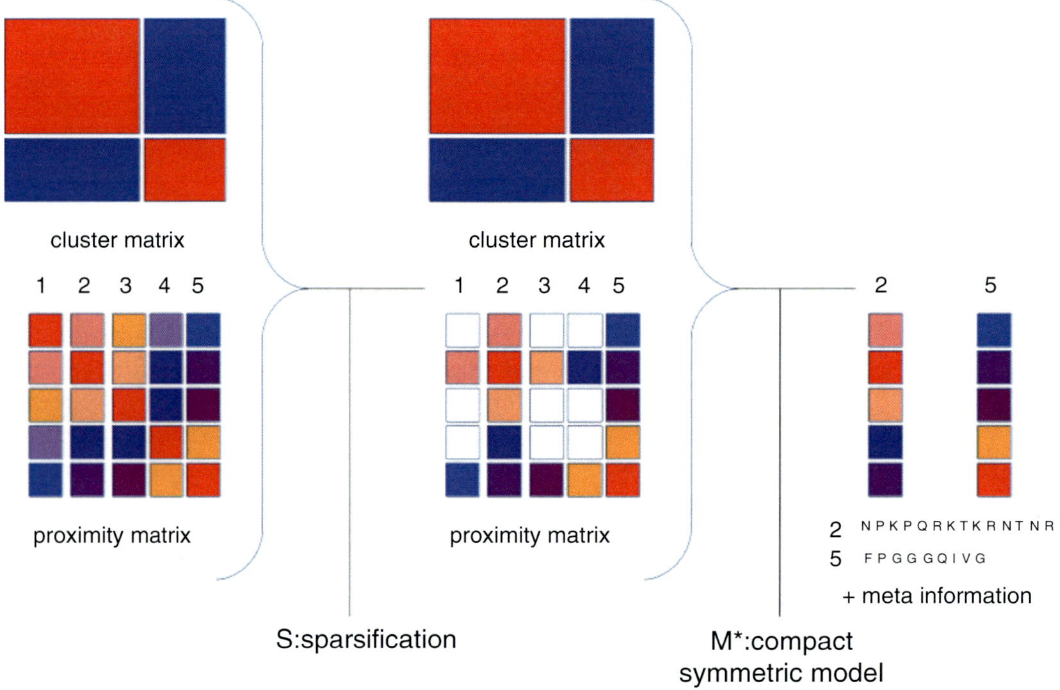

Fig. 2 Schematic view of the sparsification procedure from a given proximity model. The originally rather dense proximity matrix is reduced to a small set of reference points (cluster centers or prototypes) which are sufficient to describe the data characteristics

subsequently sparsified such that only few model parameters are needed in the decision function. This not only shrinks the model but in general also leads to a better generalization and easy out-of-sample extension with respect to new items. The steps are schematically illustrated in Fig. 2.

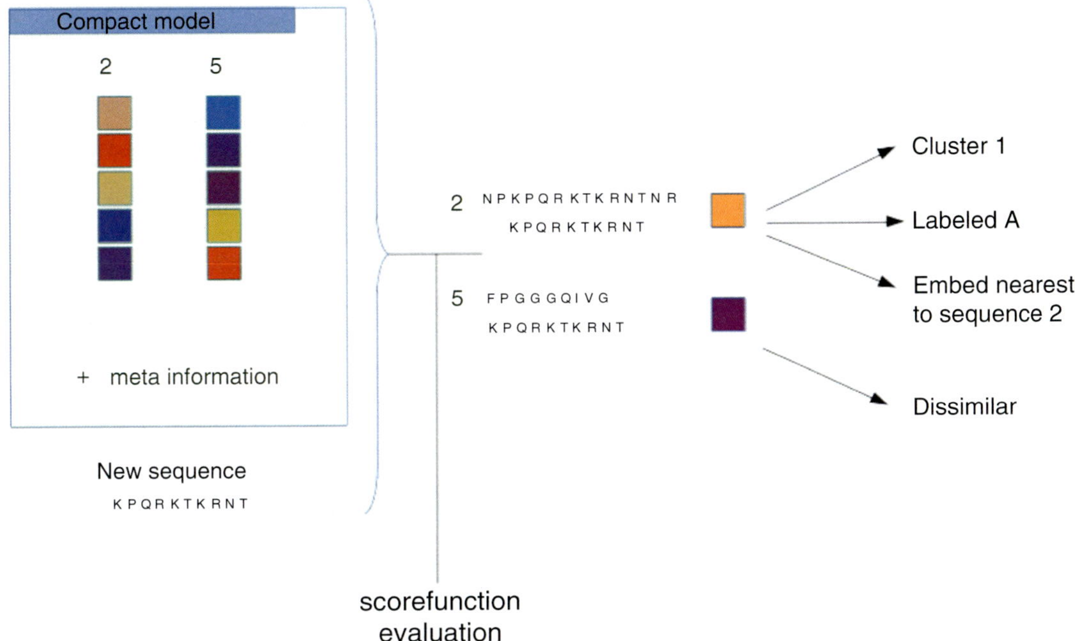

Fig. 3 Schematic view of an out-of-sample extension. The new sequence can be mapped into the model by calculating very few alignment scores to the reference points used in the model. Depending on the model objective the decision function may assign the new item to a cluster, label it by a class label, represent it in a low-dimensional display, or can just decide that the object is dissimilar

New data points can be mapped to these models by calculating proximities between the query object and in general small number of proximities, used in the modeling step. Using this procedure one can identify to which submodel, e.g., cluster or class the new item is closest to and guide the decision or representation process accordingly. This procedure is shown in Fig. 3.

In this sense the models are smart, identifying objects in the training data which are informative and useful to define generic predictive models. While this procedure appears to be very natural and could be also used in a variety of kernel algorithms, it is in fact quite challenging for nonmetric proximities at large scale. Most data analysis algorithms are based on the assumption that the given data are available in a vector space and most often the Euclidean distance, or its corresponding inner product is used to identify the proximities. Due to the aforementioned reasons this is not available for protein sequence data and rather few methods have been proposed for model learning with nonmetric proximities [7, 8]. Most of these approaches are however limited to problems at moderate volume, e.g., multiple thousand objects, and require either on-the-fly or on-block calculations of a large number of pairwise proximities, making them unattractive for large to very large problems. The comparison of protein sequences is often done at large scale by the Smith-Waterman algorithm [9]. Although widely accepted and successful in biology and biochemistry, this measure violates the triangle inequality and is costly to calculate if we consider squared input matrices of proximities. Alternative encodings for sequence data, e.g., for text documents, which are often used for classical kernel algorithms appear to be very unattractive for protein sequence data and are to limited in comparison to, e.g., Smith-Waterman. Hence, classical kernel algorithms are in general inapplicable, due to the used proximity measures, the expectation of metric input matrices or scaling issues. In the following we provide a compact overview on approaches available for the considered type of data and provide guidance in applying them in the analysis of protein sequence data.

2 Data

As already mentioned we consider protein sequence data which have been used for proximity based data analysis already very early in [10], although at a much smaller scale. For simplicity we take the sequences available in the SwissProt data base [11] of October, 2010 and we restrict our analysis to the sequences with prosite label information. Further we only consider sequences which belong to prosite classes with at least 1000 entries. This leads to approximately 80,000 sequences. The subsequent described techniques do also scale to datasets with more, e.g., 1,000,000 entries but we restrict our experiments in favour of clear results.

2.1 Encoding

We represent a large set of protein sequence data by using proximity scores only. Therefore one calculates pairwise proximities between the sequences using, e.g., the Smith-Waterman algorithm. The Smith-Waterman algorithm is a nonmetric similarity measure to compare protein sequences of varying length. The more similar to sequence are the higher the score returned by the algorithm, with a minimum score (ideally 0) for dissimilar objects. It makes internal use of weighting matrices, which describe exchange weights for the sequence symbols based on biological and biochemical pre-knowledge. Here we use the blosum62 matrix. The algorithms presented here are not restricted to a specific type of similarity measure. However we will assume in the following that the proximity measures are symmetric, e.g. we may symmetrize pairwise scores assuming that the proximity between a sequence A to a sequence B is almost the same to the score as obtained by comparing sequence B to sequence A. For a large number of sequences this would lead to a high computational load which can be avoided by two strategies. First, due to the symmetry assumption we only need to calculate $(N-1)/2$ scores with N as the number of sequences and if we assume that the self-similarities are constant (e.g., are set to the maximum observed value). As a second approach we use the Nystroem approximation, as a very efficient

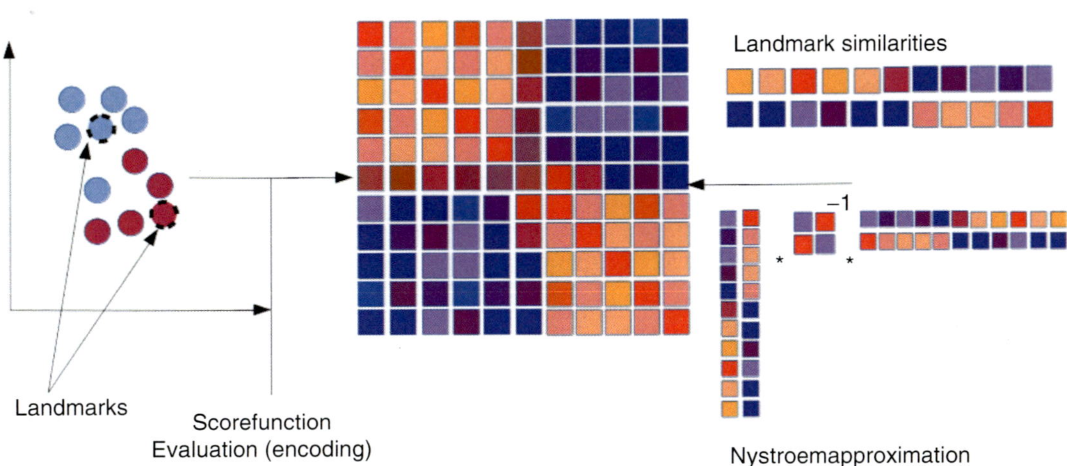

Fig. 4 Schema of the Nystroem approximation. Only few so-called landmark points (e.g., cluster centers) and their proximities to all other points are needed to represent the whole $N \times N$ proximity matrix (in the center)

technique used in kernel and prototype based learning [12, 13] with good error bounds also at large scale (although proofs are only given for metric proximity data) [14, 15]. The Nystroem approximation is a mathematical concept to approximate an arbitrary (potentially asymmetric) matrix (*see* **Note 5**, Fig. 4). In our study we consider the proximities between sequences. Lets briefly assume that our data are given in a two-dimensional vector space, a plane and we compare N pairwise points given on the plane, using our favorite proximity measure (like the Euclidean distance). Then we obtain a $N \times N$ proximity matrix which is quite costly to calculate and to store. But as we know our data are much less complex (intrinsically low dimensional) located only in two dimensions. The Nystroem approach takes this into account. The idea is to identify a number of so called landmarks (e.g., the cluster centers of our data) and to represent the proximity matrix only by scores of these landmarks to all other points (*see* **Note 1**). If we assume that our data are intrinsically not so complex we can choose the number of landmarks M, much smaller then N and approximate the full proximity matrix by a linear number of $M \times N$ proximity values. As recently shown in [16] this concept can be extended to nonmetric similarities and also dissimilarities.

Instead of using the given proximities directly one may also consider various ways of correcting them to make them metric. As shown recently in [16] this approach can be done with rather low computation costs on approximated matrices using a low rank eigenvalue decomposition. The score matrix $\mathbf{S}$ as obtained from the sequence alignments is in general nonmetric. Assuming the input matrix is a similarity matrix (this is the case if smith waterman is used) one can decompose the matrix as $\mathbf{S} = \mathrm{U\ L\ U'}$ using an eigenvalue decomposition. The eigenvalues in the matrix L show negative values only if the **S** is nonmetric. Besides various other approaches *see*, e.g., [17] one may simple correct the eigenvalues in L to get **S** metric. The most simple way is to set negative values to 0, which is called clipping or to take the absolute values, referred to as flipping. As shown in [16] this can be coupled with the Nystroem approximation at linear costs instead of otherwise cubic complexity in the number of points. The so corrected proximity matrix **S** can now be used in any standard kernel algorithm. Accordingly prominent kernel algorithms like kernel-PCA, kernel k-means, the support vector machines (SVM), or Laplacian eigenmaps for embedding become accessible [18].

3 Methods

By use of the steps addressed in the former section one has now a positive semi-definite or for short metric similarity matrix which may have been approximated by use of the Nystroem approximation. Depending on the experimental objective one may now consider an unsupervised (e.g., clustering, embedding) or supervised (classification, regression) task. In the following we focus on embedding and classification.

3.1 Embedding

A variety of embedding approaches have been proposed in the last decades. The goal is to represent a complex, potentially high dimensional dataset on a two- or three-dimensional plane. One can distinguish global approach and local approaches. While the first one try to explain the global structure of the data relations, permitting a loss of accuracy for point relations in close proximity, the later approaches try to preserve local structure, accepting that global relations are broken in the low-dimensional view. The majority of these approaches is directly based on proximity data and in general even permits the input of a proximity matrix, e.g., a distance matrix. Some require metric proximities, and others are more flexible allowing even for asymmetric relations [19]. A major drawback of many of these methods is or was the scalability to larger datasets. In the last years various extensions of classical approaches have been proposed, not only improving the proximity preservation in the low dimensional embedding but also to improve scalability. The most common techniques are subsampling, triangulation techniques, N-body approximation schemes, and low rank approximations [20–22]. However such approaches can often only be used for metric data representation such that the steps mentioned before are mandatory in preprocessing the inputs. Laplacian eigenmap is an interesting technique which can be easily extended to incorporate eigenvalue correction and Nystroem approximation [18, 20]. An application of this technique to our data is shown in Fig. 5 where we show an eigenmap for the top 21 prosite-labeled classes in the SwissProt database.

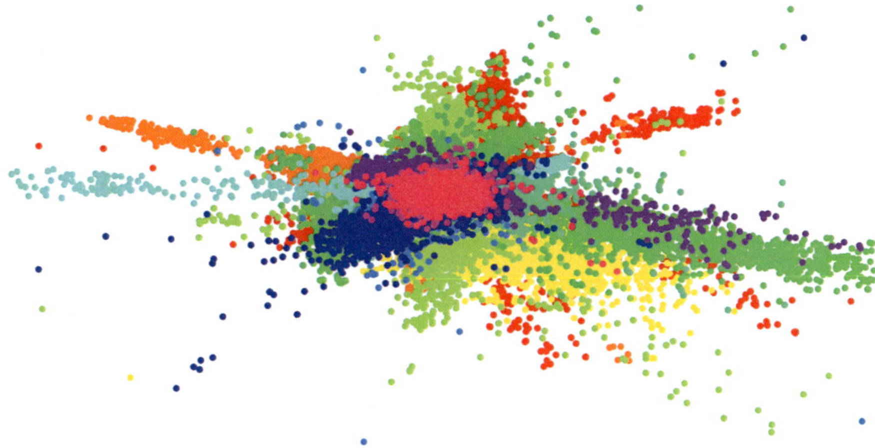

Fig. 5 Zoom into the embedding of a large part of the SwissProt database using a laplacian eigenmap approach. Laplacian eigenmaps require as input a valid kernel matrix; hence the discussed preprocessing steps are needed. Here we also used the Nystroem approximation reducing the computation costs from cubic to linear

The colors indicate the various classes and we can identify relationships within the data. The approach is obviously very interesting for sequence data sets where the label information may not yet known and an interactive view is relevant to visually explore the given data.

3.2 Classification

As pointed out before kernel classifiers are very effective tools and can also deal with sequence data by dedicated kernel like global alignment kernel [23] or others. However our objective is to keep the domain expert the freedom to use its own well accepted measure like the Smith-Waterman score. Saying this we are then forced to use a kernel method on a large set of nonmetric similarity scores. This is challenging in two ways. First of all we cannot (or want not) calculate 265166^2 scores. Given each score is represented by 8 bytes we end up with roughly 524 GB. Further it is also not very time efficient. If we assume that each score can be calculated in 1/1000 s we have to wait for a half year. Parallelization will help of course because the queries can be nicely parallelized [24] but still the handling of a big matrix becomes very unhandy. Modern kernel classifiers like the core-vector machine (CVM) [25] do not need to evaluate all the kernel similarities but our similarities are nonmetric and an online eigenvalue correction is not immediate available. However we can use our presented approach approximating the score matrix as well as performing an eigenvalue correction. To do this we need to specify a reasonable set of landmarks. As mentioned before the landmarks should capture the major data characteristics. Taking this into account we will assume that picking a large number of landmarks from different ProSite classes should be sufficient. Here we use 3354 landmarks, selected in accordance to the most prominent classes with up-to 3 landmarks per class and minimal class overlap due to the multilabeling. To simplify the experiment we reduce the multilabel of the sequences to a single label by using only the most prominent label for each multilabeled object. Accordingly we need only around 7 GB to store the matrix and some days to calculate the scores. The approximation and eigenvalue correction by the presented approach takes only some minutes. Here we use a flipping approach to correct the eigenvalues, which was found often effective for different other, smaller datasets used in former studies [16]. The obtained approximated and now positive semi definite similarity matrix is used by a CVM in a fivefold cross validation to generate a classification model. The results are very promising with a prediction accuracy of 76.85 ± 0.36 % on the hold out test data in a model with 521 classes. An additional benefit of the CVM approach is that it naturally leads to very sparse models (*see* **Note 2**). Accordingly the out of sample extension to new sequences requires only few score calculations to the sequences of the training set (*see* **Note 4**). The majority of the sequence database can be ignored from now on. By parallelizing the queries to the individual class models we get another speedup of around 500 (*see* **Note 3**).

4 Notes

1. Landmark selection: The Nystroem approximation requires the selection of a number of landmarks to capture the rank of the proximity matrix. In the literature multiple approaches have been proposed to address the landmark selection but are in general based on the availability of the proximity matrix as a whole.

   If meta information (e.g., the prosite label information) is available and a full calculation of the proximity matrix is no option we suggest to use a random subset from each class. This widely ensures that each part of the data space is reasonable covered by the landmarks and the eigenvalue structure of the proximity matrix is reasonable kept. A more expensive alternative is to calculate a full proximity matrix on a subsample of the sequence data and apply traditional approaches [20].

2. Model sparsification: Depending on the objective the used proximity (kernel) method can lead to quite complex decision models. This means the number of basis functions is rather large. Beside of using so called sparse approaches like the applied Core Vector Machine one can also employ k-approximation schemes [6, 17]. This can be especially interesting for a variety of clustering algorithms. While sparsification and approximation is in general a good idea it can have a

negative impact if the data distributions are already were sparse. This happens for the SwissProt data in the way that a larger number of prosite-labeled classes have only a small number of entries. It is reasonable to assume that for these classes any kind of approximation may have a negative impact and one better keeps these entries individual in the decision step. Therefore the proposed approach can be likely improved in the prediction accuracy by focusing on the densely populated classes. This will still cover the large majority of the databases widely keeping the intended speedup during the identification of new sequences.

3. Speed up of calculations: Calculations can be accelerated not only by means of approximation schemes but also by using parallelization approaches. The calculation of a pairwise score can be easily parallelized over the whole set of sequences. Due to the inherent overhead costs of parallelization we found it most efficient to split the parallelization by the number of landmarks. In this way each processor or cluster node calculates all proximities to one landmark. Also the model calculations can be parallelized. The CVM with a one-versus-all scheme calculates C different models, where C is the number of classes. This can be perfectly parallelized to C nodes. The usage of shared memory operations can further speed up the process by avoiding to load the Nystroem approximated kernel matrix for each of the C models from hard disk. Due to the Nystroem approximation a test sequence can be used by the model very easy. One only has to calculate the proximities to the M landmark points instead of all N sequences, then the Nystroem approximated matrix can be used to calculate the proximity values to all other points with some basic algebraic operations. A sparse model can lead to additional speed-up's during the usage of the model. If the number of basis functions used in the decision function is very small compared to N, e.g., for CVM, the number of relations to the new test point which have to be considered is very small and the decision function evaluates quickly.

4. Out-of-sample extension: For the models considered in this paper the out of sample extension is directly available. Using the approaches presented in [17] each new sequence can be easily converted in a vector of valid proximities. For the used or cited clustering approaches this proximity vector can be directly used in the decision function. For the Laplacian eigenmaps the embedding is done by a projection function and hence given a vector of proximities extended to all N training points. For other embedding approaches one may use techniques as suggested in [26].

5. General comments: Those interested on the implementation of the approach outlined in this paper are strongly encouraged to read the cited papers which describe the mathematical formalism which has been widely omitted here due to obvious reasons.

Acknowledgment

I would like to thank my former colleagues Andrej Gisbrecht, Barbara Hammer and Alexander Grigor'yan, University of Bielefeld for supporting parts of this work. I am grateful to the Max-Planck-Institute for Physics of Complex Systems in Dresden and Michael Biehl, Thomas Villmann and Manfred Opper as the organizer of the Statistical Inference: Models in Physics and Learning-Workshop for providing a nice working atmosphere leading to some of the ideas outlined in this manuscript. The author was supported by a Marie Curie Intra-European Fellowship (IEF): FP7-PEOPLE-2012-IEF (FP7-327791-ProMoS). I would like to thank R. Duin and E. Pekalska for providing various information about pseudo-Euclidean spaces and access to distools and prtools.

References

1. Shawe-Taylor J, Cristianini N (2004) Kernel methods for pattern analysis and discovery. Cambridge University Press, Cambridge
2. Pekalska E, Duin R (2005) The dissimilarity representation for pattern recognition. World Scientific, Amsterdam
3. Hastie T, Tibshirani R, Friedman J (2013) The elements of statistical learning. Springer, New York
4. Hofmann D, Schleif F-M, Hammer B (2014) Learning interpretable kernelized prototype-based models. Neurocomputing 131:43–51
5. Graepel T, Obermayer K (1999) A stochastic self-organizing map for proximity data. Neural Comput 11:139–155
6. Hammer B, Hasenfuss A (2010) Topographic mapping of large dissimilarity data sets. Neural Comput 22:2229–2284
7. Chen Y, Gupta MR, Recht B (2009) Learning kernels from indefinite similarities. In: Danyluk AP, Bottou L, Littman ML (eds) Proceedings of the 26th annual international conference on machine learning, Montreal, Quebec, Canada, June 14–18, 2009. ACM international conference proceeding series, Madison, WI, USA, pp 145–152
8. Pekalska E, Haasdonk B (2009) Kernel discriminant analysis for positive definite and indefinite kernels. IEEE Trans Pattern Anal Mach Intell 31:1017–1032
9. Gusfield D (1997) Algorithms on strings, trees, and sequences: computer science and computational biology. Cambridge University Press, Cambridge
10. Kohonen T, Somervuo P (2002) How to make large self-organizing maps for nonvectorial data. Neural Netw 15:945–952
11. Boeckmann B, Bairoch A, Apweiler R et al (2003) The Swiss-Prot protein knowledgebase and its supplement TrEMBL. Nucleic Acids Res 31:365–370
12. Gisbrecht A, Mokbel B, Schleif F-M et al (2012) Linear time relational prototype based learning. Int J Neural Syst 22:1250021
13. Williams CKI, Seeger M (2000) Using the Nystrom method to speed up kernel machines. In: Todd KL, Dietterich TG, Tresp V (eds) Advances in neural information processing

systems 13, papers from neural information processing systems, Denver, CO, USA, pp 682–688

14. Kwok M, Li JT, Lu B-L (2010) Making large-scale Nystrom approximation possible. In: Furnkranz J, Joachims T (eds) Proceedings of the 27th international conference on machine learning (ICML-10), June 21–24, 2010, Haifa, Israel, Omnipress, Madison, WI, USA, pp 631–638
15. Drineas P, Mahoney MW (2005) On the Nystrom method for approximating a gram matrix for improved kernel-based learning. J Mach Learn Res 6:2153–2175
16. Schleif F-M, Gisbrecht A (2013) Data analysis of (non-)metric proximities at linear costs. In: Hancock ER, Pelillo M (eds) Similarity-based pattern recognition—second international workshop, York, UK, July 3–5, 2013. Proceedings, pp 59–74
17. Chen Y, Garcia EK, Gupta MR et al (2009) Similarity-based classification: concepts and algorithms. J Mach Learn Res 10:747–776
18. Schleif F-M (2014) Proximity learning for non-standard big data. In: Verleysen M (ed) 22th European symposium on artificial neural networks, 2014, Bruges, Belgium, April 23–25, 2014
19. Strickert M, Bunte K, Schleif FM et al (2014) Correlation-based embedding of pairwise score data. Neurocomputing 141:97–109
20. Zhang K, Kwok JT (2010) Clustered Nystrom method for large scale manifold learning and dimension reduction. IEEE Trans Neural Netw 21:1576–1587
21. Lee J, Verleysen M (2007) Nonlinear dimensionality reduction, Information science and statistics. Springer, New York
22. Yang Z, Peltonen J, Kaski S (2013) Scalable optimization of neighbor embedding for visualization. In: Proceedings of the 30th international conference on machine learning, Atlanta, GA, USA, 16–21 June 2013. JMLR proceedings, vol 28. pp 127–135. JMLR.org
23. Cuturi M (2011) Fast Global Alignment Kernels. In: Getoor L, Scheffer T (eds) Proceedings of the 28th international conference on machine learning, Bellevue, Washington, USA, June 28–July 2, Omnipress 2011, Madison, WI, USA, pp 929–936
24. Rognes T (2011) Faster smith-waterman database searches with inter-sequence SIMD parallelisation. BMC Bioinformatics 12:221
25. Tsang IW, Kocsor A, Kwok JT (2007) Simpler core vector machines with enclosing balls. In: Ghahramani Z (ed) Machine learning, proceedings of the twenty-fourth international conference, Corvallis, Oregon, USA, June 20–24. 227. ACM, pp 911–918
26. Gisbrecht A, Lueks W, Mokbel B et al (2012) Out-of-sample kernel extensions for nonparametric dimensionality reduction. In: Verleysen M (ed) 20th European symposium on artificial neural networks, 2012, Bruges, Belgium, April 25–27

Part IV

Data Integration

Chapter 13

Statistical Method for Integrative Platform Analysis: Application to Integration of Proteomic and Microarray Data

Xin Gao

Abstract

To perform integrative analysis on multiple genomic data sources, we propose to use Fisher's combined probability test for consolidated inference. The method combines the individual *p*-values from different data sources and constructs a chi-square test statistics for the overall significance. This method is valid to combine results across independent data sources. We further improve the method to accommodate the scenario that the data sources are dependent or the data samples are too small to obtain valid *p*-values through exact distributions. The proposed method is convenient to use in practice and is robust to distributional assumptions and small sample sizes.

Key words Data integration, Fisher's combined probability test, Permutation, Microarray, Proteomics

1 Introduction

With the advent of new biotechnologies, researchers perform experiments on various experimental platforms to study the biological changes in organisms. Different types of experiments offer quantitative measurements of the underlying physical process on different aspects. Combining information across platforms can help validate the results and increase the sensitivity and selectivity of the analysis. Data integration has become increasingly important due to the large amount of data available in different data and knowledge repositories, including clinical data, laboratory data, database of genotypic/phenotypic associations, ontologies, and knowledge bases and rule bases.

To select the significant biomarkers which are predictive to some physical conditions such as disease development, scientists can carry both microarray experiment and proteomic experiment on thousands of biomarkers. In contrast to traditional approach of selecting biomarkers based on one type of experiment, the data

Klaus Jung (ed.), *Statistical Analysis in Proteomics*, Methods in Molecular Biology, vol. 1362,
DOI 10.1007/978-1-4939-3106-4_13, © Springer Science+Business Media New York 2016

integration of proteomic and microarray experiments is to combine the results together from both experiments. Many biomarkers behave differently in disease versus control groups and they can act as predictors or classifiers for the underlying disease status. This phenomenon is often manifested consistently across two platforms. Therefore, it is often desirable that results from the two experiments can be combined in a statistical way to correctly identify the significant biomarkers.

Reif et al. [1] pointed out that integrative analysis of multiple data types would improve the identification of biomarkers of clinical end points. However, there are a number of challenges to integrate data from different sources. First, genomic data provide a wide variety of physical measurements [2]. The mRNA expression data are continuous measurements, and proteomic data usually consist of discrete counting variables. The underlying distribution densities vary across different measurements. Second, the biological samples used in different platforms can be taken from the same or related subjects. Therefore, results across different platforms could be correlated.

In recent literature, classifications on biological samples were performed by combining the measurements from different experimental platforms [3–5]. Correlation analysis has been proposed to integrate diverse data types by examining the correlations between the microarray and proteomic measurements [6–9]. However, correlation analysis only reveals the concordance or dis-concordance among measurements across different platforms. The presence of strong correlation does not directly translate to significant involvement in the underlying biological process. Furthermore, correlation analysis cannot be applied on complicated data sets, such as factorial experiments, times series, or longitudinal data.

The rank aggregation method [10] has been proposed to aggregate ranking lists across different platforms to get the final ranking of the genes. However, top ranking in the list does not always relate to statistical significance to the underlying process. Rhodes et al. [11] combined multiple microarray data sets to identify genes deregulated in prostate cancer by comparing individual test statistics with thresholds and select the top genes in terms of the number of times the gene exceeds the thresholds for each individual platform. However, this method cannot offer a direct measure of how significant that gene is statistically. Wu et al. [12] proposed to combine statistical evidence across different platforms through summary statistics. For each experimental platform, a null hypothesis is formulated and a summary test statistic is formed. By randomization, the null distribution of the vector of statistics across different platforms can be performed. The integrative test statistics is a weighted sum of the individual test statistics inversely weighted by their own variability. The significance of each integrative test statistic is assessed by permutation method. The method can integrate different types of test statistics from various platforms. However, the implementation of this algorithm is complicated and thus limits its wide use in practice.

2 Methods

To consolidate the statistical significance of each biomarker across different experimental platforms, we propose a simple algorithm which is convenient for practitioners to use. The objective of our multi-platform integration method is to select a set of biomarkers that act differently in the treatment group versus the control group. Our method is conducted by formulating individual hypothesis for each platform. Each null hypothesis specifies that there is no difference in the biomarker's level comparing the two groups in that particular experimental setting. For example, for platform i, $i = 1,\ldots K$, the null hypothesis is the following: the gene does not behave differently on platform i comparing the control group versus the treatment group.

Based on the set of K null hypotheses for K experimental platform data, we construct a set of K test statistics, which can be different and constructed to summarize the specific experiments on the corresponding platforms. For instance, if the proteomics experiment generates discrete and non-normal data, the chosen test statistic can be a nonparametric rank test; if the microarray experiment has an unbalanced two samples, the appropriate test statistic can be a Welch's student t test. For each platform and each gene, the p-value of the gene on that platform is obtained by performing the chosen test. We will employ Fisher's combined probability test [13] to combine the p-values and obtain a consolidated new statistic. This test has been used in data fusion field to aggregate statistical significance [14].

Let p_1, $p_2,\ldots p_K$ denote the individual p-values obtained from the K experimental platforms for one gene. We construct the integrative test statistic:

$$T = \sum_{i=1}^{K} -2\log(p_i).$$

As the minus two log transformation on the p-value leads to a chi-square statistic of two degrees of freedom, the integrative test statistic T has a chi-square distribution of 2K degrees of freedom. Then the data integration analysis yields a consolidated p-value for this gene by calculating the probability that a chi-square random variable with 2K degrees of freedom exceeds the observed test statistic T. Such consolidation of statistical significance leads to correct inference and offers an overall p-value by combining the significance from each individual source.

Here we present the integration of microarray and proteomic data as an example. Suppose there are microarray expression data measured on l_1 diseased samples and l_2 control samples and proteomic data measured on m_1 diseases samples and m_2 control samples. The aim is to select biomarkers that behave differently in disease group and control group.

Step 1: Formulate two individual null hypotheses: H_{01}: the gene's mRNA level is the same across diseased and normal populations; H_{02}: the protein level is the same across diseased and normal populations.

Step 2: For the two hypotheses, choose two test statistics, T_{m} and T_{p}. The test statistics can be different and tailored to each experimental platform. The two p-values are denoted as p_{m} and p_{p}.

Step 3: The integrative test statistic takes the form

$$T = -2\left\{\log\left(p_{\mathrm{m}}\right) + \log\left(p_{\mathrm{p}}\right)\right\}.$$

The consolidated p-value is the probability of a chi-square random variable with four degrees of freedom exceeding the observed value of the statistic T. This method aggregates the p-values through log transformation and obtains a chi-square test statistic for the overall data.

2.1 Multiplicity Adjustment

There are a variety of methods to perform the multiplicity adjustment. In order to control the familywise type I error rate at α, we can perform Bonferroni's correction and the individual level should be adjusted to be $\frac{\alpha}{n}$, where n is the total number of biomarkers. When n is large, the Bonferroni's correction provides very stringent threshold. For each biomarker, the consolidated p-value is compared with the Bonferroni-corrected significance level. A gene is selected to be significant if the consolidated p-value is below that threshold.

Alternatively, one can control the number of false discoveries. To control the estimated number of false discoveries to be equal to or less than f, we can adjust the individual p-value threshold to be $\frac{f}{n\hat{\pi}}$, where $\hat{\pi}$ denotes the estimated proportion of non-differentially expressed biomarkers. If there is no $\hat{\pi}$ available, we use $\hat{\pi} = 1$ and that gives a conservative threshold. Compared to Bonferroni's correction, as f is usually set to tens in the view of thousands of biomarkers, this adjusted threshold is much less stringent. Furthermore, one can apply more sophisticated false discovery rate control methods proposed by Benjamini and Hochberg [15]. Readers can be referred to the references therein.

2.2 Direction of the Hypotheses

The null hypotheses can be one sided or two sided. Even when the directions of the null hypotheses are different across different platforms, it would not affect the validity of the integrative test statistic T.

2.3 Dependent Experimental Platforms

The integrative test statistic T has a chi-square distribution under the assumption that the individual test statistics are independent across all the platforms. This assumption may be violated in practice when the samples for different experiments are taken from the same subject or related subjects. For example, the same tissue from a mouse can be divided and used in both microarray and proteomic experiments. This leads to dependent test statistics across platforms. The test statistics tend to be correlated because of the fact that samples are obtained from the same or correlated sources.

Under such dependency situation, the integrative test statistic T no longer has a simple-to-use chi-square distribution. To address this problem, we propose to use the same T statistic, but make inference and obtain the overall p-value of T based on its empirical null distribution through permutation method. The permutation method offers valid overall p-value for dependent experimental results across multiple platforms [16].

The permutation method is also advantageous when each individual test statistic has unknown distribution. This often occurs when some distributional assumptions are violated or sample sizes are too small so that the theoretical distribution for the individual test statistic differs from the actual underlying null distribution. If this is the case, one can still proceed to obtain individual p-values based on the theoretical distribution. The overall inference will be based on the empirical null distribution of the integrative statistic T.

In order to perform permutation, we randomly shuffle the samples between the control groups and the treatment groups and generate permuted data. In order to preserve the dependency structure, if one sample is shuffled, then the corresponding experiment results for the sample should be shuffled in the same way across all the platforms. For each permuted data set, we generate the integrative test statistic T_i^*, $i = 1,\ldots B$. The empirical p-value for the observed T statistic would be calculated as $\sum_{i=1}^{B} I\left(T_i^* > T\right) / B$, where $I(.)$ is an indicator function.

To alleviate the discretization of the permuted p-value, we can obtain more precise p-value evaluation by collapsing all the T_i^* from all the permuted data and from all the genes. For each permuted data set, we generate the integrative test statistic for gene n as T_{in}^*, $i = 1,\ldots B$, $n = 1,\ldots, N$.

The empirical p-value for the observed T statistic would be calculated as $\sum_{i=1}^{B}\sum_{n=1}^{N} I\left(T_{in}^* > T\right) / (BN)$.

3 Data Analysis

In this section, we apply our proposed integrative platform analysis (IPA) to a dataset in Jayapal et al. [17]. This data was used to study growth and stationary phase adaption in Streptomyces coelicolor. The multi-platform experiment obtained both isobaric stable isotope labeled peptide (iTRAQ™)-derived shotgun proteomic result and DNA microarray transcriptome data. In order to investigate different growth stages of *S. coelicolor* M145 cells, cell samples were collected at eight time point: 7, 11, 14, 16, 22, 26, 34, and 38 h. As the iTRQA™ system can only analyze four distinct samples in a single experiment, three runs of mass spectrometric (MS) analysis were conducted for a total of eight samples. In each of the three MS experiments, the protein sample from 11 h was included as an internal reference. Therefore, protein abundance ratios were computed as the ratio of the amount of the protein in each sample at time j over the sample at 11 h denoted as $r_{j/11\text{h}}$. Jayapal et al. [17] carried out protein identification and quantification by comparing the raw spectral data against a theoretical proteome of *S. coelicolor* using proteinPilot™ software and the inbuilt Paragon™ search engine. They conducted further analysis for proteins identified more than 99 % confidence. Last, they formulated a protein abundance ratio with respect to the first time point (7 h) sample using $r_{j/11\text{h}}/r_{7/11}$ for all identified proteins. In the end, a total of 886 proteins identified in the 7 h sample are available for our platform integrative analysis. For microarray data, they isolated total mRNA from the same eight time point samples and conducted spotted DNA microarray experiments. Genomic DNA (gDNA) was used as a reference in the hybridization. The mRNA abundance was calculated as the ratio of $\log_2 \frac{\text{cDNA}}{\text{gDNA}}$. To follow the same reference time point as the protein data, mRNA abundance data for each time point t is compared with that of the 7 h example: $\log_2 \frac{cDNA}{gDNA} \mid t \log_2 \frac{cDNA}{gDNA} \mid 7h$. There are 886 genes with both gene expression values and protein values available for the analysis. The samples were partitioned into growth phase (7, 11, 14, and 16 h) and stationary phase (22, 26, 34, and 38 h).

We applied the integrative platform analysis (IPA) to select the biomarkers that are differentially expressed between the two growth phases. We considered two null hypotheses as H_{01}: the mRNA expression level is the same between the two phases and H_{02}: the protein level is the same between the two phases. We computed Welch's student t-statistics for each platform, which are denoted as T_m and T_p. We formed the integrative chi-square statistic T based on the log transformations of p_m and p_p. The overall p-value for each gene was obtained by comparing T with the chi-square statistic with four degrees of freedom. Alternatively, we also performed permutation analysis and collapsed all the permuted T statistics from 1000 permutations and 886 genes to form the empirical null distribution of T. The overall empirical p-value is obtained by the proportion of permuted T exceeding the observed T.

The Bonferroni threshold for controlling the familywise type I error rate of 0.05 is 5.6433×10^{-5}, whereas the less stringent threshold for controlling the estimated overall false discoveries to be less than equal to 20 is 0.0011. Using Bonferroni's threshold, the protein data and microarray data both lead to no significant genes; whereas the IPA permutation method gives zero and IPA chi-square method gives two significant genes. If we adopt the threshold of 0.0011, the protein data identifies two and the microarray data identifies three genes, where IPA permutation method identifies nine genes and IPA chi-square method identifies eight genes. Table 1 provides the result of the nine genes identified by IPA permutation method together with the *p*-values generated by single-platform analysis and the IPA *p*-values provided by chi-square method. Figure 1 also provides the plot of IPA permutation *p*-value against IPA chi-square *p*-value. It is shown that for this dataset IPA permutation *p*-values agree with IPA chi-square *p-values* very well (*see* **Note 1**).

4 Conclusion

In this chapter, we outline a multiple platform integration method based on Fisher's combined probability test. The method is simple to use and is able to consolidate statistical significance across multiple sources. It allows different platforms to have specific tests tailed to each platform and it only requires output of *p*-values from each source. The method provides valid statistical inference with a consolidated *p*-value, which is useful for the control of multiple testing errors and false discovery rates. Furthermore, we propose to use permutation method to generate empirical null distribution of the Fisher's combined probability test statistic. The proposed

Table 1
Analysis results of the integrative platform analysis (IPA) using chi-square distribution and using empirical null distribution. Each platform's student *t* statistic and individual *p*-value are provided

Gene names	*t*-stat protein	*p*-value protein	*t*-stat mRNA	*p*-value mRNA	IPA *p*-value chi-square	IPA *p*-value permutation
SCO2262	1.3213	0.2489	8.3156	0.0004	0.0011	0.0008
SCO2390	7.1427	0.0009	2.9601	0.0963	0.0009	0.0006
SCO2578	4.0878	0.0118	4.2189	0.0093	0.0011	0.0008
SCO3669	1.7881	0.1493	8.4897	0.0005	0.0008	0.0005
SCO3909	3.2170	0.0272	5.6657	0.0041	0.0011	0.0008
SCO4231	2.4746	0.1156	8.0202	0.0006	0.0007	0.0005
SCO4917	6.0833	0.0018	3.1749	0.0689	0.0013	0.0010
SCO5737	5.1900	0.0036	4.6492	0.0056	0.0002	0.0001
SCO7000	7.2779	0.0023	5.2582	0.0034	0.0001	0.0001

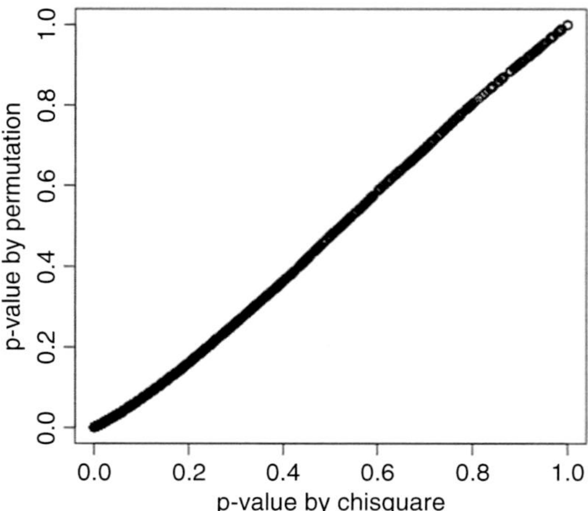

Fig. 1 Comparison of IPA p-values by chi-square and permutation method

permutation method is able to deal with dependency structure across different platforms. The permutation based Fisher's test is distribution free and robust to small sample sizes.

5 Notes

1. In this data analysis, the p-values generated by permutation method agrees with the p-values generated by Fisher's combined probability test. This may not always hold true for other real datasets. If the underlying distributional assumptions are violated and the dependency is very strong across different platforms, the two types of p-values can be different. In those situations, permutation p-values will be more trustworthy due to its distribution free property and its ability to accommodate the dependency structure.

References

1. Reif D, White B, Moore J (2004) Integrated analysis of genetic, genomic and proteomic data. Expert Rev Proteomics 1:67–75
2. Hamid J, Hu P, Roslin M et al (2009) Data integration in genetics and genomics: methods and challenges. Human Genomics and Proteomics 2009, Article ID 869093
3. Lanckriet G, De Bie T, Cristianini N et al (2004) A statistical framework for genomic data fusion. Bioinformatics 20:2626–2635
4. Daemen A, Gevaert O, De Bie T et al (2008) Integrating microarray and proteomics data to predict the response on cetuximab in patients with rectal cancer. Pac Symp Biocomput 13:166–177
5. Buness A, Ruschhaupt M, Kuner R et al (2009) Classification across gene expression microarray studies. BMC Bioinformatics 10:453
6. Tian Q, Stepaniants S, Mao M et al (2004) Integrated genomic and proteomic analyses of gene expression in mammalian cells. Mol Cell Proteomics 3:960–969
7. Bussey K, Chin K, Lababidi S et al (2006) Integrating data on DNA copy number with

gene expression levels and drug sensitivities in the NCI-60 cell line panel. Mol Cancer Ther 5:853–867

8. Adourian A, Jennings E, Balasubramanian R et al (2003) Correlation network analysis for data integration and biomarker selection. Royal Soc Chem 4:249–259
9. Ma Y, Ding Z, Qian Y et al (2009) An integrative genomic and proteomic approach to chemosensitivity prediction. Int J Oncol 34:107–115
10. Aerts S, Lambrechts D, Maity S et al (2006) Gene prioritization through genomic data fusion. Nat Biotechnol 24:537–544
11. Rhodes D, Yu J, Shanker K et al (2004) Large-scale meta analysis of cancer microarray data identifies common transcriptional profiles of neoplastic transformation and progression. Proc Natl Acad Sci U S A 25:9309–9314
12. Wu S, Xu Y, Feng Z et al (2012) Multiple-platform data integration method with application to combined analysis of microarray and proteomic data. BMC Bioinformatics 13:320
13. Fisher RA (1925) Statistical methods for research workers. Oliver and Boyd, Edinburgh
14. Brierley N, Tippetts T, Cawley P (2014) Data fusion for automated non-destructive inspection. Proc R Soc A 470, Issue 2167
15. Benjamini Y, Hochberg Y (1995) Controlling the false discovery rate: a practical and powerful approach to multiple testing. J Royal Stat Soc Series B 57:289–300
16. Gao X (2006) Construction of null statistics in permutation based multiple testing for multi-factorial microarray experiments. Bioinformatics 22:1486–1494
17. Jayapal K, Philp R, Kok Y et al (2008) Uncovering genes with divergent mRNA-protein dynamics in *Streptomyces coelicolor*. PLoS One 7:e2097

Chapter 14

Data Fusion in Metabolomics and Proteomics for Biomarker Discovery

Lionel Blanchet and Agnieszka Smolinska

Abstract

Proteomics and metabolomics provide key insights into status and dynamics of biological systems. These molecular studies reveal the complex mechanisms involved in disease or aging processes. Invaluable information can be obtained using various analytical techniques such as nuclear magnetic resonance, liquid chromatography, or gas chromatography coupled to mass spectrometry. Each method has inherent advantages and drawbacks, but they are complementary in terms of biological information.

The fusion of different measurements is a complex topic. We describe here a framework allowing combining multiple data sets, provided by different analytical platforms. For each platform, the relevant information is extracted in the first step. The obtained latent variables are then fused and further analyzed. The influence of the original variables is then calculated back and interpreted.

Key words Chemometrics, Discriminant analysis, PLS-DA, eCVA, Latent variable, Variable selection

1 Introduction

Metabolomics and proteomics have been intensively developed over the last decade(s) as a mean to understand the complexity of biological systems and their flexibility when facing either disease or new environmental conditions. Both domains carry valuable information about the biochemical pathway involvement in, e.g., human diseases. However, the description of these biochemical pathways clearly shows that metabolites and proteins are not two disjoint systems but elements of a single one. Typical the abundance of a particular enzyme will directly impact in the concentration of its substrate(s) and product(s). The corresponding increase or decrease in metabolic concentrations will then propagate along the biochemical pathways. Paradoxically both fields are frequently studied separately.

Joint analysis, also known as data fusion/data concatenation, appears as a logical solution. This intuitive solution comes with challenges, principally at the statistical level. The first challenge is

Klaus Jung (ed.), *Statistical Analysis in Proteomics*, Methods in Molecular Biology, vol. 1362,
DOI 10.1007/978-1-4939-3106-4_14, © Springer Science+Business Media New York 2016

Fig. 1 Flow chart of the fusion procedure proposed for proteomics and metabolomics data

common to any data analysis in omics; the generated data is very high dimensional. Fusing two or more set of measurements obtained from the same samples makes this problem more severe. Moreover, multiple data fusion strategies have been proposed.

Our objective here is to detail and justify a data fusion framework applicable to metabolomics and proteomics. Figure 1 presents the general outline of this framework. The first step concerning sample acquisition, sample handling and the actual measurements will not be detailed here since they are very specific to each study. Once the measurements are performed, the data from each analytical platform need to be preprocessed adequately. nuclear magnetic resonance (NMR) metabolomics data [1] may, for example, require a different type of alignment than the data generated by gas chromatography coupled to mass spectrometry (GC-MS) [2, 3].

After preprocessing, a preliminary explorative analysis is performed. The objective here is to detect potential outliers and unexpected trends present in the data. When detected, the artifacts should be appropriately taken care of, either via correction or deletion of unreliable data. The data fusion is then performed in two steps. The first one corresponds to a supervised compression of the data. Here, the most relevant information are extracted from the separate metabolomics and proteomic platform. The second part of step 2 is the fusion itself where the concatenation of various platforms takes place. This step may also involve missing values and need to be adapted to handle incomplete data. Finally, the interpretation of these complex models is quickly discussed.

1.1 Sample Handling

The importance of sample collection and preparation is largely discussed in all branches of analytical chemistry [4–7]. Obviously, the employed procedure depends strongly on the instrumentation and the research question. In the data fusion framework, only one extra point should be emphasized on: The same sample should be divided between all analytical platforms. The samples and data handling need to be carefully organized to avoid any confusion between sub-samples that would lead to an incorrect analysis.

1.2 Data Preprocessing

Ideally the data obtained from either proteomics or metabolomics measurements should be summarized in tables reporting a concentration value for each protein and each metabolite in all samples. Unfortunately, this would assume an accurate and complete identification of all chemical entities. This level of precision is not necessarily possible especially in untargeted studies. Unidentified compounds could be left out of the analysis. However, they might be the most relevant ones. A disease may give rise to a new metabolite (hence difficult to identify); this metabolite would be, thus, an ideal biomarker.

An alternative is to use the relative intensity of the peaks observed in LC-MS or NMR without explicitly associate them to chemical compounds. This approach requires an adequate preprocessing of the data [8–10]. The preprocessing protocol is highly dependent on the type of instrumentation used and the nature of the samples analyzed. Preprocessing may include (but the list is not exhaustive) baseline correction [11–13], alignment [14–17], denoising [18, 19], scaling [20], and binning or peak picking [21]. Numerous methods are available for each of these tasks. Choosing the most appropriate one is a challenge in itself, and combining them properly is another concern [22].

The data is organized in tables, one per analytical platform (*see* **Note 1**), where each line represents a sample and each column a variable (either a metabolite, a protein or a peptide). By convention, these data matrices are call $\mathbf{X}_i$ and the number of samples is designated by n while the number of variable by p_i. We assume that all analytical platforms received the same samples hence n is equal for all data matrices. These data tables are schematically represented in Fig. 2, this type of representation will be used throughout this chapter.

2 Explorative Analysis and Outlier Detection

A step often performed but rarely described in data mining is the explorative analysis and the corresponding outlier detection. Analytical chemists are perfectly aware that all measurements do not always go flawlessly smoothly and that some errors can be introduced in the resulting data. While most deviations are minors,

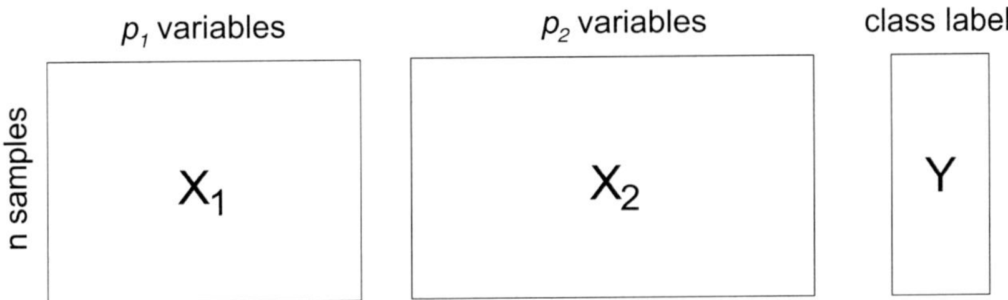

Fig. 2 Schematic representation of the data obtained from two analytical platforms ($\mathbf{X}_1$ and $\mathbf{X}_2$) both measuring n samples and, respectively, p_1 and p_2 variables. **Y** represents the response matrix containing the class labels (e.g., control vs. disease)

this has little impact on the analysis; however, larger errors or batch effects can dramatically disturb the data analysis (*see* **Note 2**). It is, therefore, advisable to perform an explorative analysis of the data using robust unsupervised method such as robust principal component analysis (ROBPCA) [23, 24]. This approach allows one to check for the presence of outliers and possible groupings unrelated to the investigated class. Figure 3 presents two examples of outlier map produced by ROBPCA. The horizontal axis in Fig. 3 is called score distance, while the vertical axis orthogonal distance. The score distance represents the distance to the origin in the modeled space whereas the orthogonal distance corresponds to residuals from the model. Any sample exceeding the critical values (indicated by red lines in Fig. 3) in the score and/or orthogonal distances is likely to be an outlier. Detected outliers should be removed before proceeding to the rest of the analysis.

Groupings of samples would typically happen when they are measured in multiple batches over a lengthy period or on different instruments. Sub-groupings in the data must ideally be avoided, but they may also reflect some real underlying effects such as gender, age, and diet differences. The decision to correct sub-groupings is, therefore, case dependent.

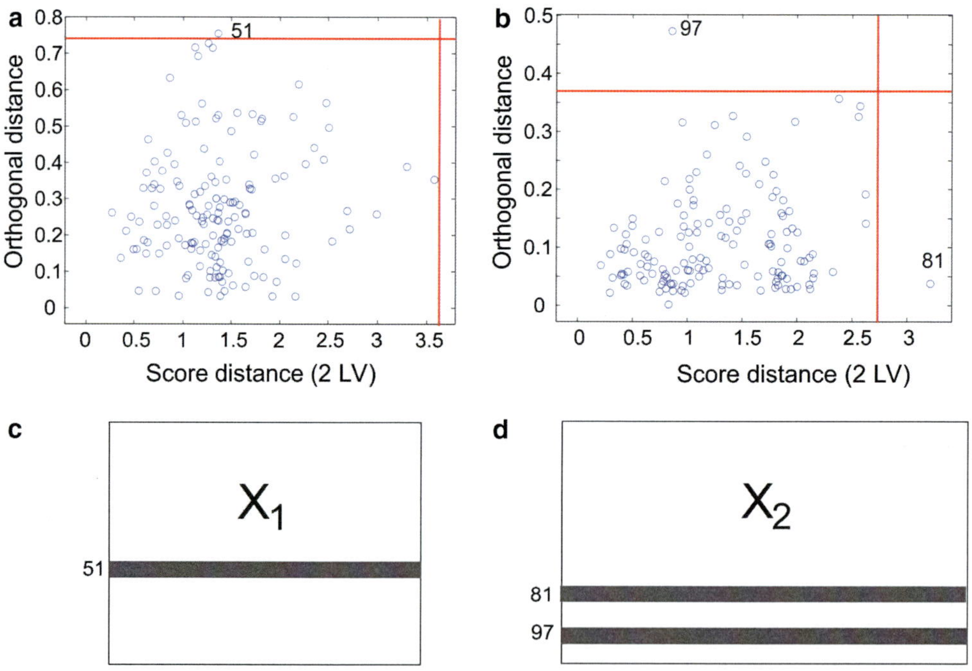

Fig. 3 Outlier maps of (**a**) $\mathbf{X}_1$ and (**b**) $\mathbf{X}_2$ permit to detect the sample 51 in $\mathbf{X}_1$ and 81 and 97 in $\mathbf{X}_2$ as outliers since they exceed the cutoff value in either score or orthogonal distance. The corresponding values are removed from the two data matrices (**c**) $\mathbf{X}_1$ and (**d**) $\mathbf{X}_2$

3 Data Fusion Architectures

Combining multiple types of measurements into a single analysis have been the subject of research in numerous scientific fields. In analytical chemistry, data fusion is usually described in three forms: namely the low-, high- and mid-level fusion [25]. All three approaches mostly rely on linear methods hence a fourth category, named kernel data fusion, has been proposed [26–28].

The low-level fusion is the most intuitive approach since it simply consists of concatenating the two data blocks into a single one and then proceeds to the statistical analysis [29]. The challenge in omics is that each block contains many variables and that they represent different amount of information. The term variables refer here to either metabolites, protein, or peptide concentrations. Variables can also refer to more abstract signals such as value at a given ppm range in NMR or a specific MS feature. One option is to give equal weight to all variables using autoscaling (also known as unit variance scaling) [20]. However, the influence of a platform measuring a large number of variables would then dominate one with fewer variables. A proteomics data usually contains more variables and would, thus, dominate the analysis if their influence is not appropriately scaled down compared to the metabolomics data block. The usual approach is to block scale [30] in function of the total variance of each block. However, the total variance is a bad approximation of the amount of relevant information contained in the two data blocks. The main advantage of low-level fusion is that the original variables are conserved along the analysis, which facilitates the biological interpretation.

The high-level fusion consists of constructing two separate multivariate models and to fuse only the prediction i.e., the decision taken by the model. The two models being independent the block scaling problem is not present. This approach is powerful when the primary objective is the application of the classifier but not its interpretation.

As indicated by its name, mid-level fusion takes an intermediate route where each block is first analyzed separately in order to select the most relevant variables or to construct latent variables. In turn, the selected or latent variables of both blocks are concatenated together and reanalyzed as a single data set. Using selected (original) variables facilitates biological interpretation. Since the number of selected variables per block does not have to be equivalent, thus the scaling might be applied before the final analysis. Using latent variables space for both blocks reduces problem of scaling in the final analysis. However, such approach increases the complexity of the biological interpretation since the original variable importance need to be evaluated through the one of the latent variables.

4 First Layer: Compression of Relevant Information

The preprocessed data obtained from both metabolomics and proteomics consist of hundreds or thousands of variables per samples. One can reasonably expect that part of these variables is unrelated to the problem of interest. Hence, it is relevant to filter out irrelevant and redundant information and focus on the relevant one. To do so, we first analyze each data block separately.

The idea is here to reduce the size of the data sets ($\mathbf{X}_1$ and $\mathbf{X}_2$) while conserving the relevant information, as depicted in Fig. 4. This reduction can be done in two ways, either by selecting the most relevant variables or by constructing latent variables that summarize multiple original variables at once (*see* **Note 3**). A number of supervised methods are available to perform this first analysis. The most popular latent variable approach in metabolomics is PLS-DA [31, 32] (and affiliated method such as O-PLS-DA [33]) but also PCA followed by LDA [34], Fisher LDA [35] could as well be used (*see* **Note 4**).

These supervised methods explicitly use a priori information, such as the belongingness to the class. Hence, it is crucial to validate the resulting models. To do so a number of samples must be left out from the analysis: the test set (aka validation set). The class labels of the test set samples are then predicted using the constructed multivariate models. This prediction provides an objective evaluation of the quality of the multivariate model. The samples use

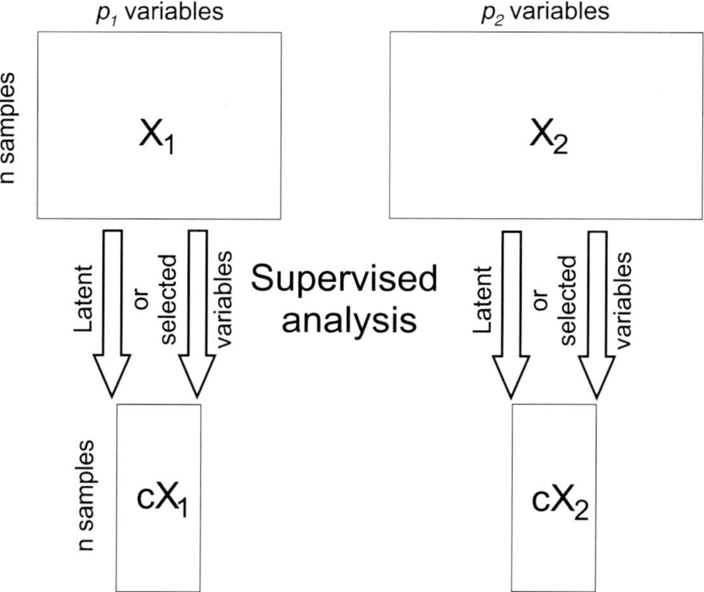

Fig. 4 Compression of the information contain in each platform ($\mathbf{X}_1$ and $\mathbf{X}_2$), into smaller sets ($\mathbf{cX}_1$ and $\mathbf{cX}_2$) by supervised methods such as PLS-DA or eCVA

to construct the models, the training set, must ideally be equally distributed between the different classes. This aspect is often referred as a balanced training set; this is done to avoid artificial bias toward one of the classes.

4.1 Latent Variables

4.1.1 PLS-DA

Here we focus first on the most popular methods in chemometrics PLS (adapted to the classification problem, i.e., PLS-DA). The underlying concept of PLS is to define new (latent) variables as a linear combination of the original variables measured. The linear combination is chosen as to maximize the variance explained in both the **X** block (the data) and the response **y** block.

Partial least squares (PLS) is in a family of latent regression methods where a new (latent) space with a reduced number of dimensions is constructed via a decomposition of the original data matrix, **X** (*see* **Note 5**). This new latent space, composed of *a* latent variables (principal components), is represented by the score matrix **T** and orthogonal loadings **P** according to Eq. 1 [31, 36]:

$$\mathbf{X} = \mathbf{T}\mathbf{P}^{\mathrm{T}} + \mathbf{E} \tag{1}$$

The scores **T** and loadings **P** are obtained through an iterative procedure such as the NIPALs algorithm. For each component $a = 1,2,\ldots$, the weights vectors are obtained via Eq. 2:

$$\mathbf{w}_a = \frac{\mathbf{X}_a^T \mathbf{y}_a}{\left\| \mathbf{X}_a^T \mathbf{y}_a \right\|} \tag{2}$$

$$\mathbf{t}_a = \mathbf{X}_a \mathbf{w}_a \tag{3}$$

$$\mathbf{p}_a = \mathbf{X}_a \frac{\mathbf{t}_a}{\mathbf{t}_a' \mathbf{t}_a} \tag{4}$$

$$\mathbf{q}_a = \mathbf{y}_a' \frac{\mathbf{t}_a}{\mathbf{t}_a' \mathbf{t}_a} \tag{5}$$

The data matrix **X** and the response **y** are then deflated according to Eqs. 6 and 7, so the next iteration will focus on the remaining variance:

$$\mathbf{X}_{a+1} = \mathbf{X}_a - \mathbf{t}_a \mathbf{p}_a' \tag{6}$$

$$\mathbf{y}_{a+1} = \mathbf{y}_a - \mathbf{t}_a \mathbf{q}_a' \tag{7}$$

Instead of performing the least square regression directly in the original space, PLS projects the response variable **y** on the new latent variables, *a*, resulting in the regression coefficients, $\mathbf{b}_{\text{score}}$, in Eq. 8 where the final PLS regression coefficients are given by Eq. 9:

$$\mathbf{b}_{\text{score}} = \left(\mathbf{T}'\mathbf{T}\right)^{-1} \mathbf{T}'\mathbf{y} \tag{8}$$

$$\hat{\mathbf{b}}_{\text{PLS}} = \mathbf{R}\mathbf{b}'_{\text{score}} = \mathbf{W}\left(\mathbf{P}'\mathbf{W}\right)^{-1}\mathbf{b}'_{\text{score}} \tag{9}$$

where $\mathbf{W}$ is a weight matrix, defined by $\mathbf{w}_a = \mathbf{X}_a^T\mathbf{y}_a$, with the deflated matrices $\mathbf{X}_a$ and $\mathbf{y}_a$, corresponding to the original space as defined by $\mathbf{X}$ and $\mathbf{y}$, respectively.

The usual form of $\mathbf{y}$ is a vector of continuous values, this corresponds to the regression problem. One possible extension is to operate a regression of multiple responses. In such case the response is designated as $\mathbf{Y}$ since it is now a matrix. The response can be a list of labels, a binary classification problem will use a vector $\mathbf{y}$ consisting of −1 and 1 a multiclass problem is instead encoded into a matrix containing only 0 and 1. A three-class problem can be encoded as 1 0 0 for class 1, 0 1 0 for class 2, and 0 0 1 for the last class.

The more latent variables are used (high a) the better is captured the information in both blocks. However, if too many latent variables are used, the resulting model overfits the data. Overfitting implies that the generated model is only able to describe this (training) data and loses its applicability to new samples. In simpler terms, it won't be able to describe a new set of measurements correctly.

A number of strategies have been proposed to set the number of latent variables in PLS-DA, the most popular being using cross-validation. However, this decision often remains rather subjective. Moreover, in the data fusion framework describe here two models need to be created. They do not necessarily have the same number of latent variables that leads to scaling issues in the later stage of the data fusion.

4.1.2 eCVA

An alternative approach, extended canonical variate analysis (eCVA) [37, 38], bypasses this issue. eCVA also focus only on the information related to the specified groups (e.g., patients versus healthy control). This method also creates latent variables, but their number is directly related to the number of groups. eCVA, similarly to PLS-DA, has been developed to cope with multi-collinear data (*see* **Note 5**). However, in its concept eCVA resembles more the Fisher LDA technique. Given a data matrix $\mathbf{X}$ $(n \times p)$ where g groups of n_i samples are present, the within-group covariance matrix $\mathbf{S}_{\text{within}}$ is defined as

$$\mathbf{S}_{\text{within}} = \frac{1}{(n-g)}\sum_{i=1}^{g}\sum_{j=1}^{n_i}\left(\mathbf{x}_{ij} - \bar{\mathbf{x}}_i\right)\left(\mathbf{x}_{ij} - \bar{\mathbf{x}}_i\right)' \tag{10}$$

and the between-group $\mathbf{S}_{\text{between}}$ is

$$\mathbf{S}_{\text{between}} = \frac{1}{(g-1)}\sum_{i=1}^{g}\left(\mathbf{x}_i - \bar{\mathbf{x}}\right)\left(\mathbf{x}_i - \bar{\mathbf{x}}\right)' \tag{11}$$

where $\mathbf{x}ij$ is the jth sample in the ith group, $\bar{\mathbf{x}}_i$ is the mean vector in the ith group, and $\bar{\mathbf{x}}$ is the overall mean vector. The best discrimination between the groups is obtained by defining a direction w maximizing the ratio of the between-group on the within-group covariances:

$$J(w) = \frac{w'\mathbf{S}_{\mathbf{between}}w}{w'\mathbf{S}_{\mathbf{within}}w} \tag{12}$$

When $\mathbf{S}_{\mathbf{within}}$ is non-singular, Eq. 3 can be rewritten in the form of an eigenvalue problem:

$$\mathbf{S}_{\mathbf{between}}\mathbf{w}_a = \lambda_a \mathbf{S}_{\mathbf{within}}\mathbf{w}_a \tag{13}$$

Note that here a represents the number of directions, whereas in PLS-DA it corresponds to the number of LVs. The two concepts are related however in eCVA the number of directions is directly related to the number of groups whereas in PLS-DA it is not the case. λ_a are the eigenvalues and $\mathbf{w}_a$ the eigenvectors. Equation 4 has a maximum dimensionality of $a = \min(p, g-1)$. Therefore, even in the case of high-dimensional data, the number of extracted latent variables is equal to the number of groups minus one ($g-1$). Equation 4 can be turned into a regression problem [38]. PLS method is then used to solve Eq. 5 corresponding to this regression problem:

$$\mathbf{Y} = \mathbf{S}_{\mathbf{between}}\mathbf{B} + \mathbf{F} \tag{14}$$

where $\mathbf{Y}$ contains the differences $(x_i - \bar{x})$ the columns of $\mathbf{B}$ are $\mathbf{w}_a$ and $\mathbf{F}$ is a residual matrix. The scores $\mathbf{T}$ can then be calculated projecting them along the directions $\mathbf{w}_a$, i.e., the canonical variates (**CV**). The classification model is then constructed using these scores.

4.1.3 Selected Variables

As mentioned earlier, the use of latent variables is not mandatory. Instead, the results of either PLS-DA or eCVA could be used to select the most interesting variables from each data set. The selected variables are then concatenated instead of the latent variables. The main advantage is that the interpretation of the final results is much clearer since one can directly see the impact of a given metabolites or protein on the final result. The risk is to leave out part of the correlation structure by selecting the most prominent variables only.

In practice, modern data contain tens of thousands of variables for only a few samples. Handling this type of data is computationally expensive. Moreover, most multivariate methods suffer from an inaccurate estimation of the covariance matrix (as it is always the case when $p \gg n$) [39]. This partly explains why numerous studies use univariate testing as a screening approach. The variables are evaluated using a Student t-test or its nonparametric equivalent.

Variables associated to a *p*-values under a given threshold (traditionally 0.05) are retained for further analysis. Mathematically speaking, this approach makes no sense if the second step is multivariate. Informative patterns spread across multiple variables are likely to be eliminated from the analysis because individually none of the implicated variables appears to bring information. The strength of multivariate analysis is its ability to capture this pattern. Hence, any variable selection/ filtering should be also done in a multivariate fashion.

4.2 Second Layer: Combination of the Latent Variables

The aim of the first step was to compress the two data blocks ($\mathbf{X}_1$ and $\mathbf{X}_2$) into smaller or compressed versions ($\mathbf{cX}_1$ and $\mathbf{cX}_2$) containing only the relevant information. It is now necessary to combine these two "clean" matrices together. This second analysis could be done again using supervised analysis; however, this would require a new validation set and, therefore, a new test set. Most studies can only afford a limited number of samples hence it might be more practical to use an unsupervised approach here. The application of unsupervised analysis is possible based on the assumption that the first step successfully extracted the relevant information in both data blocks. The main sources of variances should be related to the studied problem.

Regardless of the method used, specific challenges must be addressed. Due to samples handling, measurements specificities, and outliers detection it is likely that some measurements would be missing. In the previous step, if the measurements are missing then the corresponding samples are simply ignored. In this step, if the measurements for a sample are missing only in one of the two data blocks, it will be a waste to ignore the part successfully measured.

The problem of missing values has been abundantly studied. Multiple imputations are generally considered as the most optimal solution [40] but are complex and computationally intensive. Expectation Maximization (EM) methods are a seducing alternative in term of speed and performance. Note that both unsupervised (PCA) and supervised (PLS-DA, eCVA) methods can be adapted to include an EM step [41]. The concept underlying the EM algorithm is that it is easy to estimate the missing values when the covariance structure of the data is known. Similarly, it is straightforward to determine the covariance structure is no values were missing (assuming that enough data is available). The Expectation step basically estimates the covariance structure given an estimate of the missing values. This covariance structure will be used in the Maximization step to reestimate the missing values and fill the updated values in the data matrix. The procedure iterates until the system stabilizes.

After dealing with missing values, the combination of the two compressed data blocks $\mathbf{cX}_1$ and $\mathbf{cX}_2$ can be further analyzed using standard tools such as PCA. By design, the matrices $\mathbf{cX}_1$ and $\mathbf{cX}_2$ should

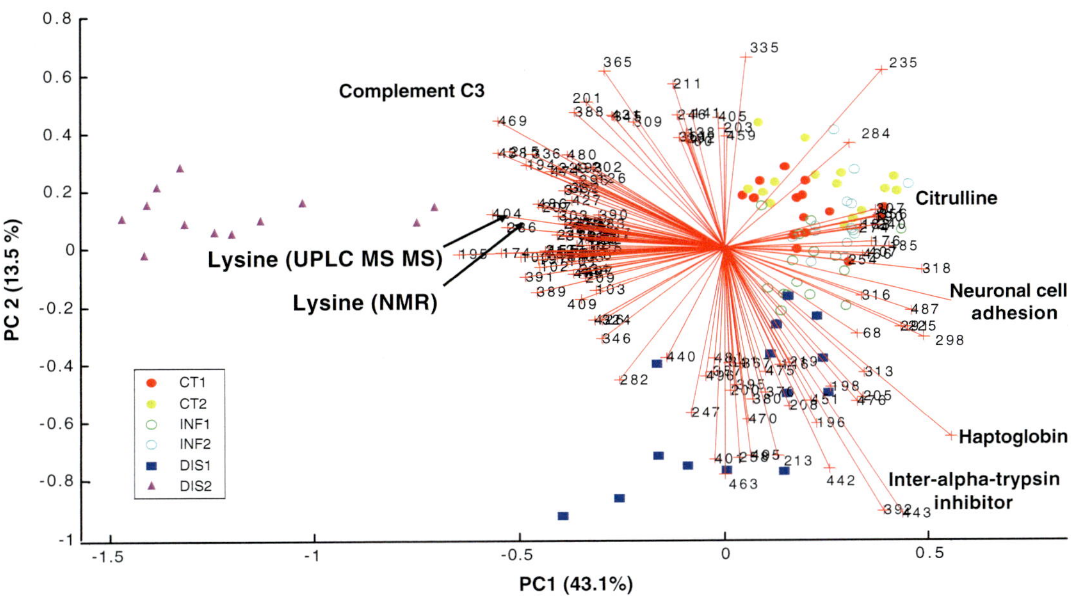

Fig. 5 Biplot obtained after data fusion of five analytical platforms, the names of selected variables are displayed

only provide relevant information. Hence, clear separation should be visible on the first PCs for binary problems; more complex problems will require more complex PCA models. In all cases, the projection of the original variables in the PCA model can be recalculated. A biplot combining information from all platforms can they be created and used to define the specific pattern associated to, e.g., a disease. Figure 5 presents an example based on two proteomics and three metabolomics platforms. The complete data include over 20,000 variables that are reduced to approximately 500 variables relevant to the separation of the six groups present in the data. Those six groups represent cohort of control inflamed and diseased animals at different ages. This example demonstrates the feasibility of data fusion on complex problem involving large number of platforms. The influence of proteins and metabolites (represented by red arrows) can be evaluated simultaneously in this representation. The case of lysine is interesting here. Lysine was measured independently by two analytical platforms and the two variables are lying close to each other on this biplot, as they should. This confirms the quality of the data used here.

4.3 When Linear Model Fails, Nonlinear Ones Can Be Tested

The methods described in the previous sections are suitable to investigate linear relationships between variables and groups. However, these methods are less efficient to study the nonlinear relations that may be present in complex biological systems. Nonlinear associations might originate from various metabolic backgrounds of hidden disease evolution, variability in disease stages or individual patient conditions or medication. A special case of data fusion takes into account nonlinear relations between variables is multiple kernel learning (MKL). Originally pioneered by Lanckriet et al. [26] and Bach et al. [27], MKL is an extension of single kernel (i.e., transformed original data via a specific function) to combine multiple kernels in support vector machine. The principle of MKL is to concatenate kernel matrices into a single kernel using their parameterized linear combination. Recently this approach was extended to k-PLS-DA [28]. In MKL each kernel matrix, delivered from different data blocks, gets a weight (i.e., importance) optimized by various regularization techniques (e.g., the L_1 or L_2 norm). The weights are next utilized in the linear combination of kernel matrices. The bigger the weight the more important is a platform associated with the weight. Clearly, the platform with the higher weight in the MKL procedure delivers the most optimal information about a studied problem (e.g., difference between disease and healthy groups).

5 Interpretation

The biological interpretation is performed by assessing the importance of the original variables. This importance can be evaluated by various means. One can either recalculate the pseudo regression coefficient of the original variables by multiplying the canonical variates (or PLS-DA regression coefficients) from the first step by the loadings of the PCA (second step). It is possible to introduce more advanced variable selection methods in both the construction and interpretation steps of the supervised models. Variable importance in projection [42] is commonly used together with PLS-DA, but it suffers from some drawbacks [43, 44]. More recently significant multivariate correlation has been proposed based on a more solid statistical background [45]. The objective here is to reduce the amounts of variable to consider in order to make it more manageable for human interpretation.

The most important variables can be put into context by projecting them into a correlation network. This network can be constructed *ab novo* by calculating all pair-wise Pearson or Spearman correlations. However, complete networks are too large and complex to be used. Therefore, only part of the network is considered by using the most important variables as a seed that is expanded to all variables with a correlation superior to 0.8 (in absolute value).

Besides their increasing popularity, such networks only partially represent the complexity of the data. First, only pair-wise relations are included, while more complex relations are likely to be present. Second, this type of correlation analysis is solely based on the data at hand. Biochemical atlases like KEGG [46] or GO [47] compile numerous studies and contain more information.

A number of methods have been proposed to join these two worlds by allowing to project the results of data analysis on known pathways [48, 49]. The available ontology enriches the analysis yet one should remember that they can only bring information confirmed by previous studies and hence do not allow for the discovery of new mechanism or interactions.

6 Conclusion

Capturing all information describing a biological system is the implicit objective of all the omics methods. Powerful by themselves, genomics, transcriptomics, proteomics and metabolomics need to be combined to approach this goal. The framework presented here was successfully applied to the combination of multiple proteomics and metabolomics data sets. This procedure can (and should) be extended to include all possible source of information. As hinted before, biological relations are rarely strictly linear; hence, nonlinear methods are potentially more suited for this type of analysis. Regardless the choice of linear on nonlinear method, the main bottleneck, in our opinion, is the biological interpretation of the resulting models. Intuitively researchers tend to split complex questions into simpler elements. Here the multivariate approach brings more power to the analysis but it might also hinder interpretation since highly complex multivariate model need to be understood as a whole pattern rather than simple elevation or decrease in concentration of a particular molecule.

7 Notes

1. Correct organization of the data is always crucial, even more in data fusion. Make sure that the correct samples are matched across the different platforms. Any permutation (even within the same group) can have dramatic consequences for the analysis.
2. Appropriate preprocessing and outlier removal are cumbersome tasks yet they are conditioning the rest of the analysis. A single outlier or a single artifact might perturbate the construction of the latent variables.
3. Multiclass problems, i.e., more than two groups of samples, might be too difficult to solve at once. If unsuccessful, a global model can be replaced by a combination of simpler binary models.
4. The assumption that the separation can be obtained using linear combinations of the original variables does not necessarily hold true. Nonlinear models are, however, more complex and should, therefore, be used only if the linear one failed.

5. PLS-DA, eCVA and PCa are all able to cope with matrices containing more variables than samples, yet if the number of variables greatly exceed the number of samples, the estimation of the covariance structure will be incorrect and thus the multivariate model based on it.

References

1. Smolinska A, Blanchet L, Buydens LMC et al (2007) NMR and pattern recognition methods in metabolomics: from data acquisition to biomarker discovery: a review. Anal Chim Acta 750:82–97
2. Koek MM, Jellema RH, van der Greef J et al (2011) Quantitative metabolomics based on gas chromatography mass spectrometry: status and perspectives. Metabolomics 7:307–328
3. Almstetter MF, Oefner PJ, Dettmer K (2012) Comprehensive two-dimensional gas chromatography in metabolomics. Anal Bioanal Chem 402:1993–2013
4. Álvarez-Sánchez B, Priego-Capote F, Luque de Castro MD (2010) Metabolomics analysis I. Selection of biological samples and practical aspects preceding sample preparation. TrAC Trends Anal Chem 29:111–119
5. Álvarez-Sánchez B, Priego-Capote F, de Castro MDL (2010) Metabolomics analysis II. Preparation of biological samples prior to detection. TrAC Trends Anal Chem 29: 120–127
6. Vuckovic D (2012) Current trends and challenges in sample preparation for global metabolomics using liquid chromatography-mass spectrometry. Anal Bioanal Chem 403: 1523–1548
7. Bodzon-Kulakowska A, Bierczynska-Krzysik A, Dylag T et al (2007) Methods for samples preparation in proteomic research. J Chromatogr B Anal Technol Biomed Life Sci 15:1–31
8. Smolinska A, Hauschild A-C, Fijten RRR (2014) Current breathomics--a review on data pre-processing techniques and machine learning in metabolomics breath analysis. J Breath Res 8:027105
9. Ebbels TMD, Lindon JC, Coen M (2011) Processing and modeling of nuclear magnetic resonance (NMR) metabolic profiles. Methods Mol Biol 708:365–388
10. Dallinga J, Smolinska A, van Schooten F-J (2014) Analysis of volatile organic compounds in exhaled breath by gas chromatography-mass spectrometry combined with chemometric analysis. In: Raftery D (ed) Mass spectrometry in metabolomics: methods and protocols. Springer, New York, pp 251–263
11. Eilers PHC (2003) A perfect smoother. Anal Chem 75:3631–3636
12. Eilers PHC, Marx BD (1996) Flexible smoothing with B-splines and penalties. Stat Sci 11: 89–121
13. Xu Z, Sun X, Harrington PB (2011) Baseline correction method using an orthogonal basis for gas chromatography/mass spectrometry data. Anal Chem 83:7464–7471
14. Bloemberg TG, Gerretzen J, Wouters HJP et al (2010) Improved parametric time warping for proteomics. Chemom Intell Lab Syst 104: 65–74
15. Nielsen NPV, Carstensen JM, Smedsgaard J (1998) Aligning of single and multiple wavelength chromatographic profiles for chemometric data analysis using correlation optimised warping. J Chromatogr A 805:17–35
16. Tomasi G, Van Den Berg F, Andersson C (2004) Correlation optimized warping and dynamic time warping as preprocessing methods for chromatographic data. J Chemom 18: 231–241
17. Wei X, Shi X, Merrick M (2013) A method of aligning peak lists generated by gas chromatography high-resolution mass spectrometry. Analyst 138:5453–5460
18. Walczak B (2000) Wavelets in chemistry, 1st edn. Elsevier, Amsterdam
19. Trygg J, Gabrielsson J, Lundstedt T (2009) Background estimation, denoising, and preprocessing. In: Phan-Tan-Luu R, Leardi R, Sarabia L (eds) Comprehensive chemometrics. Elsevier, Amsterdam, pp 1–8
20. Van den Berg RA, Hoefsloot HCJ, Westerhuis JA (2006) Centering, scaling, and transformations: improving the biological information content of metabolomics data. BMC Genomics 7:142
21. Pluskal T, Castillo S, Villar-Briones A et al (2010) MZmine 2: modular framework for processing, visualizing, and analyzing mass spectrometry-based molecular profile data. BMC Bioinformatics 11:395

22. Engel J, Gerretzen J, Szymańska E et al (2013) Breaking with trends in pre-processing? TrAC Trends Anal Chem 50:96–106
23. Hubert M, Rousseeuw P, van der Branden K (2005) ROBPCA: a new approach to robust principal component analysis. Technometrics 47:64–79
24. Daszykowski M, Serneels S, Kaczmarek K et al (2007) TOMCAT: a MATLAB toolbox for multivariate calibration techniques. Chemom Intell Lab Syst 85:269–277
25. Roussel S, Bellon-Maurel V, Roger JM et al (2003) Fusion of aroma, FT-IR and UV sensor data based on the Bayesian inference. Application to the discrimination of white grapes varieties. Chemom Intell Lab Syst 65:209–219
26. Lanckriet GRG, Cristianini N, Bartlett P (2004) Learning the Kernel Matrix with semi-definite programming. J Mach Learn Res 5:27–72
27. Bach FR (2008) Consistency of the Group Lasso and Multiple Kernel Learning. J Mach Learn Res 9:1179–1225
28. Smolinska A, Blanchet L, Coulier L (2012) Interpretation and visualization of non-linear data fusion in kernel space: study on metabolomic characterization of progression of multiple sclerosis. PLoS One 7, e38163
29. Bro R, Nielsen HJ, Savorani F et al (2013) Data fusion in metabolomic cancer diagnostics. Metabolomics 9:3–8
30. Skov T, Honoré AH, Jensen HM (2014) Chemometrics in foodomics: handling data structures from multiple analytical platforms. TrAC Trends Anal Chem 60:71–79
31. Wold S, Sjostrom M, Eriksson L et al (2001) PLS-regression: a basic tool of chemometrics. Chemom Intell Lab Syst 58:109–130
32. Barker M, Rayens W (2003) Partial least squares for discrimination. J Chemom 17:166–173
33. Trygg J, Wold S (2002) Orthogonal projections to latent structures (O-PLS). J Chemom 16:119–128
34. Tominaga Y (1999) Comparative study of class data analysis with PCA-LDA, SIMCA, PLS, ANNs, and k-NN. Chemom Intell Lab Syst 49:105–115
35. Fisher RA (1936) The use of multiple measurements in taxonomic problems. Ann Eugen 7:79–89
36. De Jong S (1993) SIMPLS: an alternative approach to partial least squares regression. Chemom Intell Lab Syst 18:251–253
37. Blanchet L, Smolinska A, Attali A (2011) Fusion of metabolomics and proteomics data for biomarkers discovery. BMC Bioinformatics 12:254
38. Nørgaard L, Bro R, Westad F (2006) A modification of canonical variates analysis to handle highly collinear multivariate data. J Chemom 20:425–435
39. Haury AC, Gestraud P, Vert JP (2011) The influence of feature selection methods on accuracy, stability and interpretability of molecular signatures. PLoS One 6:e28210
40. Nielsen SF (2003) Proper and improper multiple imputation. Int Stat Rev 71:593–607
41. Andersson CA, Bro R (1998) Improving the speed of multi-way algorithms: part I. Tucker 3. Chemom Intell Lab Syst 42:93–103
42. Wold S, Johansson E, Cocchi M (1993) PSL - partial least-squares projections to latent structures. Escom, Leiden, pp 523–550
43. Wehrens R, Franceschi P (2012) Thresholding for biomarker selection in multivariate data using Higher Criticism. Mol Biosyst 8:2339–2346
44. Wehrens R, Franceschi P, Vrhovsek U (2011) Stability-based biomarker selection. Anal Chim Acta 705:15–23
45. Tran TN, Afanador NL, Buydens LMC et al (2014) Interpretation of variable importance in Partial Least Squares with Significance Multivariate Correlation (sMC). Chemom Intell Lab Syst 138:153–160
46. Kanehisa M, Goto S, Sato Y et al (2012) KEGG for integration and interpretation of large-scale molecular data sets. Nucleic Acids Res 40:D109–D114
47. Harris MA, Clark J, Ireland A et al (2004) The Gene Ontology (GO) database and informatics resource. Nucleic Acids Res 32:D258–D261
48. Posma JM, Robinette SL, Holmes E et al (2014) MetaboNetworks, an interactive Matlab-based toolbox for creating, customizing and exploring sub-networks from KEGG. Bioinformatics 30:893–895
49. Kaever A, Landesfeind M, Feussner K (2015) MarVis-Pathway: integrative and exploratory pathway analysis of non-targeted metabolomics data. Metabolomics 11(3):764–777

Part V

Special Topics

Chapter 15

Reconstruction of Protein Networks Using Reverse-Phase Protein Array Data

Silvia von der Heyde, Johanna Sonntag, Frank Kramer, Christian Bender, Ulrike Korf, and Tim Beißbarth

Abstract

In this chapter, we describe an approach to reconstruct cellular signaling networks based on measurements of protein activation after different stimulation experiments. As experimental platform reverse-phase protein arrays (RPPA) are used. RPPA allow the measurement of proteins and phosphoproteins across many samples in parallel with minimal sample consumption using a panel of highly target protein-specific antibodies. Functional interactions of proteins are modeled using a Boolean network. We describe the Boolean network reconstruction approach *ddepn* (dynamic deterministic effects propagation networks), which uses time course data to derive protein interactions based on perturbation experiments. We explain how the method works, give a practical application example, and describe how the results can be interpreted. Furthermore prior knowledge on signaling pathways is essential for network reconstruction. Here we describe the use of our software *rBiopaxParser* to integrate prior knowledge on protein signaling available in public databases. All applied methods are freely available as open-source R software packages. We describe the preparation of RPPA data as well as all relevant programming steps to format the RPPA data, to infer the prior knowledge, and to reconstruct and analyze the protein signaling networks.

Key words Reverse-phase protein arrays, Proteomics, Protein signaling, DDEPN, Network reconstruction, Boolean modeling

1 Introduction

Complex diseases such as cancer are caused by aberrant protein interactions due to mutations or overexpression of specific proteins. These interactions imply activations and inhibitions, e.g., via phosphorylation events after external stimuli, within a signaling network. A deeper understanding of the proteomic signaling networks could help to specifically target individual components and abort corresponding signaling pathways within a cell. It could further improve patient stratification according to their network profiles. Above that, simulations on the network models allow predictions on the system behavior under diverse conditions, supporting personalized medicine approaches.

Klaus Jung (ed.), *Statistical Analysis in Proteomics*, Methods in Molecular Biology, vol. 1362,
DOI 10.1007/978-1-4939-3106-4_15,

In the era of efficiently producible proteomic large-scale data, appropriate modeling techniques are required to reconstruct the networks of interest. Qualitative Boolean models consider the networks as directed graphs with either active or passive proteins, i.e., nodes, connected via activating or inhibiting edges. They have already been successfully applied in reverse engineering of proteomic signaling networks, and their reduced complexity is advantageous for large-scale systems [1, 2]. The basis of network reconstruction is measurements of the involved proteins, but approaches which additionally include literature-based prior knowledge can lead to improved results [3, 4]. Comprehensive prior knowledge is available in public databases like the *Pathway Interaction Database* [5], *Biocarta* [6], or *Reactome* [7]. The National Cancer Institute (NCI) exports the data in the BioPAX [8] format. The R [9] package *rBiopaxParser* [10] can be used to process the prior knowledge for further analyses.

Three major procedures are involved in the protocol presented here.

The first part describes RPPA data. Here we focus on phosphorylation measurements of growth factor-stimulated proteins in a breast cancer cell line expressing epidermal growth factor receptor (EGFR, also termed ErbB-1) and heregulin receptor (ErbB-3/HER3). Protein phosphorylation was monitored at ten different time points after ligand-induced stimulation of growth factor receptors [11].

The second subsection introduces the *rBiopaxParser* software to derive a prior network describing the wiring of ErbB signaling components.

In the third part, the *ddepn* method [3] is presented, which combines RPPA data and prior knowledge within the network reconstruction approach. Finally the actual reconstruction procedure within the R environment is explained in a detailed stepwise manner.

2 Materials

2.1 Data

The example RPPA data is part of a set which is available via the Gene Expression Omnibus (GEO) with accession number GSE50109. It includes protein phosphorylation measurements which were carried out as described by Henjes et al. [11]. Additional information on RPPA can be found in literature [12–14].

In brief, starved human breast cancer cell lines (e.g., SKBR3) were stimulated with the growth factors EGF and HRG alone and in combination. These growth factors activate the MAPK and PI3K pathways. Application of the stimuli defined time point 0 min. Lysate preparation was carried out after 0, 4, 8, 12, 16, 20, 30, 40, 50, and 60 min. The measurements were repeated as five biological replicates of which each replicate was spotted in triplicate on the nitrocellulose-coated glass slides.

To illustrate the modeling approach, data of seven phosphoproteins (pEGFR Y1068, pERBB2 Y1248, pERBB3 Y1289, pAKT S473, pERK12 T202Y204, pMEK S217/S221, and pPDK1 S241) were selected. The RPPA data has been normalized to total protein concentration using the dye Fast Green FCF [14].

2.2 R Session Information

The network reconstructions were run in R (version 3.1.1), applying *ddepn* (version 2.2.1) and *rBiopaxParser* (version 1.3.3).

3 Methods

3.1 RPPA Data

To address the network topology downstream of ErbB-1/EGFR and ErbB-3/HER3, adherent cells were stimulated with specific ligands (EGF, HRG) in a time-resolved fashion. To reduce the experimental error, all stimulation experiments were carried out five times. To reduce the analytical noise, that might be derived from the analytical procedure itself and add to the biological noise, all experimental samples—time points, perturbations, and biological replicates generated throughout the experimental part of this project—were analyzed in parallel in a single analytical run. Currently only array-based formats guarantee a sufficiently high sample capacity to analyze several hundred samples in parallel. We used RPPA, a targeted proteomics approach, that requires deposition of sample lysates on nitrocellulose-coated slides as solid-phase carrier and detection of the proteins of interest with highly specific antibodies. Serial dilutions of suitable calibrator samples are co-printed along with lysate samples to assess the dynamic range of a particular detection antibody. Our RPPA approach which relies on near-infrared (NIR) fluorescence detection is employed since this detection approach allows measurements over a three- to fourfold dynamic range [14].

1. Prepare cellular lysates for all experimental samples and determine total protein concentration in μg/μl.
2. Adjust total protein concentration in all samples with lysis buffer to a range between 1 and 3 μg/μl. Take care that the number of outliers is less than 5 %.
3. Prepare serial dilutions (6–12 steps) of suitable control samples, e.g., the cell line(s) used in this particular experiment. If required, prepare positive controls with high levels of protein phosphorylation to guarantee that phosphoproteins of interest can be detected.
4. Transfer controls and lysate samples to a multi-titer plate compatible with the printing robot.

5. Print all samples in an addressable format on nitrocellulose-coated slides. Just before printing, add SDS/DTT buffer to all samples and boil samples briefly to denature proteins. After completing the print run, allow lysates to bind overnight at RT and store slides at –20 °C in a sealed box with a desiccator pad.
6. Determine total protein content per spot to calculate a spot-specific correction factor that will be required during data analysis (*see* **Note 1**). This can be made by staining slides with an NIR-detectable fluorescent dye, e.g., Fast Green FCF [14, 15].
7. Detect proteins of interest using highly specific primary antibodies at a dilution of 1:300 and appropriately labeled secondary antibodies or Fab fragments as described [14, 15].
8. Dry slides protected from light and obtain images using a suitable NIR scanner with a resolution of at least 20 μm. Store slides as *tif* files and determine signal intensities for all spots on all slides using a suitable software package.
9. Normalize raw data using the R package *RPPanalyzer* [16] as presented in **Note 1**.

3.2 Prior Knowledge

Vast amount of knowledge on molecular interactions, for example protein signaling, has been accumulated within the last decades. This knowledge has been compiled into numerous pathway databases. Currently, the website *pathguide.org* lists over 500 databases which host pathway knowledge in various formats and specializations [17]. Well-renowned examples for pathway databases are the *Kyoto Encyclopedia of Genes and Genomes* (*KEGG*) database [18], *Reactome* [19], and *WikiPathways* [20].

The integration of prior knowledge for network reconstruction is recommended to increase the robustness and power of bioinformatic approaches [21, 22]. Different strategies for integrating prior knowledge can be applied, depending on the method for network reconstruction, pathway databases, and computing environments [23, 24].

Within this subsection, the *KEGG* database will be parsed into R using the *rBiopaxParser* [10] to provide prior knowledge for the *ddepn* package [3]. The first step describes installation of the package, *KEGG* data download, and parsing into R. Afterwards we specify how to access the ErbB signaling pathway and how to perform necessary ID mapping and pathway reduction steps.

1. Install the *rBiopaxParser* package and load it into R via the following commands:

```
source("http://bioconductor.org/biocLite.R")
biocLite("rBiopaxParser")
library(rBiopaxParser)
```

2. Download *KEGG* data in BioPAX level 3 format (KEGG.bp3.owl.gz) from www.mmnt.net/db/0/0/ftp1.nci.nih.gov/pub/PID/BioPAX_Level_3.

3. Import the Biopax *.owl* file and generate an internal data.frame format. This step can take a while. So we recommend to store the data for further analyses as shown below:

```
biopax = readBiopax(file="KEGG.bp3.owl.gz")
save(biopax, file="KEGG_biopax.RData")
```

4. List IDs and names of the instances of class "pathway" in an ordered data.frame. Select the ErbB signaling pathway (pathway ID "pid_10379"):

```
pw_list = listInstances(biopax, class="pathway")
id = 'pid_10379'
erbb_graph = pathway2RegulatoryGraph(biopax, id,
splitComplexMolecules=T, verbose=T, useIDasNodenames=T)
```

5. Adapt internal IDs to common protein names, e.g., *HUGO Gene Nomenclature Committee* (*HGNC*) [25] nomenclature. Therefore we provide the useful function *getIDs_pid* in **Note 2** which is applied in the following commands. This step can be adapted corresponding to the identifiers used within specific pathway databases, for example using the *biomaRt* package [26]:

```
library(biomaRt)
mart = useMart('ensembl')
ensembl = useMart("ensembl",
dataset="hsapiens_gene_ensembl")
idtab = getIDs_pid(erbb_graph, biopax)
```

6. Reduce the ErbB graph to nodes of interest (with MAP2K as MEK, MAPK1 as ERK, and NRG1 as HRG) and apply *HGNC* nomenclature:

```
targets = c("^AKT1/AKT3/AKT2$", "^EGF$", "^EGFR$",
"^ERBB2$", "^ERBB3$", "^MAP2K1/MAP2K2$",
"^MAPK3/MAPK1$", "^MTOR$", "^NRG1$")
targetIDs <- nameTargets <- c()
for(i in 1:length(targets)){
  targetIDs = c(targetIDs, rownames(idtab)
  [grep(targets[i], idtab[,"hgnc"])])
  nameTargets = c(nameTargets, rep(targets[i],
  length(grep(targets[i], idtab[,"hgnc"]))))
}
```

```
names(targetIDs) <- nameTargets <- sub("\\$", "",
sub("\\^", "", sub("AKT1/AKT3/AKT2", "AKT",
sub("MAP2K1/MAP2K2", "MEK1/2", sub("MAPK3/MAPK1",
"ERK1/2", sub("NRG1", "HRG", nameTargets))))))
nameTargets[which(duplicated(nameTargets))] =
paste(nameTargets[which(duplicated(nameTargets))], 2,
sep="_")
erbb_subgraph = subGraph(nodes(erbb_graph)
[match(targetIDs, nodes(erbb_graph))], erbb_graph)
nodes(erbb_subgraph) = nameTargets
```

7. Merge duplicate nodes, which can occur when parsing database information. Duplicate nodes can occur during ID conversion, due to mapping between internal identifiers, gene names, *RefSeq* [27] numbers, or protein families. Within this example graph the conversion led to duplicated information on MTOR, EGFR, ERBB3, MEK1/2, and ERK1/2. This is fixed within the next commands:

```
erbb_sub_comb = erbb_subgraph
for(i in which(duplicated(names(targetIDs)))){
   erbb_sub_comb = combineNodes(nameTargets[(i-1):i],
   erbb_sub_comb, nameTargets[i-1])
}
```

8. Add missing edges manually. Create the adjacency matrix "prior" with 1 denoting activation and −1 denoting inhibition from row to column target. Figure 1 shows the graph resulting from applying the following commands:

```
prior_graph = addEdge("EGFR", "MEK1/2", erbb_sub_comb,
1)
prior_graph = addEdge("ERBB2", "MEK1/2", prior_graph,
1)
prior_graph = addEdge("ERBB2", "AKT", prior_graph, 1)
prior_graph = addEdge("ERBB3", "AKT", prior_graph, 1)
prior = as(prior_graph, "matrix")
plotRegulatoryGraph(prior_graph)
```

These steps were a straightforward way to retrieve a reduced ErbB signaling pathway with *HGNC* symbol identifiers for further use as prior knowledge within the *ddepn* functionality. These steps can be adapted to specific needs of other available methods for network reconstruction, pathways, or pathway databases.

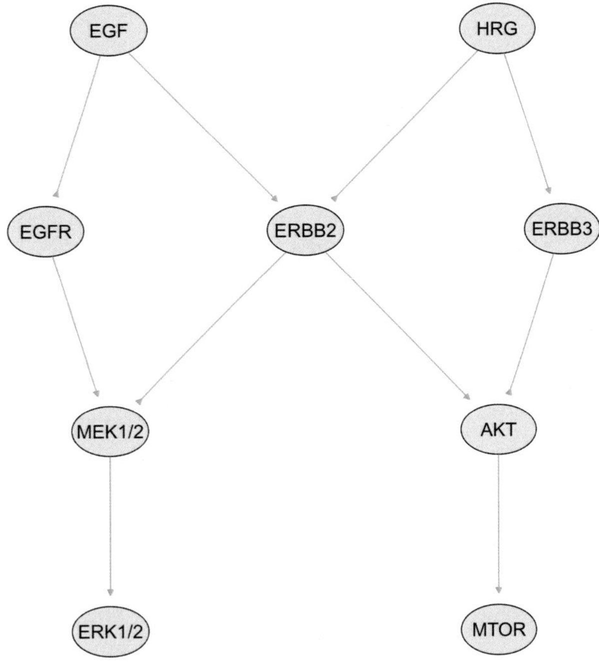

Fig. 1 Graph incorporating the prior literature knowledge on ErbB signaling

3.3 Dynamic Deterministic Effects Propagation Networks (ddepn)

The goal of the *ddepn* method is to reconstruct signaling networks from time-resolved data under specific stimulation and inhibition of input nodes for the network. Starting at the stimuli, a signal propagation scheme is applied to model the dynamics of a signal across a network of proteins or genes, which are connected by either activating or inhibiting edges. The method was developed specifically for phosphoprotein signaling networks but is, in principle, applicable to all types of networks that describe a dynamically changing system of interconnected nodes.

The protein interaction networks are modeled as directed and possibly cyclic graphs with proteins and stimuli as nodes and interactions as edges. Activating edges are represented by 1 and inhibiting ones as −1, respectively, in an adjacency matrix. A missing edge between two nodes is represented by 0. The measurement data are stored in a matrix format, too, as explained in more detail in the following.

To infer an optimal network regarding the measurements and the prior knowledge, we apply the stochastic Markov Chain Monte Carlo (MCMC) approach of *ddepn*, called *inhibMCMC*. It considers all possible network structures and evaluates them according to their posterior probabilities: the posterior distribution of a network given the data is proportional to the product of its prior probability distribution and the likelihood of the data given the network. The network is defined by a system state matrix with active (1) and passive (0) states per protein and time point. The entries result from a Boolean signal propagation rationale. All nodes, except the active stimuli, are initialized with inactive states before the signal propagation starts. Nodes get activated if at least one activating parent node is active and all inhibiting parents are inactive. Using this scheme, a series of all reachable system states for a given network structure with N nodes is calculated, leaving less than the maximum of $2N$ states that can be realized. From this "repository" of possible system states an optimized state matrix is then fitted to the measured time points by a hidden Markov model (HMM), identifying a most likely sequence of state transitions over time. According to the activity states, the measurements are assumed to result from an active or passive normal distribution. The corresponding parameters are estimated as empirical mean and standard deviation for each protein in the corresponding class across all time points. The prior probability distribution implies penalization of differences between the inferred network and a defined prior network. The latter is represented as an adjacency matrix, as described before. Although *ddepn* allow to weight the confidence in an edge (*see* **Note 3**), we chose maximum confidence, i.e., entries of 0, −1, and 1.

Network sampling is done within an MCMC network structure sampling framework, adapted for usage with two different kinds of edges (activating and inhibiting). To be able to reach each possible edge type and direction from any edge in a single step, five possible transitions are defined for the edge sampling procedure: add activation, add inhibition, revert, switch type, simultaneously revert and switch type. The MCMC sampler is then calculating an acceptance ratio for each step based on the posterior of the current and the new network structure, which is the basis for accepting or rejecting a change in the network. This sampling is done for thousands of iterations to infer a well-defined and converging sampling of the posterior distribution of all network structures.

Within this subsection, the *ddepn* package will be applied to the example data which has to be replaced by user-specific data and stored as "data.RData" to be loaded successfully with the commands shown below. The corresponding data format will be explained and further information can be inferred from the *ddepn* vignette. After some preprocessing steps, networks will be reconstructed via MCMC sampling. Afterwards the convergence will be checked and a consensus network will be created. Finally the consensus network will be plotted and analyzed.

1. Install the *ddepn* package and load it into R via the following commands:

```
install.packages("ddepn", repos =
"http://R-Forge.R-project.org")
library(ddepn)
```

2. Load the RPPA data matrix with protein (node) names as row names and stimuli/time points as column names. Stimuli and time points should be separated by underscore. If you have combined stimuli, the software will recognize it by "&". In our example, cells were exposed to the stimuli EGF and HRG, either alone (EGF, HRG) or in combination (EGF&HRG). In the network EGF and HRG are modeled as separate nodes. Time points are represented as numeric values. Replicates have the same column names. Table 1 shows a reduced set of our example data. For data preprocessing steps as normalization and export to table format we recommend the R package *RPPanalyzer* [16, 28] as presented in **Note 1**.

   Furthermore, the *ddepn* package itself provides example RPPA data and corresponding formatting instructions.

   The following command stores the data in a variable called "data":

```
data = get(load("data.RData"))
```

3. Add the stimuli defined in the column names of the data matrix to the data matrix as dummy nodes. This results in a list of the stimuli and a modified data matrix. The column names stay the same, but the matrix will be extended by two rows named "EGF" and "HRG" with zero entries after applying the *addstimuli* function as follows:

```
stim.dat = addstimuli(dat=data)
```

4. Use the commands shown below to define an additional prior adjacency matrix "phiorig" which equals the prior but with 2 denoting inhibition instead of –1. This reference network will be used for comparison to the inferred network. Here we do not use extra edge weights within the prior, defining the trust in an edge.

Table 1
Excerpt of the example data with signal intensities per stimulus and time point

	EGF&HRG_0	EGF&HRG_0	EGF_0	EGF_0	HRG_0	HRG_0
pMEK12_S217S221	2494.66	2505.87	2801.97	2888.26	3472.66	3443.23
pERBB3_Y1289	1623.24	1606.16	1636.40	1676.40	2598.70	2283.89
pEGFR_Y1068	1678.96	1645.55	1527.87	1544.19	1489.53	1506.92
pERBB2_Y1248	4141.60	4175.52	7847.07	8152.05	5673.07	5507.46
pAKT_S473	2359.72	2450.71	2196.77	2290.46	3789.89	3636.11
pERK12_T202Y204	2957.40	3083.52	3084.82	3148.46	12150.63	11906.87
pmTOR_S2448	1919.05	1983.95	2411.44	2305.56	2163.50	2194.71

Take care that your "prior" and "stim.dat$dat" matrices have the same row names in the same order. The same holds for the column names of the prior. In our example we adapted the row names of the data matrix according to the prior and reordered the matrix rows to be in the same order ("EGF", "HRG", "MTOR", "MEK1/2", "ERBB3", "EGFR", "ERBB2", "AKT", "ERK1/2"):

```
phiorig = prior
phiorig[phiorig == -1] = 2
```

5. Reconstruct networks via MCMC sampling with ten parallel and independent runs (parameter *cores*). We define 10,000 maximum iterations per run with 5000 as burn-in phase, taking approximately 1 h of computation time. To penalize deviation of the networks from the prior knowledge, we use the Laplace prior model (*laplaceinhib*) with the parameters *lambda* and *gamma*. Here we choose a strong influence of the prior knowledge, as the EGFR network is well described. The parameter *lambda* is set to 0.0001 and *gamma* is set to 1. For the motivation of the parameter choice *see* **Note 4**. The following commands perform the network reconstructions and store the results:

```
library(parallel)
nets = ddepn(dat=stim.dat$dat, phiorig=phiorig,
inference="mcmc", outfile="mcmc.pdf", multicores=T,
maxiterations=10000, usebics=FALSE, cores=10,
priortype="laplaceinhib", lambda=0.0001, B=prior,
gam=1, burnin=5000, always_sample_sf=T)
save(nets, file="mcmc.RData")
```

6. Convergence is validated via Gelman diagnostic [29] applying the commands shown below. Approximate convergence is achieved when the upper confidence limit of the "potential scale reduction factor" is close to 1. In our example runs we achieved a value of 1.06:

```
nets = get(load("mcmc.RData"))
gelman_diag(nets)
```

Apart from the convergence check, one can further validate the reconstructed edges by plotting the confidences for each edge obtained from the independent MCMC chains. An explanation is given in **Note 5**.

Furthermore the *ddepn* package offers a functionality to create a time profile for each experiment and protein in the data set (*see* **Note 6**).

7. Using independent MCMC chains, we can identify edges that occur significantly more often than expected by chance and merge them into a consensus network. A Wilcoxon rank sum test is used to test for the difference between the numbers of

sampled activating and inhibiting edges. If significantly more activating or inhibiting edges are observed, the corresponding edge is finally included into the network. If not, no edge is assumed. The procedure can be embedded in a leave-one-out cross-validation approach, in which each of the ten MCMC chains is left out once, and the testing algorithm is applied to the remaining runs. Adjustment of *p*-values due to multiple testing is done according to Benjamini and Hochberg [30]. An edge is included in the consensus network if it occurs in all of the cross-validation runs. The corresponding R code looks as follows:

```
cons.net = create_signetwork_cv(ret=nets, adj.
method="BH")

save(cons.net, file="consensusNet.RData")
```

8. Create a plot including the prior and the reconstructed network via the commands shown below. This requires a layout matrix defining the positions of vertices in the plot. The row names are the targets in the same order as in the consensus network, and the column names are "x" and "y" as coordinates. Saving the output as *pdf* document allows easy editing in vector graphic programs afterwards. Figure 2 shows our edited example output:

```
plot_overlap(adjcore=phiorig, adjX=cons.net,
overlap=rownames(phiorig), nameA="Prior",
nameB="Reconstruction", main="SKBR3",
symbols.plot=rownames(phiorig),ig.layout=layoutMatrix,
stimnodes=c("EGF","HRG"))
```

As displayed in Fig. 2, we inferred six novel edges which were not included in the prior network. All of them are activating ones which is represented by solid blue lines in Fig. 2. None of the prior edges has been deleted or switched, which is due to our strong belief in the prior knowledge represented by the corresponding parameters.

The novel edges EGF → MTOR, EGF → MEK1/2, EGF → AKT, and HRG → ERK1/2 can be interpreted as indirect activating effects induced by the growth factors EGF and HRG. Likely, these edges also indicate that the model does not consider the fact that the plasma membrane presents a physical barrier for growth factor ligands (EGF, HRG). Furthermore, growth factor receptor-mediated signals are propagated with a high velocity and that is what these new edges directly emerging from EGF or HRG are actually showing. The newly inferred interaction AKT → MEK1/2 can be interpreted as cross talk between the PI3K and MAPK pathway. Finally, the novel edge ERBB3 → ERBB2 might represent close interaction of the ErbB family members ErbB-2 and ErbB-3 as well-characterized heterodimers [31].

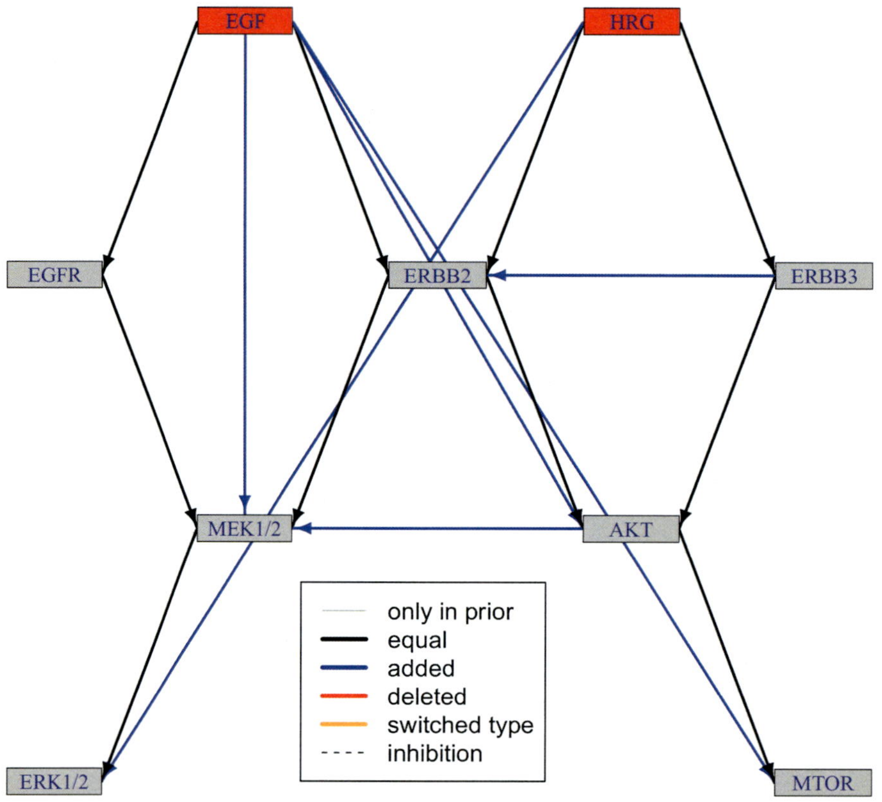

Fig. 2 Reconstructed graph in comparison to the prior network. Newly inferred edges are marked *blue. Black* denotes no change compared to the prior

4 Notes

1. The R package *RPPanalyzer* [28] offers useful functions for RPPA data preprocessing, statistical analyses, and data visualization. A software upgrade was published recently [16] providing supplementary R code to preprocess and analyze RPPA example data. The stepwise instructions start with data preprocessing. This first step simultaneously performs data import, background noise correction, generation of plots to assess data quality, and spot-specific data normalization based on the total protein concentration determined by the dye Fast Green FCF [14].

   The raw data are imported from files storing the spot-specific signal intensities. These files have to be in *gpr* file format as provided by applying the commercial image analysis software GenePixPro. The background noise correction performs subtraction of an intercept estimated for a total protein concentration of 0 μg/μL from the signal intensities. The intercept estimation is based on a representative dilution series.

For data normalization based on the total protein concentration, one slide has to be stained with Fast Green FCF to determine the total protein concentration per lysate spot.

The signal intensities from slides probed with target-specific detection antibodies are divided by these FCF factors. Finally, the corrected values are multiplied by the median of the FCF slide to scale data back to the original signal intensity range.

The next step in the R script performs data selection and calculation of the median of the technical replicates. It is further demonstrated how to export data to a file in table format. Finally a boxplot and a time course plot are created for the example data.

The *RPPanalyzer* offers methods to test and visualize correlations, to visualize data as heatmap or time courses, and to apply statistical rank sum tests for differences between sample groups which can be visualized as boxplots.

2. Function to adapt prior network IDs to common protein names: `getIDs_pid <- function (graph, biopax) {`

```
   class<-entrez<-uniprot<-hgnc<-xref_name<-instprop<-
   instprop.<-c()
   for(nod in nodes(graph)){
         message(paste("Getting annotation of", nod,
"node"))
         if(length(getInstanceProperty(biopax,
id=nod))>1){
               instprop[nod] =
   getInstanceProperty(biopax,
               id=nod)[1]
               instprop.[nod] =
getInstanceProperty(biopax,
               id=nod)[2]
         }
         if(length(getInstanceProperty(biopax,
id=nod))==1){
   instprop[nod] = getInstanceProperty(biopax,
               id=nod)[1]
               instprop.[nod] = NA
         }
         if(length(getInstanceProperty(biopax,
id=nod))==0){
               instprop[nod]<-instprop.[nod]<-NA
         }
```

```
        annt = data.frame(getXrefAnnotations(biopax,
            id=nod))
        class[nod] = getInstanceClass(biopax, nod)
        if (length(annt)==0){
            entrez[nod]<-uniprot[nod]<-
xref_name[nod]<-
            hgnc[nod]<-NA
        }else{
            xref_name[nod] = annt[1, "name"]
            # get entrez
            if (length(annt[grep("EntrezGene:",
            annt[,6]),6])==0){
                entrez[nod] = NA
            }
            if (length(annt[grep("EntrezGene:",
            annt[,6]),6])==1){
                entrez[nod] = sub("EntrezGene:", "",
                annt[grep("EntrezGene:",
annt[,6]),6])
                getsym =
                getBM(attributes=c("hgnc_symbol"),
                filters='entrezgene',
values=entrez[nod],
                mart=ensembl)[1]
            }
            if (length(annt[grep("EntrezGene:",
            annt[,6]),6])>1){
                entrez[nod] =
paste(sub("EntrezGene:", "",
                annt[grep("EntrezGene:",
annt[,6]),6]),
                collapse="/")
                gs = c()
                for(e in sub("EntrezGene:", "",
                annt[grep("EntrezGene:",
annt[,6]),6])){
                    gs[e] =
    getBM(attributes=c("hgnc_symbol"),
                    filters='entrezgene', values=e,
                    mart=ensembl)[1]
                    if(length(gs[e])==0){gs[e] =
```

```
NA}
                    }
                    gss = gs[!is.na(gs)]
                    getsym = paste(gss, collapse="/")
              }
              # get uniprot
              if (length(annt[grep("UniProt:",
              annt[,6]),6])==0){
                    uniprot[nod] = NA
              }
              if (length(annt[grep("UniProt:",
              annt[,6]),6])==1){
                    uniprot[nod] = sub("UniProt:", "",
                    annt[grep("UniProt:", annt[,6]),6])
                    getsym =
                    getBM(attributes=c("hgnc_symbol"),
   filters='uniprot_swissprot_accession',
                    values=uniprot[nod], mart=ensembl)[1]
              }
              if (length(annt[grep("UniProt:",
              annt[,6]),6])>1){
                    uniprot[nod]=paste(sub("UniProt:",
"",
                    annt[grep("UniProt:",annt[,6]),6]),
                    collapse="/")
                    gs = c()
                    for(e in sub("UniProt:", "",
                    annt[grep("UniProt:", annt[,6]),6])){
                         gs[e] =
       getBM(attributes=c("hgnc_symbol"),


   filters='uniprot_swissprot_accession',
                         values=e, mart=ensembl)[1]
                         if(length(gs[e])==0) {gs[e] = NA
}
                    }
                    gss = gs[!is.na(gs)]
                    getsym = paste(gss, collapse="/")
              }
```

```
                if(is.na(entrez[nod]) &
is.na(uniprot[nod])){
                    hgnc[nod] = NA
                    getsym = NULL
                }
                if(length(getsym)==0){
                    hgnc[nod] = NA
                }else{
                    hgnc[nod] = getsym
                }
            }
        }
        cbind(hgnc, entrez, uniprot, instprop,
        instprop.,
        xref_name, class)
    }
```

3. The confidence in an edge of the prior network correlates with its absolute value ranging from zero to one. A value of one (zero) means absolute (no) confidence. Negative values denote inhibition, while positive values denote activation.
4. Bender et al. describe how to choose the prior parameters *lambda* and *gamma* [3]. For a constant parameter *gamma* (1), small values of *lambda* penalize small differences between the reconstruction and the reference network less strongly than large differences. Large values of *lambda* almost penalize all differences equally, which is not recommendable. Small values of *gamma* penalize differences to the reference in a stronger way than large values. Hence, *gamma* should be small (≤ 1) if the reference is trustworthy, with a corresponding *lambda* $\ll 0.1$ to retain a sensitivity towards the absolute difference between reference and prior network.

   After running the *ddepn* algorithm, for each MCMC chain plots of the network reconstructions are obtained ("mcmc_1_stats.pdf", "mcmc_2_stats.pdf", ..., "mcmc_10_stats.pdf"). These plots further reveal how much your network reconstruction resembles the prior network in the form of an ROC curve. An AUC close to one indicates a reconstructed network similar to the prior. You further see the posterior, likelihood, and prior traces (absolute or their ratios between two subsequent iterations) as well as graphical representations of the prior network and the inferred network within the corresponding MCMC run.

5. The *ddepn* function plot_edgeconfidences creates a plot of edge confidences or frequencies, i.e., occurrences, for the conducted MCMC runs. If the inferred network contains N nodes, each panel in the plot contains N boxes of activation and inhibition confidences/frequencies each. In our example N equals nine. Activation boxes are colored red, while inhibition boxes are colored blue. The column name in each panel defines the source node from which an edge originates. The panel name denotes the destination node to which the edge is pointing.

   The following commands create a *pdf* file as displayed in Fig. 3 showing the edge confidences for the example networks:

```
pdf("confidences.pdf", width=12, height=10)
plot_edgeconfidences(nets, act="conf.act",
  inh="conf.inh",
cex.axis=0.8)
dev.off()
```

   As expected, our six novel edges are less confidential than our a priori-assumed ones. Large differences between activating and inhibiting edge confidences indicate that a particular type of edge is preferred over the other. To test this, a Wilcoxon test is suggested and can be conducted within the package.

6. The *ddepn* function plot_profiles can be applied to create a time profile for each experiment and protein in the data set. The data points are plotted as boxplots per time point. A smoothing spline fit represents the trend of the intensities over time.

   Additionally, estimations of the inferred model parameters for the computed Gaussian active and passive distributions are laid over the plots to denote the active (red) and passive (blue) states. The boxplots are colored using a diverging color panel from green to red, indicating the state of the node at each time point.

   The following commands create a *pdf* file as displayed in part in Fig. 4 showing the time profiles for the example data. This requires additionally the *gam* R package by Trevor Hastie for generalized additive models:

```
library(gam)
pdf("profiles.pdf")
plot_profiles(nets, mfrow=c(3,3))
dev.off()
```

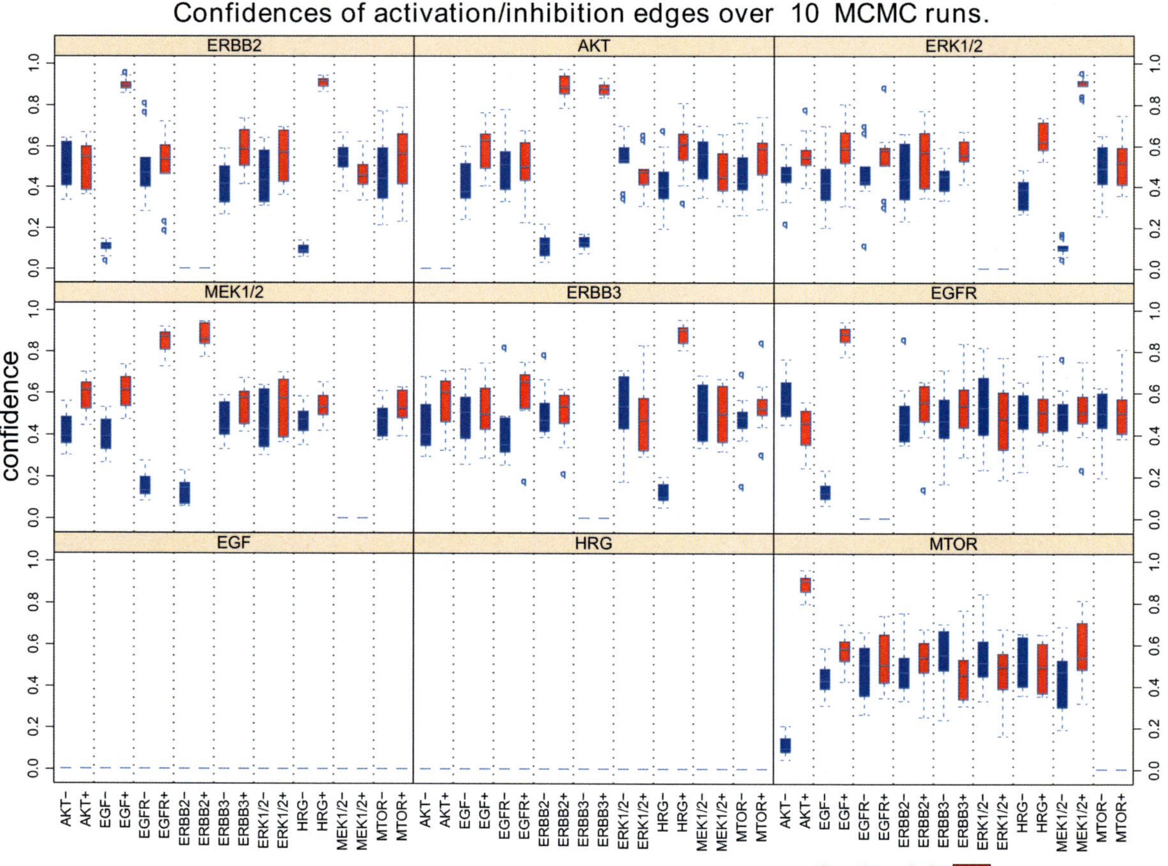

Fig. 3 Edge confidences for the reconstructed networks. Confidence strength is illustrated as denote confidence strength for activating edges while denote confidence strength for inhibiting edges. The column names in each panel define the source nodes from which an activating ("+") or inhibiting ("−") edge originates. The denotes the destination node to which the edge is pointing

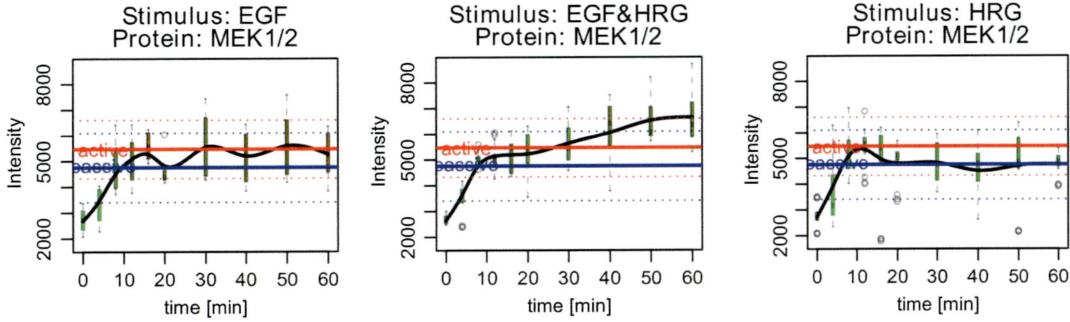

Fig. 4 Time profiles for the phosphorylation of MEK1/2 under the diverse stimuli combinations. The data are represented as boxplots per time point. A smoothing spline fit represents the trend of the intensities. The *red* and *blue lines* denote the estimated model parameters for the active (*red*) and passive (*blue*) Gaussian distributions

References

1. Wang R-S, Saadatpour A, Albert R (2012) Boolean modeling in systems biology: an overview of methodology and applications. Phys Biol 9:055001
2. von der Heyde S, Bender C, Henjes F et al (2014) Boolean ErbB network reconstructions and perturbation simulations reveal individual drug response in different breast cancer cell lines. BMC Syst Biol 8:75
3. Bender C, vd Heyde S, Henjes F et al (2011) Inferring signalling networks from longitudinal data using sampling based approaches in the R-package “ddepn”. BMC Bioinformatics 12:291
4. Eduati F, De Las Rivas J, Di Camillo B et al (2012) Integrating literature-constrained and data-driven inference of signalling networks. Bioinformatics 28:2311–2317
5. Schaefer CF, Anthony K, Krupa S et al (2009) PID: the Pathway Interaction Database. Nucleic Acids Res 37:D674–D679
6. Nishimura D (2001) BioCarta. Biotech Softw Internet Rep 2:117–120
7. Croft D, O’Kelly G, Wu G et al (2011) Reactome: a database of reactions, pathways and biological processes. Nucleic Acids Res 39:D691–D697
8. Demir E, Cary MP, Paley S et al (2010) The BioPAX community standard for pathway data sharing. Nat Biotechnol 28:935–942
9. R Development Core Team (2013) R: a language and environment for statistical computing. R Foundation for Statistical Computing, Vienna, Austria. http://www.R-project.org. Accessed 1 Feb 2015
10. Kramer F, Bayerlová M, Klemm F et al (2013) rBiopaxParser—an R package to parse, modify and visualize BioPAX data. Bioinformatics 29:520–522
11. Henjes F, Bender C, von der Heyde S et al (2012) Strong EGFR signaling in cell line models of ERBB2-amplified breast cancer attenuates response towards ERBB2-targeting drugs. Oncogenesis 1, e16
12. Korf U (ed) (2011) Protein microarrays—methods and protocols. Springer Science+Business Media, LLC, New York
13. Akbani R, Becker K-F, Carragher N et al (2014) Realizing the promise of reverse phase protein arrays for clinical, translational, and basic research: a workshop report: the RPPA (Reverse Phase Protein Array) society. Mol Cell Proteomics 13:1625–1643
14. Loebke C, Sueltmann H, Schmidt C et al (2007) Infrared-based protein detection arrays for quantitative proteomics. Proteomics 7:558–564
15. Sonntag J, Bender C, Soons Z et al (2014) Reverse phase protein array based tumor profiling identifies a biomarker signature for risk classification of hormone receptor-positive breast cancer. Transl Proteomics 2:52–59
16. von der Heyde S, Sonntag J, Kaschek D et al (2014) RPPanalyzer toolbox: an improved R package for analysis of reverse phase protein array data. Biotechniques 57:125–135
17. Bader GD, Cary MP, Sander C (2006) Pathguide: a pathway resource list. Nucleic Acids Res 34:D504–D506

18. Ogata H, Goto S, Sato K et al (1999) KEGG: Kyoto encyclopedia of genes and genomes. Nucleic Acids Res 27:29–34
19. Joshi-Tope G, Gillespie M, Vastrik I et al (2005) Reactome: a knowledgebase of biological pathways. Nucleic Acids Res 33:D428–D432
20. Kelder T, van Iersel MP, Hanspers K et al (2012) WikiPathways: building research communities on biological pathways. Nucleic Acids Res 40:D1301–D1307
21. Werhli AV, Husmeier D (2007) Reconstructing gene regulatory networks with bayesian networks by combining expression data with multiple sources of prior knowledge. Stat Appl Genet Mol Biol 6, Article15
22. Fröhlich H, Tresch A, Beissbarth T (2009) Nested effects models for learning signaling networks from perturbation data. Biom J 51:304–323
23. Mukherjee S, Speed TP (2008) Network inference using informative priors. Proc Natl Acad Sci U S A 105:14313–14318
24. Kramer F, Bayerlová M, Beißbarth T (2014) R-based software for the integration of pathway data into bioinformatic algorithms. Biology (Basel) 3:85–100
25. Gray KA, Yates B, Seal RL et al (2014) Genenames.org: the HGNC resources in 2015. Nucleic Acids Res 43:D1079–D1085
26. Durinck S, Moreau Y, Kasprzyk A et al (2005) BioMart and Bioconductor: a powerful link between biological databases and microarray data analysis. Bioinformatics 21:3439–3440
27. Pruitt K, Brown G, Tatusova T et al (2002) The Reference sequence (RefSeq) database. In: McEntyre J, Ostell J (eds) The NCBI handbook [Internet], Bethesda (MD): National Library of Medicine (US), National Center for Biotechnology Information, Chapter 18. http://www.ncbi.nlm.nih.gov/books/NBK21091. Accessed 1 Feb 2015
28. Mannsperger HA, Gade S, Henjes F et al (2010) RPPanalyzer: analysis of reverse-phase protein array data. Bioinformatics 26: 2202–2203
29. Brooks SP, Gelman A (1998) General methods for monitoring convergence of iterative simulations. J Comput Graph Stat 7:434–455
30. Benjamini Y, Hochberg Y (1995) Controlling the false discovery rate: a practical and powerful approach to multiple testing. J R Stat Soc Series B 57:289–300
31. Junttila TT, Akita RW, Parsons K et al (2009) Ligand-independent HER2/HER3/PI3K complex is disrupted by trastuzumab and is effectively inhibited by the PI3K inhibitor GDC-0941. Cancer Cell 15:429–440

Chapter 16

Detection of Unknown Amino Acid Substitutions Using Error-Tolerant Database Search

Sven H. Giese, Franziska Zickmann, and Bernhard Y. Renard

Abstract

Recent studies have demonstrated that mass spectrometry-based variant detection is feasible. Typically, either genomic variant databases or transcript data are used to construct customized target databases for the identification of single-amino acid variants in mass spectrometry data. However, both approaches require additional data to perform the identification of SAAVs. Here, we discuss the application of an error-tolerant peptide search engine such as BICEPS for identifying variants exclusively based on standard Uniprot databases. Thereby, unnecessary and redundant extensions of the search space are avoided. The workflow provides an unbiased view on the data; the search space is not limited to known variants and simultaneously does not require additional data. In a subsequent step a second identification search is performed to verify the initially identified variant peptides and aggregate information on the protein level.

Key words Mass spectrometry, Variant peptide identification, Error-tolerant peptide identification, Single-amino acid variations, Single-nucleotide variants, Proteomics

1 Introduction

In recent years, mass spectrometry (MS)-based proteomics has become more and more popular in numerous applications [1, 2]. Initially only capable to characterize proteins in a sample, the set of use cases has expanded to, i.e., the analysis of posttranslational modifications (PTMs) [3, 4], protein-protein interactions, single-amino acid variations (SAAVs) [5, 6], quantification [7], and combinations of the respective fields [8, 9].

The general workflow for standard proteomic experiments includes sample preparation, digestion of proteins into peptides, and mass spectrometric analysis of the samples [2]. The fragment spectra are then associated to peptide sequences via database searches, spectral library searches, de novo identification, or hybrid methods. For a general review of peptide identification algorithms and statistical analysis the reader is referred to [10, 11]. MS-based methods are not only employed for proteomic analysis but also in

Klaus Jung (ed.), *Statistical Analysis in Proteomics*, Methods in Molecular Biology, vol. 1362,
DOI 10.1007/978-1-4939-3106-4_16, © Springer Science+Business Media New York 2016

the context of proteogenomics [12, 13]. Of particular interest are single-nucleotide variations (SNV) that can be located in coding or noncoding parts of the DNA. Disease-associated SNVs are challenging to find and to interpret because complex diseases can be caused by multiple mutations in different genes [14, 15]. The proteomic perspective can help to understand the consequences of genomic variations at the protein level by looking for SAAVs.

However, the practical application of SAAV identification is challenging. First, not all peptides of a protein fly equally well in the mass spectrometer because of their physicochemical properties. Thus, identifying all of them might not be possible since not all peptides are ionized. Generally, only non-synonymous mutations can be detected because they result in visible sequence variations. In contrast, nonsense mutations invoke the protein degradation mechanism [16] and may thus only be present in low-abundance and truncated proteins. Therefore, nonsense mutations are difficult to detect [17]. In addition, standard spectral database search algorithms are not able to detect amino acid variations from standard reference databases. By definition, the exact sequence of a peptide containing a SAAV is not in the database; hence, it is not possible to detect it with standard searches such as Mascot, OMSSA, MyriMatch, X!TANDEM, InsPect or Sequest [18–23]. Also, spectral library searches can only detect what has been previously stored in the corresponding libraries and are thus not suited for detecting novel variants.

Further, critical issues in the detection of SAAVs are search time and quality control of search results. A common way for quality control in standard database searches is the so-called target-decoy approach. Here, in addition to the proteomic target database known false positives (decoys) are added to the databases [24, 25]. It is infeasible to generate all possible combinations of amino acid substitutions as single-database entries [26] without increasing the false-positive rate and the search time dramatically.

Further, not all organisms have comprehensive protein databases available, making standard database searches not sensitive enough. Here, de novo transcriptome assembly [27] can help to construct a database exclusively based on RNA-Seq reads. However, transcriptome assembly is far from trivial and the resulting transcripts are prone to contain false positives that can bias the identification search.

To overcome the complex challenges of SAAV detection, customized reference databases can be used. There, knowledge from the genome level is integrated into proteomic analysis, for instance in the form of online databases that store so far discovered sequence variants, e.g., dbSNP [39]. MS databases are then constructed by the translation of the annotated genome variants into their respective proteins. Following the general dogma of gene expression, here genomic information, for instance derived in next-generation sequencing experiments, can be combined with MS data.

Thereby, a better understanding of gene regulation, genome annotation, or disease-associated variants is achieved. In this way, Li et al. [40] constructed a database containing cancer-related variations. Thereby, it becomes feasible to detect common cancer variations on the proteome level. Obviously, a drawback of this approach is that it does not allow searching for unknown variants. Moreover, different mutations on the same amino acid and SAAV that are not present in the sample increase the database size unnecessarily.

To avoid unnecessary expansion of the database size RNA-Seq [41] can be used to construct customized protein databases [9, 42, 43]. Thereby, an unbiased view at the proteomic sample is achieved. However, in many settings it is not (yet) cost effective to construct personalized proteome databases for individuals. In addition, current sequencing technologies are imperfect and produce sequencing errors. Thus, on the one hand incorrectly aligned reads and consequently incorrect variant calls pose already a challenge for the construction of the database [44]. High-coverage experiments yield more reliable results but are also more expensive. On the other hand, missed variant calls are still possible. In summary, building a customized SAAV database with prior knowledge from genome databases only permits a biased and limited view. In contrast, supporting RNA-Seq data is unbiased but not yet cost effective to use in all settings. However, both methods are valuable to enable the usage of standard peptide identification algorithms for proteomic MS data when the exact target sequence is not in the database.

These limitations can be approached by using error-tolerant searches on the amino acid sequence level. Generally, error-tolerant peptide identification can be categorized as (1) de novo approaches [28], (2) iterative search procedures [29, 30], (3) tag-based algorithms [31–33], and (4) hybrid approaches [34–37]. (1) De novo approaches perform well on high-quality data. However, low-quality spectra do often not provide enough evidence to sequence the complete peptide from a single spectrum. Iterative search procedures (2) limit the number of proteins that are considered for the error-tolerant search step. First, a standard database search is done on an error-free model. Then, only proteins that were identified in the initial search are extracted for an exhaustive search of SAAV in an error-tolerant mode. Tag-based approaches (3) manage to extract candidates even with low sequence coverage by creating short amino acid sequence tags from the spectrum. These tags are used to extract candidate peptides from the reference database. The achieved error tolerance is only limited because either the tag or the flanking sequences of the tag are allowed to represent sequence variations to a peptide in the database [38]. Finally, hybrid approaches (4), e.g., de novo sequencing and related protein sequence databases, are used to transfer the error-tolerant

MS search to a sequence alignment problem. Again, the quality of the tags is decisive for the success of the extraction of de novo tags. While numerous examples are available, here we focus on the error-tolerant approach of BICEPS [38], which integrates elements of tag-based and hybrid approaches.

Error-tolerant identification algorithms share the inherent advantage that no specialized target database is necessary that contains a list of possible SAAVs adapted from, e.g., a database of known single-nucleotide polymorphisms, such as dbSNP. Instead, an error-tolerant identification algorithm is able to identify peptides in the sample that have amino acid substitutions in their sequence compared to the reference database. However, error-tolerant search algorithms usually suffer from higher run times.

In order to detect variant peptides with MS data we introduce a workflow that combines the ideas of error-tolerant peptide identification with BICEPS, customized variant peptide databases from RNA-Seq [5, 13], and a bioinformatics workflow for variant peptide detection [45]. Instead of using RNA-Seq data, the introduced variant detection workflow uses BICEPS in an initial search to create a customized variant peptide database. In a second step a verification search on the customized database is performed with standard MS search engines. Instead of using the full list of so far discovered SAAV from online databases the target database becomes much smaller. The inherent advantage of the BICEPS implementation is the control of the search space by a Bayesian information criterion. The dynamic regulation of the search space leads to fast discontinuance of the search space extension when the optimal solution has been found [38].

2 Materials

2.1 Software Requirements

To conduct the described variant detection workflow the list of software in Table 1 is recommended. In addition working installations of Python (v.2.7, http://python.org) and R (v.3.03, www.r--project.org) are necessary. BICEPS [38] is the only software that is mandatory to follow the overall strategy in our variant detection workflow and can be run on Linux or Windows systems. All other software solutions are exchangeable. However, for ease of use and high reproducibility the use of the workflow system KNIME [46] along with the OpenMS plug-in [47, 48] is recommended. Furthermore, a collection of scripts for BICEPS (bictools) is available from sourceforge (http://sourceforge.net/projects/bictools/) that automates the individual steps of the workflow. The bictools repository also includes the KNIME workflow and further instructions to set up the analysis pipeline.

Table 1
List of software for the detection of single-amino acid variants on proteomic MS data

Software	Website	Publication	Comment
BICEPS	http://software.steenlab.org/	[38]	Error tolerant MS/MS identification
KNIME	https://www.knime.org	[46]	Workflow engine
OpenMS	https://www.openms.de	[48]	MS suite
X!Tandem	http://www.thegpm.org/tandem/	[21]	MS/MS identification

Note that in principle X!Tandem can be replaced with any other MS/MS identification search engine that is supported by OpenMS to use the workflow introduced here

2.2 Data Requirements

To successfully run the variant detection workflow further resources are required. These resources include a target database, e.g., a complete proteome from Uniprot [49]. The target database is usually available as FASTA file and contains sequence and protein information. Furthermore, MS/MS recordings are required in Mascot Generic Format (MGF) and mzML format (*see* **Note 1**). Optionally, a genomic derived variant database such as CanProVar [40] can be used to evaluate the identified variant peptides versus previously annotated SAAVs.

3 Methods

3.1 BICEPS: Algorithmic Overview

BICEPS was developed to identify peptides in cross-species experiments when it cannot be assumed that the database entries at hand for a related species necessarily correspond to the sequence of the species regarded in a proteomic experiment. To overcome differences from the database sequence available, BICEPS applies a multistep procedure. First, short sequence tags are generated by established de novo approaches to obtain sequence candidates [50, 51]; these are then error-tolerantly searched against the existing sequence database to obtain candidate sequences. Candidate sequences are then scored against spectra at hand using a Bayesian Information Criterion (BIC)-motivated score. The idea is to weight the potential increase in the quality of a peptide spectrum match by allowing a SAAV by the increase in the size of the search space the SAAV is associated with. Since allowing larger search spaces increases not only run time but also the risk of false-positive identification, this trade-off is highly crucial. With an underlying probabilistic statement BICEPS adaptively decides whether increasing the search space is justified on a spectrum basis. It has an internal early stopping criterion which checks whether a higher score is still theoretically achievable for a spectrum at hand given the probabilistic nature of the scoring function and otherwise terminates the search for this spectrum. Thereby, run time is reduced and false positives are avoided.

Besides typical peptide identification settings specifying the mass tolerance or modifications, BICEPS allows specifying the number of tags which are generated (more tags thereby imply a higher likelihood of finding the correct sequence). Potential mutations within the tag can be allowed by setting the mutation parameter to 1 (default 0 corresponds to no mutations) to further increase the robustness and the logarithmic size of the search space to be considered (penaltyvec) can be indicated. The search space increases with standard proteomic database search parameters such as variable modifications or non-tryptic ends and more substantially with SAAVs considered. BICEPS weights SAAVs by a PAM matrix to capture the likelihood of the SAAV at hand. Setting the penaltyvec value to 2.0 allows that all possible single SAAVs not coinciding with other modifications within a peptide are detectable for BICEPS. Larger values allow the detection of multiple SAAVs per peptide or combinations of modifications and single SAAVs. Starting from values of 4.0, it is ensured that two SAAVs within a peptide are detectable even for uncommon SAAVs.

3.2 Variant Detection Workflow

We introduce a variant detection workflow exclusively based on proteomic MS data. The key step in our workflow is to use BICEPS to create customized variant databases from MS data. Our BICEPS centric variant detection workflow consists of three steps (Fig. 1) and one optional step for further evaluation of the variant peptides. A practical guide is attached to the end of this section. In addition, Subheading 4 provides two case studies that demonstrate the usage of BICEPS with different parameterization effects of the proposed workflow. The individual steps can be divided as follows:

1. Initial search—the MS data is searched with BICEPS against a standard Uniprot database.

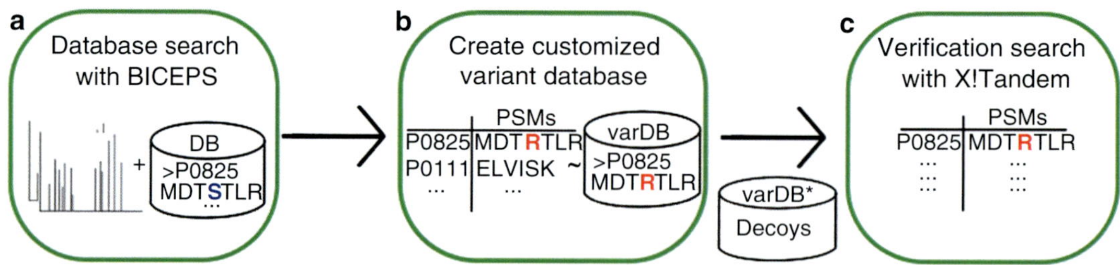

Fig. 1 Variant detection workflow using BICEPs. (a) BICEPS is used to search the spectra data set against a standard Uniprot reference database. (b) Identified peptides that show a single-amino acid variation (SAAV) are used to create a customized variation database. (c) A standard database search is performed versus the newly created database with corresponding decoy sequences and the initial input spectra

2. Database creation—BICEPS results are used to create a customized database of variant containing candidate peptides.
3. Verification search—a verification search with a standard database search engine is performed.
4. Post-processing: interpreting identified variants.

3.3 Initial Search: Finding Variant Peptide Candidates

A strength of our approach is that the initial search can be performed with any target database. The error-tolerant nature of BICEPS makes a database containing the exact sequence variants dispensable. However, the peptide scoring mechanism in BICEPS makes reliable variant detection challenging. The reason is that peptide identifications containing a substitution show in general a distinct, lower score distribution than peptides that do not contain a substitution (Fig. 2). Along with standard FDR estimations in BICEPS, e.g., 1 % FDR, a large number of identifications with substitutions would thus not pass the confidence threshold. Therefore, BICEPS is only used as a tool to generate a targeted SAAV library. Thus, higher error levels or even all peptide identifications can be used. Once the library is created and standard database tools are used the score distributions of variant peptides and normal peptides are not distinguishable anymore.

The creation of the customized database is based on the search results with BICEPS. BICEPS requires as input data spectra files in MGF format and a FASTA reference database. Natively, BICEPS relies on decoy-free estimation of false positives [52]. The BICEPS peptide identifications can either unrestrictedly added to the candidate database or according to an equivalent peptide initial FDR

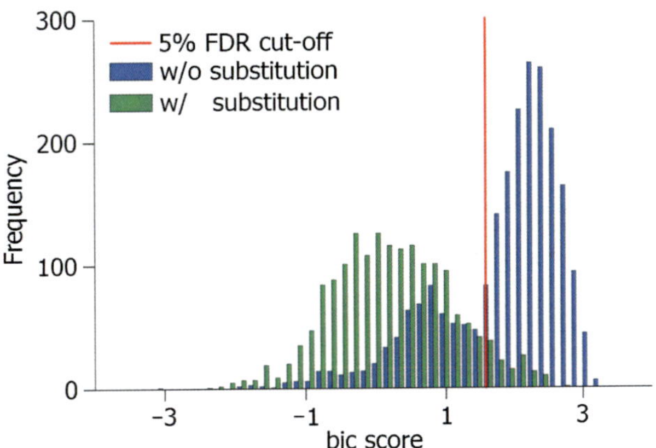

Fig. 2 Score distribution of peptides identified with BICEPS with and without substitutions, respectively. The *red line* indicates the confidence threshold at approximately 5 % false discovery rate, estimated with CurveFDP. The data shows the results from 8457 HeLa cell spectra searched with BICEPS. *See* Table 1, RunID 2, for detailed information on the search parameters

(iFDR) threshold (*see* **Notes 2** and **3**). The former creates more candidate peptides and relies on confident FDR estimation in the verification search, whereas the latter largely influences the number of candidate variants that are generated. Therefore, the parameterization of BICEPS plays an important role for the subsequent analysis. The workflow is designed such that multiple settings can easily be tested and evaluated against each other.

3.4 Database Creation: Turning Candidates into a Target Database

The database creation serves as a tool to allow the validation search of variant peptides with any standard database search program. The reported BICEPS result file contains the matched peptide sequence (identified sequence) and the original sequence (sequence in the target database) for any peptide that was identified. The variant peptide candidate database contains all sequences that show amino acid variations between the matched peptide and the original sequence during the initial search with BICEPS. However, it is possible to influence the number of candidate peptides that are written to the candidate database. For example, a minimum mass difference filter between the original sequence and the identified sequence can be set (*see* **Note 4**). The idea behind the creation of the customized database is to use standard peptide identification programs from here on. Thus, a way to control the number of false positives is needed. This is achieved via standard FDR. Therefore, in addition to the variant peptide candidates the corresponding decoy peptides are added to the database. Then a restrict FDR estimation is performed in the verification search, i.e., peptide level FDR$\leq$0.01. The final database can be searched with every standard database search. In addition, it represents a customized search space for the particular proteomic sample that was analyzed. Therefore, the search space is not unnecessarily expanded (*see* **Note 5**).

3.5 Verification Search: Validation with a Standard Search

The verification search is comparable to standard identification workflows for proteomics. The database from Subheading 3.4 is used as target database and the spectra from Subheading 3.3 are now required in mzML format. Here, we performed the verification search within the OpenMS framework [47, 48] in KNIME. The accompanying KNIME [46] workflow is available via the bictools repository. In brief, spectra files are searched versus the above-described reference database using X!Tandem.

Carbamidomethylation on cysteine was set as fixed modification and oxidation on methionine as variable modification. The precursor mass tolerance was set to 10 ppm and the fragment tolerance to 0.5 Da. The FDR was solely computed on peptide level as *q*-value. The *q*-value threshold for accepted peptide hits was set to 0.01. The KNIME workflow allows further adjustments of the search settings via the corresponding OpenMS nodes. Furthermore, the presented verification workflow can be further extended.

For instance, different or multiple search engines can be combined for a superior consensus scoring of identified peptides [53]. In addition, first studies showed that the quantitative analysis of variant peptides is promising, too [9]. The provided OpenMS framework is with a few extensions also able to perform quantitative analysis [54].

3.6 Post-processing: Interpreting Identified Variants

Important for the subsequent analysis of variant peptides is their effect in the biological system they are found in. Interesting candidates can be determined by using variant effect prediction tools such as PROVEAN. Bictools provide an output format that can be used with the PROVEAN web service [55] for human and mouse. To narrow down the number of variants of interest it is possible to select all variant peptides that are classified as damaging by SIFT [56] and PROVEAN. However, the true effect of any variant is not necessarily bound to their predictions. Therefore, additional experiments and pathway analysis are necessary to pinpoint the effects of SAAVs. Further, to investigate protein-protein interactions network databases can also be utilized to investigate specific diseases [57].

3.7 A Practical Guide to Perform the Above-Described Workflow

The examples assume that all data files are in the current working directory.

1. Preparation:
   Download all requirements that have been introduced in Subheading 2. Further, make sure that all software requirements for bictools are met.

   Critical step: BICEPS relies on decoy-free estimation of false positives. Make sure that no decoy peptides are present in the database.
2. Initial search:
   Call BICEPS via a command line and point the MGF and target database correctly as parameters:

   Example:
   ```
   >>biceps --mgf HeLa.mgf --fasta HS.fasta --tags 10 --tol 10 --mutation 0 --penaltyvector 3.0
   ```
   Critical step: Parameter settings—confer Subheadings 3.1 and 4 of the manuscript.
3. Create variant peptide database:
   Call the script from the bictools package to create customized variant peptide databases.

   Example:
   ```
   >>bictools-bic_DB --in_bic Biceps.Results.HeLa.txt --in_db HS.fasta --out_dir results/
   ```

Critical step:

Carefully investigate the optional parameters. They can be very restrictive and have a large effect on the number of SAAV candidates in your second search. For example one can set an iFDR threshold on BICEPS (--fdr) and the maximum mutations per peptide (--max_mut).

4. Verification search:
   Start KNIME and import the "SAAV_identification_workflow" workflow. Select the correct input files and choose output files for the output nodes.

   Critical step:

   The MS/MS identification in KNIME (via OpenMS) offers a variety of parameters to fine-tune the identification. Adjustments to the search engine or false discovery rate (FDR) can be performed on the respective nodes.

5. Post-processing:
   Create lists of verified SAAVs to predict the effect on the protein function. Further, filter the OpenMS results to true variant peptides.

Example:

```
>>bictools-bic_evaluate --in_bic results/
--in_oms knime_output/ --in_db HS.fasta
--CPV_tab canprovar.fasta
--map_tab mapping.csv --out_aux eval_out/
```

4 Case Study

To evaluate the usefulness of BICEPS as a variant detection tool we analyzed two different data sets with our analysis workflow. First, a publicly available HeLa data set [58, 52] is analyzed. The 8457 HeLa cell spectra were acquired on an LTQ-Orbitrap classic (Thermo Scientific). Presumably, HeLa cells contain a large number of cancerogenous mutations which makes it a valuable resource for evaluation purposes. Second, we reanalyzed data from a publicly available colorectal cancer cell line (HCT-115) [45] and compared the identified variants. The HCT-116 data was also acquired on an LTQ-Orbitrap. Raw files were converted to MGF resulting in a total of 69,865 MS2 scans. The first case study is used to demonstrate different search settings within the BICEPS identification process. The second case study is used to give insight into the false discovery setting effects. In both studies, a reviewed Uniprot database not containing isoforms[1] was used as target database.

[1] tools/loaded on 19/09/2014.

Table 2

Variant detection workflow results using BICEPS and X!Tandem on 8457 HeLa cell spectra with different parameters for BICEPS

	BICEPS parameter			Identifications			
RunID	Tags	Tolerance	Penalty	BICEPS wo SAAV	BICEPS w SAAV	X!Tandem (FDR 0.01)	CanProVar
1	10	10	2.5	2486	1219	116 (75)	2
2	10	10	3.0	2277	1704	105 (64)	1
3	5	10	2.5	2245	1245	109 (69)	2
4	5	10	3.0	2090	1706	105 (63)	1
5	15	10	2.5	2557	1149	117 (72)	3
6	15	10	3.0	2322	1652	112 (62)	2

All searches had the mutation argument of BICEPS set to 0. BICEPS results are unfiltered with regard to false discovery rate. X!Tandem results correspond to the validation step in the variant detection workflow. The numbers in brackets show the unique number of unmodified peptides. Results are filtered to 1 % false discovery rate

Abbreviations: *w SAAV* with single-amino acid variation, *wo SAAV* without single-amino acid variation. *CanProVar* refers to a database of known variants

4.1 HeLa

To create a reasonably well-suited variant database it is important to find the best parameters for BICEPs for the initial search. The search parameters can largely influence the number of identified peptides. The error tolerance, however, is mainly dependant on the instrument that was used to acquire the data. Therefore, this parameter was fixed to 10 ppm. For a selection of parameters and the corresponding search results for the HeLa data set the results are summarized in Table 2.

The penalty argument has the largest impact on the number of SAAV-containing candidates. All biceps search results with 2.5 set as penalty argument identify approximately 1200 peptides with substitutions. Searches with 3.0 as penalty vector identify approximately 1675 SAAV candidates. The number of tags has only a minor impact on the number of identifications. However, the number of peptide identifications with substitutions is only a proxy for the number of identified SAAVs. Therefore, not the SAAV candidates but the validated SAAVs on the created variant database are important to evaluate for future analysis. The number of uniquely identified peptides in the verification step with X!Tandem is comparable in all searches, with the lowest number of identified peptides being 62 and the highest number being 75. However, lower penalty vectors achieve a constantly higher number of identifications. In addition, the highest numbers of peptide spectrum matches that are in agreement with entries from CanProVar are found with a lower penalty vector. Note that the number of variant peptides that have been previously reported is very low. Only 1–3 peptides per search show known sequence variations.

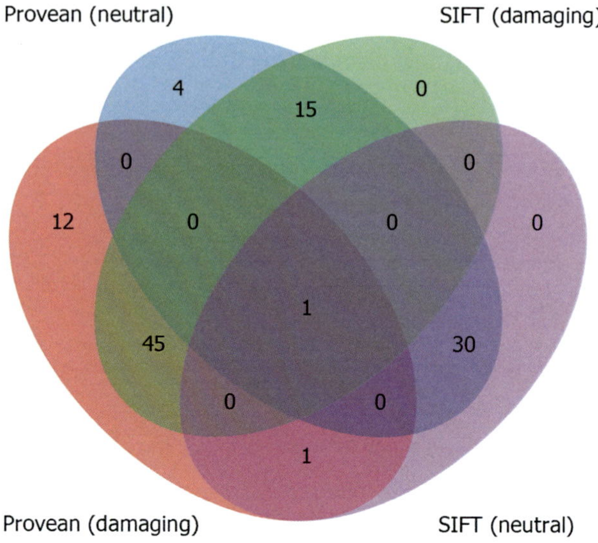

Fig. 3 Agreement of variant effect prediction on BICEPS results. The data corresponds to the search with RunID #1 from Table 1. Prediction was performed using the PROVEAN web service. Damaging corresponds to the SIFT annotation of deleterious and neutral corresponds to the PROVEAN annotation for deleterious

Further inspection of the identified variant peptides is achieved via the prediction of the effect on the protein function. Bictools provide an output format that can be used with the PROVEAN web service for human and mouse. To narrow down the number of variants of interest it is possible to select all variant peptides that are classified as damaging by SIFT [56] and PROVEAN. For instance, for the HeLa data set (run 1, Table 2) the Venn diagram in Fig. 3 summarizes the agreement of the prediction tools. Note that the number of peptides in Fig. 3 is different than in Table 2 because shared peptides are counted as one unique peptide in Table 2. For the variant effect prediction, however, the amino acid substitutions need to be placed in the context of the whole protein. Thus, a higher number of variant-containing proteins are generated than there are peptides. The highest consensus in the prediction of the variant effect is achieved in the damaging/damaging case. The total agreement between both tools (damaging/damaging or neutral/neutral predictions) is approximately 69 %. Note that the SIFT predictions are not complete because in some cases the mutations are not classified.

4.2 HCT-116

The HCT-116 cell line represents a better characterized example with more spectra and more identified SAAVs from a previous study. In the following we adjust the variant detection workflow to its more restrictive mode by setting an initial FDR (iFDR) at 20 % and 50 %, respectively, for the initial identification search with

Table 3

Variant detection workflow results using BICEPS and X!Tandem on 69,865 HCT-116 cell spectra with different parameters for BICEPS

	BICEPS parameter		Identifications			
			BICEPS	BICEPS		
Run	Tags	iFDR	wo SAAV	w SAAV	X!Tandem (FDR 0.01)	CanProVar
1	5	10	14,746	1254	258 (133)	6
2	10	10	16,104	1482	328 (162)	7
3	15	10	16,390	1561	305 (158)	7
4	5	50	16,768	5230	533 (348)	11
5	10	50	18,036	5903	607 (384)	14
6	15	50	18,158	5697	566 (362)	14

All searches had the mutation argument of BICEPS set to 0, tolerance to 10, and the same penaltyvector of 3.0. BICEPS results are filtered to two different initial false discovery rates (iFDR). X!Tandem results correspond to the validation step in the variant detection workflow. The numbers in brackets show the number of unique unmodified peptides. X!Tandem results are filtered to 1 % false discovery rate. The CanProVar column indicates the overlap to the CanProVar database

Abbreviations: *w SAAV* with single-amino acid variation, *wo SAAV* without single-amino acid variation

BICEPS. The iFDR is independent from the FDR in the verification search. Further, only three restrictive settings are used with --tol set to 10, --mutation to 0, and --penaltyvector to 3.0 and 5, 10, and 15 tags, respectively. Obviously, the iFDR threshold has a large impact on the number of candidate peptides (*see* Table 3). Subsequently, the number of identified peptides with SAAV increases with a higher iFDR.

For instance, least variant peptide candidates (1254) are found with a 10 % iFDR and five tags. In contrast, most variant peptide candidates (5903) are found with a 50 % iFDR and ten tags. In run 1 approximately 21 % of the candidate peptides are verified with X!Tandem. In run 5 the respective success rate is approximately 10 %. This indicates that the verified results with X!Tandem do not linearly increase with the number of candidate peptides, i.e., the target database. Thus, the supposed effect of arbitrarily increasing the database and getting more verified peptides is controlled by the restrictiveness in the verification search. Obviously, applying no iFDR cutoff in the initial search will allow low-scoring peptides to be included in the candidate database. However, initially low-scored peptides are expected to achieve a low score in the verification search as well. Therefore, adding more sequences in the candidate database is actually beneficial to have a robust FDR estimation in the verification step.

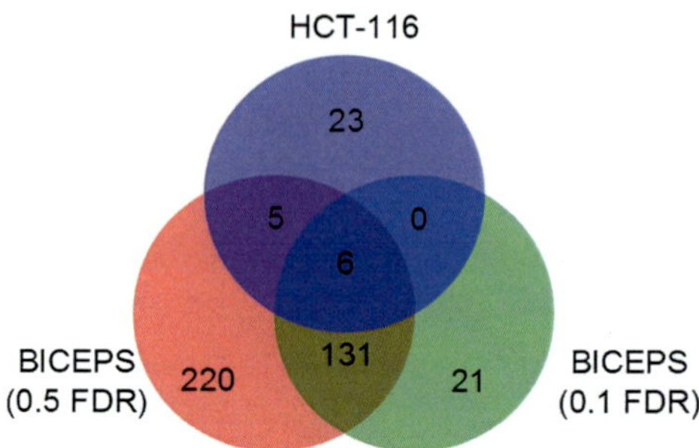

Fig. 4 Agreement of identified variant peptides with different false discovery rates (FDR) during the initial search. The data corresponds to the search with RunID #1 and #6 from Table 3. The reference, HCT-116, includes all identified variant peptides from an earlier publication [44] on the same data. The FDR for the verification search was set to 0.01 following the respective FDR searches with BICEPS

The verified variant peptides do largely agree between two runs with different iFDR thresholds (Fig. 4). 86 % of the verified peptides from run 1 are contained within the results of run 6. In addition, 11 variant peptide sequences have also been identified in the previous publication of this data set [45].

5 Notes

1. Vendor-specific formats for mass spectrometry data are not supported by OpenMS or BICEPS. Therefore, the raw files need to be converted to mzML and MGF format; *see*, e.g., MSConvert from the ProteoWizard suite [59].
2. The lack of a proper ground truth, i.e., sequencing of all identified variant peptides, is challenging for the large-scale analysis of the variants. To increase the confidence in the variant peptide identifications a time-consuming workflow needs to be performed with two obstacles: First, the initial search with BICEPS needs to identify a variant peptide at a user-defined error level. Second, the verification search engine needs to identify the same peptide in the customized database at a low-peptide-level FDR. These two steps and the error-tolerant nature of BICEPS increase the processing time of the data compared to approaches with proteogenomics databases [45]. However, only using BICEPS and its built-in FDR would result in a decreased number of identified variant peptides. Unless, the FDR estimation is adapted for the lower scoring variant peptides; for example by regarding local variant FDRs [45], a numerous variant candidates will not pass the thresh-

old. In the proposed variant detection workflow we focus on a restrictive FDR in the verification step. This allows relaxing identification conditions for the initial identification search with BICEPS.

3. Often proteins are identified by multiple peptides in a shotgun experiment. These peptides are either unique to the protein or shared by multiple proteins due to sequence homologies. In the context of this chapter we do not focus on the protein inference [60], i.e., compute protein groups that share peptides and resolve them. Rather all potential protein matches of shared peptides are used to analyze the SAAV. Thus, only the peptide-level FDR is computed and useful in this situation.

4. Increased sensitivity in the detection of SAAVs can be expected with high-resolution mass spectrometry in the fragment spectra [61]. However, substitutions such as glutamine to lysine are difficult to detect even with high-resolution instruments because the mass difference is only 0.036 Da. Another example is the mutation from serine to alanine. Though the base difference of serine to alanine is clearly larger than the machine tolerance, i.e., 15.9994 Da, it coincides with the mass of an oxidation. Therefore, a serine-to-alanine mutation needs to be evaluated with caution when an oxidation is possible in the identified peptide. These drawbacks arise with any proteomic database search tool that tries to identify SAAVs of amino acids with similar masses. High-resolution data and well-fragmented peptides can help to decide where exactly the variant (or modification) is located in the peptide, e.g., to unambiguously distinguish between modifications and amino acid substitutions.

5. Typically, either genomic variant databases or transcript data is used to construct target databases for MS-based variant detection. Since the FDR is largely influenced by the search space, i.e., the database size and allowed modifications [62], it is crucial to limit the database size to the true number of possible variant peptides. Obviously, this number remains unknown even after peptide identification. However, utilizing RNA-Seq data can partly overcome this challenge and utilize personalized transcript data for database construction.

Acknowledgment

The authors gratefully acknowledge financial support by Deutsche Forschungsgemeinschaft (DFG), grant number (RE3474/2-1 to BYR).

References

1. Yates JR, Ruse CI, Nakorchevsky A (2009) Proteomics by mass spectrometry: approaches, advances, and applications. Annu Rev Biomed Eng 11:49–79
2. Aebersold R, Mann M (2003) Mass spectrometry-based proteomics. Nature 422:198–207
3. Mann M, Ong S-E, Grønborg M et al (2002) Analysis of protein phosphorylation using mass spectrometry: deciphering the phosphoproteome. Trends Biotechnol 20:261–268
4. Ozlu N, Akten B, Timm W et al (2010) Phosphoproteomics. Wiley Interdiscip Rev Syst Biol Med 2:255–276
5. Sheynkman GM, Shortreed MR, Frey BL et al (2014) Large-scale mass spectrometric detection of variant peptides resulting from nonsynonymous nucleotide differences. J Proteome Res 13:228–240
6. Mayne SLN, Patterton H-G (2011) Bioinformatics tools for the structural elucidation of multi-subunit protein complexes by mass spectrometric analysis of protein-protein cross-links. Brief Bioinform 12:660–671
7. Bantscheff M, Schirle M, Sweetman G et al (2007) Quantitative mass spectrometry in proteomics: a critical review. Anal Bioanal Chem 389:1017–1031
8. Su Z-D, Sun L, Yu D-X et al (2011) Quantitative detection of single amino acid polymorphisms by targeted proteomics. J Mol Cell Biol 3:309–315
9. Song C, Wang F, Cheng K et al (2014) Large-scale quantification of single amino-acid variations by a variation-associated database search strategy. J Proteome Res 13:241–248
10. Nesvizhskii AI (2010) A survey of computational methods and error rate estimation procedures for peptide and protein identification in shotgun proteomics. J Proteomics 73:2092–2123
11. Nesvizhskii AI, Vitek O, Aebersold R (2007) Analysis and validation of proteomic data generated by tandem mass spectrometry. Nat Methods 4:787–797
12. Ansong C, Purvine SO, Adkins JN et al (2008) Proteogenomics: needs and roles to be filled by proteomics in genome annotation. Brief Funct Genomic Proteomic 7:50–62
13. Woo S, Cha SW, Merrihew G et al (2014) Proteogenomic database construction driven from large scale RNA-seq data. J Proteome Res 13:21–28
14. Altshuler D, Daly MJ, Lander ES (2008) Genetic mapping in human disease. Science 322:881–888
15. Sachidanandam R, Weissman D, Schmidt SC et al (2001) A map of human genome sequence variation containing 1.42 million single nucleotide polymorphisms. Nature 409:928–933
16. Brogna S, Wen J (2009) Nonsense-mediated mRNA decay (NMD) mechanisms. Nat Struct Mol Biol 16:107–113
17. McGlincy NJ, Tan L-Y, Paul N et al (2010) Expression proteomics of UPF1 knockdown in HeLa cells reveals autoregulation of hnRNP A2/B1 mediated by alternative splicing resulting in nonsense-mediated mRNA decay. BMC Genomics 11:565
18. Perkins DN, Pappin DJ, Creasy DM et al (1999) Probability-based protein identification by searching sequence databases using mass spectrometry data. Electrophoresis 20:3551–3567
19. Geer LY, Markey SP, Kowalak J et al (2004) Open mass spectrometry search algorithm. J Proteome Res 3:958–964
20. Tabb DL, Fernando CG, Chambers MC (2007) MyriMatch: highly accurate tandem mass spectral peptide identification by multivariate hypergeometric analysis. J Proteome Res 6:654–661
21. Craig R, Beavis RC (2004) TANDEM: matching proteins with tandem mass spectra. Bioinformatics 20:1466–1467
22. Tanner S, Shu H, Frank A et al (2005) InsPecT: identification of posttranslationally modified peptides from tandem mass spectra. Anal Chem 77:4626–4639
23. Eng JK, McCormack AL, Yates JRIII (1994) An approach to correlate tandem mass spectral data of peptides with amino acid sequences in a protein database. J Am Soc Mass Spectrom 5:976–989
24. Choi H, Nesvizhskii AI (2008) False discovery rates and related statistical concepts in mass spectrometry-based proteomics. J Proteome Res 7:47–50
25. Elias JE, Gygi SP (2007) Target-decoy search strategy for increased confidence in large-scale protein identifications by mass spectrometry. Nat Methods 4:207–214
26. Yates JRIII, Eng JK, McCormack AL et al (1995) Method to correlate tandem mass spectra of modified peptides to amino acid sequences in the protein database. Anal Chem 67:1426–1436

27. Evans VC, Barker G, Heesom KJ et al (2012) De novo derivation of proteomes from transcriptomes for transcript and protein identification. Nat Methods 9:1207–1211
28. Hughes C, Ma B, Lajoie GA (2010) De novo sequencing methods in proteomics. Methods Mol Biol 604:105–121
29. Creasy DM, Cottrell JS (2002) Error tolerant searching of uninterpreted tandem mass spectrometry data. Proteomics 2:1426–1434
30. Starkweather R, Barnes CS, Wyckoff GJ et al (2007) Virtual polymorphism: finding divergent peptide matches in mass spectrometry data. Anal Chem 79:5030–5039
31. Mann M, Wilm M (1994) Error-tolerant identification of peptides in sequence databases by peptide sequence tags. Anal Chem 66:4390–4399
32. Tabb DL, Saraf A, Yates JR (2003) GutenTag: high-throughput sequence tagging via an empirically derived fragmentation model. Anal Chem 75:6415–6421
33. Shilov IV, Seymour SL, Patel AA et al (2007) The Paragon algorithm, a next generation search engine that uses sequence temperature values and feature probabilities to identify peptides from tandem mass spectra. Mol Cell Proteomics 6:1638–1655
34. DiMaggio P, Floudas C, Lu B et al (2008) A hybrid method for peptide identification using integer linear optimization, local database search, and quadrupole time-of-flight or OrbiTrap tandem mass spectrometry. J Proteome Res 7:1584–1593
35. Han Y, Ma B, Zhang K (2004) SPIDER: software for protein identification from sequence tags with de novo sequencing error. Proc IEEE Comput Syst Bioinform Conf, pp 206–215
36. Searle BO, Dasari S, Turner M et al (2004) High-throughput identification of proteins and unanticipated sequence modifications using a mass-based alignment algorithm for MS/MS de novo sequencing results. Anal Chem 76:2220–2230
37. Wang X, Li Y, Wu Z et al (2014) JUMP: a tag-based database search tool for peptide identification with high sensitivity and accuracy. Mol Cell Proteomics 13:3663–3673
38. Renard BY, Xu B, Kirchner M et al (2012) Overcoming species boundaries in peptide identification with Bayesian information criterion-driven error-tolerant peptide search (BICEPS). Mol Cell Proteomics 11:M111.014167
39. Sherry ST, Ward MH, Kholodov M et al (2001) dbSNP: the NCBI database of genetic variation. Nucleic Acids Res 29:308–311
40. Li J, Duncan DT, Zhang B (2010) CanProVar: a human cancer proteome variation database. Hum Mutat 31:219–228
41. Wang Z, Gerstein M, Snyder M (2009) RNA-Seq: a revolutionary tool for transcriptomics. Nat Rev Genet 10:57–63
42. Wang X, Slebos RJC, Wang D et al (2012) Protein identification using customized protein sequence databases derived from RNA-Seq data. J Proteome Res 11:1009–1017
43. Wang X, Zhang B (2013) customProDB: an R package to generate customized protein databases from RNA-Seq data for proteomics search. Bioinformatics 29:3235–3237
44. DePristo M, Banks E, Poplin R et al (2011) A framework for variation discovery and genotyping using next-generation DNA sequencing data. Nat Genet 43:491–498
45. Li J, Su Z, Ma Z-Q et al (2011) A bioinformatics workflow for variant peptide detection in shotgun proteomics. Mol Cell Proteomics 10:M110.006536
46. Berthold MR, Cebron N, Dill F et al (2007) KNIME: the Konstanz Information Miner. Stud Classif Data Anal Knowl Organ (GfKL 2007)
47. Kohlbacher O, Reinert K, Gröpl C et al (2007) TOPP—the OpenMS proteomics pipeline. Bioinformatics 23:e191–e197
48. Sturm M, Bertsch A, Gröpl C et al (2008) OpenMS—an open-source software framework for mass spectrometry. BMC Bioinformatics 9:163
49. The UniProt Consortium (2014) Activities at the Universal Protein Resource (UniProt). Nucleic Acids Res 42:D191–D198
50. Frank A, Pevzner P (2005) PepNovo: de novo peptide sequencing via probabilistic network modeling. Anal Chem 77:964–973
51. Tabb DL, Ze-Qiang M, Martin DB et al (2008) DirecTag: accurate sequence tags from peptide MS/MS through statistical scoring. J Proteome Res 7:3838–3846
52. Renard BY, Timm W, Kirchner M et al (2010) Estimating the confidence of peptide identifications without decoy databases. Anal Chem 82:4314–4318
53. Nahnsen S, Bertsch A, Rahnenführer J et al (2011) Probabilistic consensus scoring improves tandem mass spectrometry peptide identification. J Proteome Res 10:3332–3343
54. Weisser H, Nahnsen S, Grossmann J et al (2013) An automated pipeline for high-throughput label-free quantitative proteomics. J Proteome Res 12:1628–1644

55. Choi Y, Sims GE, Murphy S et al (2012) Predicting the functional effect of amino acid substitutions and indels. PLoS One 7, e46688
56. Kumar P, Henikoff S, Ng PC (2009) Predicting the effects of coding non-synonymous variants on protein function using the SIFT algorithm. Nat Protoc 4:1073–1081
57. Franceschini A, Szklarczyk D, Frankild S et al (2013) STRING v9.1: protein-protein interaction networks, with increased coverage and integration. Nucleic Acids Res 41:D808–D815
58. Renard BY, Kirchner M, Monigatti F et al (2009) When less can yield more—computational preprocessing of MS/MS spectra for peptide identification. Proteomics 9:4978–4984
59. Kessner D, Chambers M, Burke R et al (2008) ProteoWizard: open source software for rapid proteomics tools development. Bioinformatics 24:2534–2536
60. Huang T, Wang J, Yu W et al (2012) Protein inference: a review. Brief Bioinform 13:586–614
61. Mann M, Kelleher NL (2008) Precision proteomics: the case for high resolution and high mass accuracy. Proc Natl Acad Sci U S A 105:18132–18138
62. Qian WJ, Liu T, Monroe ME et al (2005) Probability-based evaluation of peptide and protein identifications from tandem mass spectrometry and SEQUEST analysis: the human proteome. J Proteome Res 4:53–62

Chapter 17

Data Analysis Strategies for Protein Modification Identification

Yan Fu

Abstract

Mass spectrometry-based proteomics provides a powerful tool for large-scale analysis of protein modifications. Statistical and computational analysis of mass spectrometry data is a key step in protein modification identification. This chapter presents common and advanced data analysis strategies for modification identification, including variable modification search, unrestrictive approaches for modification discovery, false discovery rate estimation and control methods, and tools for modification site localization.

Key words Protein modifications, Variable modification search, Modification discovery, False discovery rate, Modification localization

1 Introduction

Most proteins in cells carry some posttranslational modifications, which often significantly alter the structures and functions of proteins and play key roles in biological and pathological processes [1]. The information of protein modifications is not directly encoded in DNA sequences and thus has to be studied at the protein level. In recent years, the technology of tandem mass spectrometry has rapidly developed and become the method of choice for large-scale analysis of proteins and the modifications on them [2, 3].

In shotgun proteomics, proteins are digested into peptides and the latter are analyzed by tandem mass spectrometry. Identification of peptides is traditionally accomplished by searching the tandem mass spectra against a database of protein sequences [4], and the peptides carrying modifications are identified by the so-called variable modification search approach [5]. However, as the research focus of the field moves from protein sequences to protein modifications, the variable modification search approach shows insuperable deficiencies. A fatal drawback of this approach is that it is restricted to modification types specified by the user only and cannot disclose unanticipated or novel modifications. Therefore, various

Klaus Jung (ed.), *Statistical Analysis in Proteomics*, Methods in Molecular Biology, vol. 1362,
DOI 10.1007/978-1-4939-3106-4_17, © Springer Science+Business Media New York 2016

unrestrictive approaches for modification identification have been proposed. These approaches are mostly based on open database search or open clustering of mass spectra, e.g., [6, 7].

Quality control of identified modifications is a very important step following database search. At present, the common way to estimate the false discovery rate (FDR) of peptide identifications is the target-decoy search strategy [8]. In current proteomic data analyses, the modified and unmodified peptides are usually identified together by a database search and are filtered using the same score threshold. In studies of protein modifications, it is often the case that a global FDR of all peptide identifications is estimated while only a subgroup of identifications with specific modifications are focused on. However, an important fact is that the subgroup FDR of modified peptide identifications is not equivalent to the global FDR. Therefore, one must be very careful when using the target-decoy strategy for validating modification identifications.

The modification sites assigned by the peptide identification software may not be correct even if the peptide sequences have been correctly identified. This is due to multiple reasons. For example, the fragmentation of peptides is often incomplete, leading to information loss in mass spectra. Even for variable modification search, in which the amino acid specificities of modifications are known, the modifications can be misplaced when there are multiple candidate modification sites on a peptide. The problem becomes even more serious for unrestrictive modification identification, in which the identities of discovered modifications are unknown and all composing amino acids of peptides are potential modification sites. Therefore, the modification sites had better be re-localized after peptide identification [9, 10].

In the following part of this chapter, the variable modification search is first introduced in Subheading 2. Then, several unrestrictive approaches for modification identification are described in Subheading 3. After that, Subheading 4 presents the methods for FDR control for modification identification. Finally, the modification site localization problem is discussed in Subheading 5.

2 Variable Modification Search

Variable modification search mimics the dynamic nature of protein modifications in cells. Given a list of modification types, each potential modification site is tested for whether it is modified or not. When multiple modification sites exist on a peptide, all possible forms of modification combinations are generated for matching with the input spectrum. All database search engines for peptide identification, e.g., SEQUEST [4], Mascot [11], X!Tandem [12], pFind [13–15], Andromeda [16], and PEAKS DB [17], support the variable modification search mode. During the database search process, candidate peptides are fragmented theoretically and are compared with the input experimental mass spectrum if their calculated precursor masses match the experimental value within a given tolerance, e.g., several Daltons (Da) or parts per million (ppm).

2.1 Refinement Search

The search space of candidate peptides would be geometrically expanded if a large number of variable modification types are specified for search. In such circumstances, only small databases can be efficiently searched. One way to reduce the database size is the two-round search strategy, in which a coarse search with few modification types specified is first performed to identify the presented proteins in the sample, and then a refinement search with many modification types considered is performed against the small database composed of identified proteins. Some search engines, such as X!TANDEM [18] or Mascot [19], support automated refinement search. In Mascot, it is called the error-tolerant search mode.

2.2 Multiple Searches

Another potential problem caused by searching many types of modifications simultaneously is the decreased sensitivity of peptide identification. As illustrated by Fig. 1, as the number of specified variable modification types increases, the search time goes up exponentially, and meanwhile the number of peptide identifications decreases as more and more competing (modified) candidate peptides are generated in the search space. If the variable modifications specified are relevant, that is, they do exist in the protein sample, then the number of modified peptide identifications will increase at first. However, as more and more irrelevant modification types are included, the number of both modified and unmodified peptide identifications will decrease. If one indeed has to search many modification types, performing multiple searches is a suggested

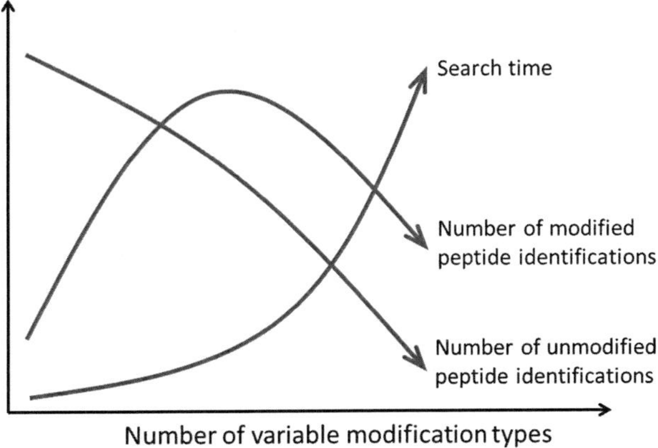

Fig. 1 Search time and sensitivity of peptide identification versus the number of specified variable modification types

choice to effectively relieve the sensitivity problem. In multiple searches, only a subset of the specified modification types is considered in each individual search, and the search results from all searches are combined finally. In this way, the number of identified spectra is expected to be maximized. Of course, the error rate of peptide identifications must be properly controlled for each search. Error rate control for modification identification is another topic and will be discussed later. In general, multiple searches do not need more search time in total than a single search that considers all modification types simultaneously.

3 Discovery of Unanticipated and Unknown Modifications

Variable modification search is also called the restrictive approach to modification identification, because it can only identify modification types that are anticipated to be present in the sample. However, there are a huge number of known modification types. For example, the Unimod database contains more than 1000 entries of modifications [20]. Moreover, many unknown modifications apparently leave to be discovered. On the other hand, the presence of unconsidered protein modifications is a major reason for the low spectral identification rate in proteomics [6, 21, 22]. A study of human proteome samples estimated that there are 8–12 modified versions for each unmodified tryptic peptide [23]. In order to identify the peptides with unanticipated or unknown modifications and increase the spectral identification rate, unrestrictive approaches to modification identification have been proposed.

3.1 Open Sequence Database Search

The first class of unrestrictive approaches is the open sequence database search, in which the precursor mass tolerance window is expanded from the normal width (e.g., several Da or ppm) to hundreds of Da, and hypothetical modifications (masses) are inferred to account for the offset of precursor masses [6, 22, 24–31]. As this approach greatly increases the number of candidate peptides, it is applicable to small databases or datasets only. Software tools for open sequence database search include MS-Alignment [6, 24, 30], Protein Prospector [22, 25], P-Mod [26], Interrogator [27], TwinPeaks [28], SeMoP [29], and PTMap [31]. To identify multiply modified peptides in an unrestricted manner will greatly reduce the search speed and identification accuracy. Therefore, it is recommended that only one unanticipated modification per peptide is allowed.

3.2 Open Spectral Clustering or Library Search

Another class of unrestricted approaches to modification identification is based on the simultaneous existence of modified and unmodified forms of the same peptide in protein samples [7, 21, 32–38]. Unlike the variable modification search or open sequence database search, in which the spectra are used independently of each other to identify modified peptides, this class of approaches uses information from related spectra to infer modifications. Suppose that two spectra were produced from the modified and unmodified forms of a peptide, respectively. The amino acid sequence of the peptide can be first identified from the spectra of the unmodified peptide, and then the modification on the modified peptide can be easily inferred. The spectral-pair-based algorithms include spectral networks [32, 33], spectral clustering [34], and library search [7, 35, 37, 38]. In the spectral library search approach, spectra are searched against a reference spectral library constructed from previously identified spectra, and the search can be carried out in the open search mode to identify peptides with unanticipated modifications. pMatch is such a tool for open spectral library search and can be used in combination with the conventional database search approach [7].

3.3 Precursor Clustering

The above approaches are all based on large-scale comparison of fragment spectra, and thus are time consuming and sensitive to the spectrum quality. DeltAMT implemented in the pCluster software is an algorithm that clusters the peptide precursors according to their mass and retention time information and can detect abundant modifications in a very fast speed [36]. DeltAMT is based on the fact that the modified and unmodified forms of a peptide often exist simultaneously in a sample and an abundantly present modification will lead to many spectral pairs with consistent precursor mass shifts and retention time shifts. By clustering the precursors based on their mass and retention time shifts using a mixture distribution model, abundant modifications can be discovered, and the modification-related spectral pairs can be detected at a given confidence level. Peptides with detected modifications can be identified in two ways. One is to incorporate the detected modifications into a variable modification search, and the other is to propagate the peptides identified by database search between paired spectra.

3.4 De Novo-Based Approach

Another approach for modification discovery is based on de novo peptide sequencing [39–47]. In this approach, the amino acid sequence of a possibly modified peptide is first located in the database from a sequence tag that is constructed by de novo sequencing, and then the potential modification is inferred according to the remaining mass difference between the full peptide sequence and the observed spectrum. Software tools of this approach are, for example, OpenSea [39], SPIDER [40], UStag [41], and MODa [47].

3.5 Joint Use of Multiple Tools for Modification Discovery

The above-described approaches for modification discovery are based on different assumptions and each of them has its own advantages and disadvantages. Low sensitivity is often a problem faced by all of the algorithms. Applying any individual algorithm to a data set is expected to disclose only a fraction of modifications

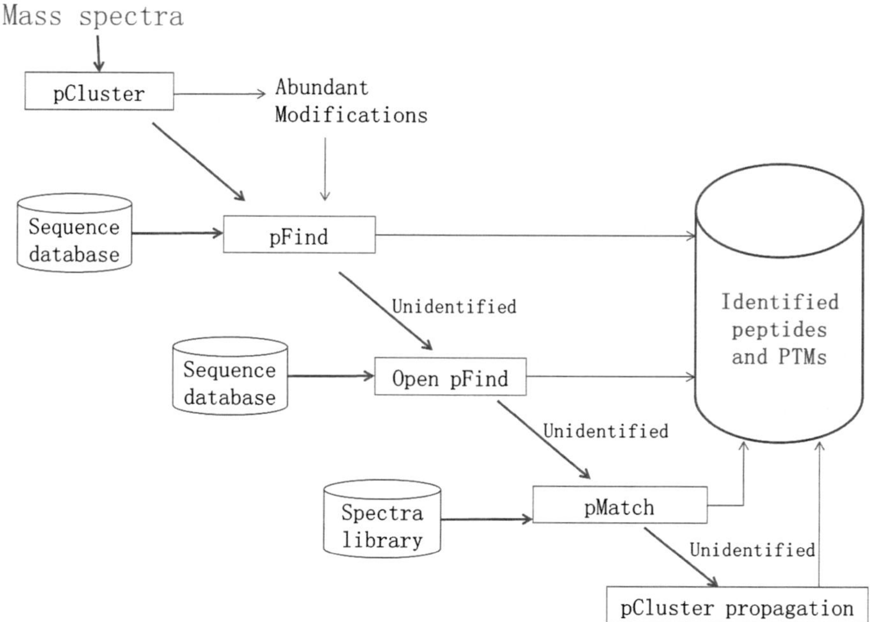

Fig. 2 A data analysis workflow for in-depth modification discovery

hidden in the data. A better way is to use them jointly. Presented in Fig. 2 is a data analysis workflow based on several software tools for in-depth discovery of protein modifications. The steps of this workflow include fast detection of abundant modifications using pCluster [36], variable modification search using pFind [15], open sequence database search using pFind (in open search mode), open spectral library search using pMatch [7], and peptide propagation using pCluster. These tools are available on the website of pFind studio (http://pfind.ict.ac.cn).

Note that this workflow can be flexibly configured in practice according to the preference of users and the features of their data. For example, these tools can be used in different order, or one or more of them can be skipped for simplicity and efficiency.

4 False Discovery Rate Control of Modification Identifications

4.1 Target-Decoy Database Search

The standard method to control the FDR of peptide identifications is the so-called target-decoy database search [8]. In this method, the mass spectra are searched against a database composed of the target protein sequences and the same number of decoy protein sequences (e.g., reverse sequences of the target proteins), as illustrated by Fig. 3. The FDR of peptide identification above a score threshold is simply estimated as the number of decoy matches divided by the number of target matches. This is based on the

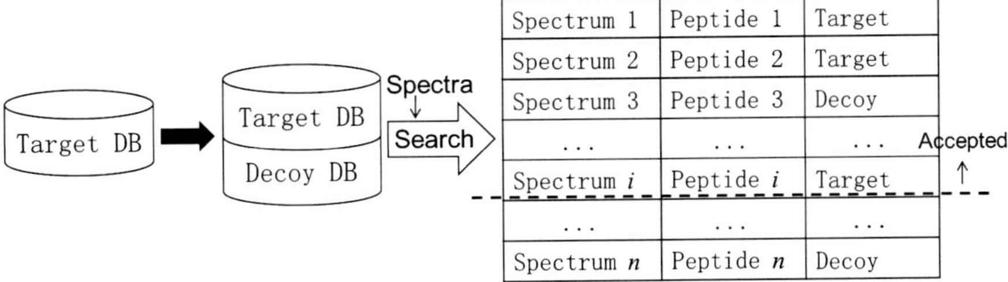

Fig. 3 Target-decoy database search for FDR estimation. The FDR is commonly estimated by N_d/N_t, where N_d and N_t are the numbers of target and decoy identifications, respectively, among the accepted identifications

assumption that an incorrect identification has an equal chance of being a match to the target sequences or to the decoy sequences. Though the target-decoy database search method has been very successful in practice, it can be misused when an experiment is focused on identifying peptides in specific forms, e.g., modified peptides [48].

4.2 Separate FDR Control Strategy

Because of multiple reasons, the subgroup FDR of modified peptides may be very different from the global FDR of all (modified and unmodified) peptides at the same score threshold. It can be theoretically shown that these two types of FDRs are not equivalent to each other (*see* **Note 1**). However, in the past it was often the case that only the identifications with specific modifications were emphasized or reported while a global FDR was estimated, leading to the risk of seriously under- or over-estimated FDR for reported results. To properly estimate the FDR of modified peptides of interest, the identifications of modified peptides should be extracted out from the identification result and subjected to separate FDR estimation, as pointed out recently [48–50]. The separate FDR can be estimated based on the target-decoy search in the same way as the global FDR except that all unwanted identifications of unmodified peptides are excluded from the data. pBuild is a tool that allows users to specify the modification types of interest and performs separate filtering and FDR estimation for modification identifications [15].

4.3 Transferred FDR for Rare Modifications

The target-decoy method for FDR estimation is effective only when the data set is large. When the number of evaluated peptide identifications is very small, the variance of the estimates made by this method becomes large [51]. For proteomic experiments, the whole data set is in general sufficiently large, but the modifications of interest may be rare in the data set. The rareness of a type of modification makes the separate FDR estimation for it inaccurate if the target-decoy strategy is applied directly. Recently, a method named transferred FDR was proposed to solve this problem [52]. The transferred FDR is also based on the target-decoy search strategy, but it does not directly use the number of decoy identifications of modified peptides to estimate the number of false target identifications of modified peptides. It infers the subgroup FDR of modified peptides from the global FDR based on the quantified relationship between the two FDRs (*see* **Note 2**).

5 Modification Localization

Many algorithms have been developed to re-localize the modification sites after peptide identification, and most of them were specifically designed for phosphorylation [53–58]. They can be basically categorized into two classes. The first class of algorithms are based on the probability of random peak matches, including AScore [53], PTMScore [54], PhosphoScan [59], SLoMo [55], phosphoRS [60], and LuciPHOr [58]. The second class of algorithms are based on the difference of peptide identification scores with different modification sites, including MD-Score [56], SLIP [57], and D-Score [61]. For the case of unrestrictive modification identification, few work has been done for modification re-localization. PTMFinder uses information of multiple overlapping peptides containing a modification site to assess the accuracy of this site [62]. Modification localization is not a satisfactorily resolved problem, especially for more generic modification types than phosphorylation and for unrestrictive modification identification.

6 Notes

1. Under trivial assumptions, the following relationship exists between the subgroup FDR of modified peptides and the global FDR of all peptides:

$$\mathrm{FDR}_k(x) = \frac{\mathrm{FDR}(x)}{\mathrm{FDR}(x) + \frac{\lambda_k(x)}{\gamma_k(x)}\left(1 - \mathrm{FDR}(x)\right)} \tag{1}$$

   where FDR(x) represents the global FDR of all identified peptides at the score threshold x, FDRk(x) represents the subgroup FDR of identified peptides with modification k, $\lambda k(x)$ is the probability that a spectrum is identified as a k -modified peptide given that the identification is true and the score is greater than x, and $\gamma k(x)$ is the probability that a spectrum is identified as a *k-modified* peptide given that the identification is false and the score is greater than x. From Eq. 1, it can be easily seen that FDRk(x) equals FDR(x) if and only if $\lambda k(x)$ equals $\gamma k(x)$, which, however, rarely happens in reality [52].

2. The idea of transferred FDR is that if we can estimate the values of $\lambda k(x)$ and $\gamma k(x)$ in Eq. 1, then FDR$k(x)$ can be derived from FDR(x) instead of being estimated from inadequate data (note that it is assumed that FDR(x) can always be accurately estimated). In fact, we have the following estimate for FDR$k(x)$:

$$\widehat{\mathrm{FDR}_k(x)} = \frac{N(x)}{N_k(x)}(ax+b)\mathrm{FDR}(x) \tag{2}$$

where $N(x)$ is the number of target identifications with scores greater than x, and $Nk(x)$ is the number of target identifications of modification k with scores greater than x, and the linear function $ax+b$ is an approximation of $\gamma k(x)$ in Eq. 1. The values of coefficients a and b can be estimated from decoy identifications as follows. Given the results of a database search and an arbitrary value of x, calculate the proportion of modified peptides among decoy identifications with scores above x. The value of x and the calculated proportion constitute a training sample of the linear function. Changing the value of x generates a training data set, from which values of the two coefficients a and b can be estimated using least squares regression [52].

References

1. Walsh CT (2005) Posttranslational modification of proteins: expanding nature's inventory. Roberts & Company Publisher, Englewood, CO
2. Mann M, Jensen ON (2003) Proteomic analysis of post-translational modifications. Nat Biotechnol 21:255–261
3. Witze ES, Old WM, Resing KA et al (2007) Mapping protein post-translational modifications with mass spectrometry. Nat Methods 4:798–806
4. Eng JK, McCormack AL, Yates JR III (1994) An approach to correlate tandem mass spectral data of peptides with amino acid sequences in a protein database. J Am Soc Mass Spectrom 5:976–989
5. Yates JR III, Eng JK, McCormack AL et al (1995) Method to correlate tandem mass spectra of modified peptides to amino acid sequences in the protein database. Anal Chem 67:1426–1436
6. Tsur D, Tanner S, Zandi E et al (2005) Identification of post-translational modifications by blind search of mass spectra. Nat Biotechnol 23:1562–1567
7. Ye D, Fu Y, Sun RX et al (2010) Open MS/MS spectral library search to identify unanticipated post-translational modifications and increase spectral identification rate. Bioinformatics 26: 399–406
8. Elias JE, Gygi SP (2007) Target-decoy search strategy for increased confidence in large-scale protein identifications by mass spectrometry. Nat Methods 4:207–214
9. Chalkley RJ, Clauser KR (2012) Modification site localization scoring: strategies and performance. Mol Cell Proteomics 11:3–14
10. Na S, Paek E (2015) Software eyes for protein post-translational modifications. Mass Spectrom Rev 34:133–147
11. Perkins DN, Pappin DJ, Creasy DM et al (1999) Probability-based protein identification by searching sequence databases using mass spectrometry data. Electrophoresis 20: 3551–3567
12. Craig R, Beavis RC (2004) TANDEM: matching proteins with tandem mass spectra. Bioinformatics 20:1466–1467
13. Fu Y, Yang Q, Sun R et al (2004) Exploiting the kernel trick to correlate fragment ions for peptide identification via tandem mass spectrometry. Bioinformatics 20:1948–1954
14. Li D, Fu Y, Sun R et al (2005) pFind: a novel database-searching software system for automated peptide and protein identification via

tandem mass spectrometry. Bioinformatics 21:3049–3050

15. Wang LH, Li DQ, Fu Y et al (2007) pFind 2.0: a software package for peptide and protein identification via tandem mass spectrometry. Rapid Commun Mass Spectrom 21:2985–2991
16. Cox J, Neuhauser N, Michalski A et al (2011) Andromeda: a peptide search engine integrated into the MaxQuant environment. J Proteome Res 10:1794–1805
17. Zhang J, Xin L, Shan B et al (2012) PEAKS DB: de novo sequencing assisted database search for sensitive and accurate peptide identification. Mol Cell Proteomics 11:M111.010587
18. Craig R, Beavis RC (2003) A method for reducing the time required to match protein sequences with tandem mass spectra. Rapid Commun Mass Spectrom 17:2310–2316
19. Creasy DM, Cottrell JS (2002) Error tolerant searching of uninterpreted tandem mass spectrometry data. Proteomics 2:1426–1434
20. Creasy DM, Cottrell JS (2004) Unimod: protein modifications for mass spectrometry. Proteomics 4:1534–1536
21. Savitski MM, Nielsen ML, Zubarev RA (2006) ModifiComb, a new proteomic tool for mapping substoichiometric post-translational modifications, finding novel types of modifications, and fingerprinting complex protein mixtures. Mol Cell Proteomics 5:935–948
22. Chalkley RJ, Baker PR, Medzihradszky KF et al (2008) In-depth analysis of tandem mass spectrometry data from disparate instrument types. Mol Cell Proteomics 7:2386–2398
23. Nielsen ML, Savitski MM, Zubarev RA (2006) Extent of modifications in human proteome samples and their effect on dynamic range of analysis in shotgun proteomics. Mol Cell Proteomics 5:2384–2391
24. Pevzner PA, Dancik V, Tang CL (2000) Mutation-tolerant protein identification by mass-spectrometry. J Comput Biol 7:777–787
25. Chalkley RJ, Baker PR, Huang L et al (2005) Comprehensive analysis of a multidimensional liquid chromatography mass spectrometry dataset acquired on a quadrupole selecting, quadrupole collision cell, time-of-flight mass spectrometer: II. New developments in Protein Prospector allow for reliable and comprehensive automatic analysis of large datasets. Mol Cell Proteomics 4:1194–1204
26. Hansen BT, Davey SW, Ham AJ (2005) P-Mod: an algorithm and software to map modifications to peptide sequences using tandem MS data. J Proteome Res 4:358–368
27. Tang WH, Halpern BR, Shilov IV et al (2005) Discovering known and unanticipated protein modifications using MS/MS database searching. Anal Chem 77:3931–3946
28. Havilio M, Wool A (2007) Large-scale unrestricted identification of post-translation modifications using tandem mass spectrometry. Anal Chem 79:1362–1368
29. Baumgartner C, Rejtar T, Kullolli M et al (2008) SeMoP: a new computational strategy for the unrestricted search for modified peptides using LC-MS/MS data. J Proteome Res 7:4199–4208
30. Tanner S, Pevzner PA, Bafna V (2006) Unrestrictive identification of post-translational modifications through peptide mass spectrometry. Nat Protoc 1:67–72
31. Chen Y, Chen W, Cobb MH et al (2009) PTMap—a sequence alignment software for unrestricted, accurate, and full-spectrum identification of post-translational modification sites. Proc Natl Acad Sci U S A 106:761–766
32. Bandeira N, Tsur D, Frank A et al (2006) A new approach to protein identification. 10th annual international conference on research in computational molecular biology, April 2–5, Venice, Italy. Springer-Verlag, Berlin Heidelberg
33. Bandeira N, Tsur D, Frank A et al (2007) Protein identification by spectral networks analysis. Proc Natl Acad Sci U S A 104: 6140–6145
34. Falkner JA, Falkner JW, Yocum AK et al (2008) A spectral clustering approach to MS/MS identification of post-translational modifications. J Proteome Res 7:4614–4622
35. Ahrne E, Masselot A, Binz PA et al (2009) A simple workflow to increase MS2 identification rate by subsequent spectral library search. Proteomics 9:1731–1736
36. Fu Y, Xiu L-Y, Jia W et al (2011) DeltAMT: a statistical algorithm for fast detection of protein modifications from LC-MS/MS data. Mol Cell Proteomics 10:M110.000455
37. Ahrne E, Nikitin F, Lisacek F et al (2011) QuickMod: a tool for open modification spectrum library searches. J Proteome Res 10:2913–2921
38. Hu Y, Li Y, Lam H (2011) A semi-empirical approach for predicting unobserved peptide MS/MS spectra from spectral libraries. Proteomics 11:4702–4711
39. Searle BC, Dasari S, Turner M et al (2004) High-throughput identification of proteins and unanticipated sequence modifications using a mass-based alignment algorithm for MS/MS de novo sequencing results. Anal Chem 76:2220–2230
40. Han Y, Ma B, Zhang K (2005) SPIDER: software for protein identification from sequence

tags with de novo sequencing error. J Bioinform Comput Biol 3:697–716

41. Shen Y, Tolic N, Hixson KK et al (2008) De novo sequencing of unique sequence tags for discovery of post-translational modifications of proteins. Anal Chem 80:7742–7754
42. Shilov IV, Seymour SL, Patel AA et al (2007) The Paragon Algorithm, a next generation search engine that uses sequence temperature values and feature probabilities to identify peptides from tandem mass spectra. Mol Cell Proteomics 6:1638–1655
43. Liu C, Yan B, Song Y et al (2006) Peptide sequence tag-based blind identification of post-translational modifications with point process model. Bioinformatics 22:e307–e313
44. Na S, Jeong J, Park H et al (2008) Unrestrictive identification of multiple post-translational modifications from tandem mass spectrometry using an error-tolerant algorithm based on an extended sequence tag approach. Mol Cell Proteomics 7:2452–2463
45. Liu J, Erassov A, Halina P et al (2008) Sequential interval motif search: unrestricted database surveys of global MS/MS data sets for detection of putative post-translational modifications. Anal Chem 80:7846–7854
46. Bern M, Cai Y, Goldberg D (2007) Lookup peaks: a hybrid of de novo sequencing and database search for protein identification by tandem mass spectrometry. Anal Chem 79: 1393–1400
47. Na S, Bandeira N, Paek E (2012) Fast multi-blind modification search through tandem mass spectrometry. Mol Cell Proteomics 11:M111.010199
48. Fu Y (2012) Bayesian false discovery rates for post-translational modification proteomics. Stat Interface 5:47–59
49. Baker PR, Medzihradszky KF, Chalkley RJ (2010) Improving software performance for peptide electron transfer dissociation data analysis by implementation of charge state- and sequence-dependent scoring. Mol Cell Proteomics 9:1795–1803
50. Marx H, Lemeer S, Schliep JE et al (2013) A large synthetic peptide and phosphopeptide reference library for mass spectrometry-based proteomics. Nat Biotechnol 31:557–564
51. Huttlin EL, Hegeman AD, Harms AC et al (2007) Prediction of error associated with false-positive rate determination for peptide identification in large-scale proteomics experiments using a combined reverse and forward peptide sequence database strategy. J Proteome Res 6:392–398
52. Fu Y, Qian X (2014) Transferred subgroup false discovery rate for rare post-translational modifications detected by mass spectrometry. Mol Cell Proteomics 13:1359–1368
53. Beausoleil SA, Villén J, Gerber SA et al (2006) A probability-based approach for high-throughput protein phosphorylation analysis and site localization. Nat Biotechnol 24: 1285–1292
54. Olsen JV, Blagoev B, Gnad F et al (2006) Global, in vivo, and site-specific phosphorylation dynamics in signaling networks. Cell 127: 635–648
55. Bailey CM, Sweet SM, Cunningham DL et al (2009) SLoMo: automated site localization of modifications from ETD/ECD mass spectra. J Proteome Res 8:1965–1971
56. Savitski MM, Lemeer S, Boesche M et al (2011) Confident phosphorylation site localization using the Mascot Delta Score. Mol Cell Proteomics 10:M110.003830
57. Baker PR, Trinidad JC, Chalkley RJ (2011) Modification site localization scoring integrated into a search engine. Mol Cell Proteomics 10:M111.008078
58. Fermin D, Walmsley SJ, Gingras AC et al (2013) LuciPHOr: algorithm for phosphorylation site localization with false localization rate estimation using modified target-decoy approach. Mol Cell Proteomics 12:3409–3419
59. Wan Y, Cripps D, Thomas S et al (2008) PhosphoScan: a probability-based method for phosphorylation site prediction using MS2/MS3 pair information. J Proteome Res 7: 2803–2811
60. Taus T, Kocher T, Pichler P et al (2011) Universal and confident phosphorylation site localization using phosphoRS. J Proteome Res 10:5354–5362
61. Vaudel M, Breiter D, Beck F et al (2013) D-score: a search engine independent MD-score. Proteomics 13:1036–1041
62. Tanner S, Payne SH, Dasari S et al (2008) Accurate annotation of peptide modifications through unrestrictive database search. J Proteome Res 7:170–181

Chapter 18

Dissecting the iTRAQ Data Analysis

Suruchi Aggarwal and Amit Kumar Yadav

Abstract

In the era of large-scale quantitative biology, mass spectrometry-based quantitative proteomics is progressively becoming indispensable for gaining insights into the biological systems at molecular level. Various quantitative study designs rely on chemical tagging approaches to study disease, stress, or drug response and temporal studies aiming at disease/developmental progression in a biological system. Isobaric tags for relative and absolute quantitation (iTRAQ) is one of the most popular chemical labeling techniques which allows four, six, or eight samples to be multiplexed in a single run. As the iTRAQ tag has a balancer group to equalize all states of a labeled peptide to same mass, the differentially labeled iTRAQ peptides are mixed before chromatography and elute as a single combined peak in MS. This enhances the peptide signal and quantitation is performed during MS/MS along with sequencing, where reporter ions of different masses are released to give relative quantitation. Known amount of a spiked-in protein can also help in absolute quantitation of the proteins in a sample.

Key words iTRAQ, Quantitative proteomics, Statistics, Relative protein quantitation, Chemical labeling

1 Introduction

With the advancement of next-generation mass spectrometric proteomic sequencing techniques [1], the detailed analysis of whole proteomes *en masse* and their dynamics in space and time have become feasible with high resolution. The analysis of whole proteomes, including identification and quantitation, for any organism at a given time or biological state has become not only possible but also faster than the earlier approaches. As the proteome may change with time, developmental conditions, and environmental effects, studies need to be designed in such a way that this dynamic behavior of a biological system can be captured and understood. The genome remains fairly static, but the expression of genes in a specific context is more interesting, for which the proteome studies need to factor in time because of its ever-changing nature. Multiplexing various states by labeling techniques is therefore of utmost necessity for quantitative proteomics. The quantitative

Klaus Jung (ed.), *Statistical Analysis in Proteomics*, Methods in Molecular Biology, vol. 1362,
DOI 10.1007/978-1-4939-3106-4_18, © Springer Science+Business Media New York 2016

information for the proteins expressed can be used for modeling the cellular and metabolic response of an organism [2] and thus can help in designing inhibitors for specific targets in a disease study [3]. With the improvement of qualitative techniques, many quantitative techniques are now in existence for high-throughput proteomics data sequencing that offer improved sensitivity, precision, and easy automation of experiments and analysis [4].

Spatiotemporal proteome dynamics can be quantitatively studied using different kinds of stable isotope labeling like metabolic (e.g., SILAC [5] and dimethyl [6]), enzymatic (^{18}O [7]), and chemical labeling strategies (like iTRAQ [8] and TMT [9]). Mass spectrometry is not inherently quantitative. The abundance/intensity of differentially labeled precursor ions or fragment ions are used as a proxy for protein quantitation. Relative expression changes are easier to monitor by chemical labeling techniques but absolute quantitation is challenging as it requires spiking-in of known amounts of a reference standard protein [10]. Quantification by chemical labeling is achieved by measuring peak intensities (or the peak areas) at either MS1 or MS2 level. In-depth review of various quantitative strategies is available elsewhere [11–13]. This chapter mainly focuses on how iTRAQ is used for multiplexed quantitative proteomics and the basic guiding principles, tools, and techniques for data analysis. This chapter does not cover the wet lab protocols for iTRAQ as these have been extensively covered earlier in various reviews and book chapters [14–22]. Combining replicates and dealing with technical and biological variations have also been covered elsewhere [23–26] and are out of scope for this chapter. This chapter covers the method of basic iTRAQ data analysis and the underlying informatics methods and their explanations. We conclude by providing the tools and software available for data analysis.

Introduced in 2004, iTRAQ was developed as a multiplexed method for studying protein quantitation using isobaric tags containing a reporter and a balancer group. In iTRAQ, the amine groups at peptide N-terminal and lysine residues are derivatized by isobaric labels (4-plex, 6-plex, or 8-plex) after digestion and mixed together before MS [8, 27, 28]. Ross et al. [8] introduced four different *reporters* having the masses of 114.1, 115.1, 116.1, and 117.1. Corresponding *balancers* with masses ranging from 28 to 31 Da add up to form the *isobaric tag* with the combined mass of 145.10 Da (Fig. 1a). It was later extended to 6-plex and 8-plex [27, 28]. In 8-plex iTRAQ, four new reporters, 113.1, 118.1, 119.1, and 121.1, were added, supplemented with a corresponding balancer mass of 192–184 Da that takes the total mass to 305.10 Da (Fig. 1b). Reporter with a mass of 120.08 Da was omitted from 8-plex masses as it is isobaric with phenylalanine immonium ion and could have contaminated the peak resulting in inaccurate quantitation. An iTRAQ reaction utilizes an N-hydroxysuccinimide

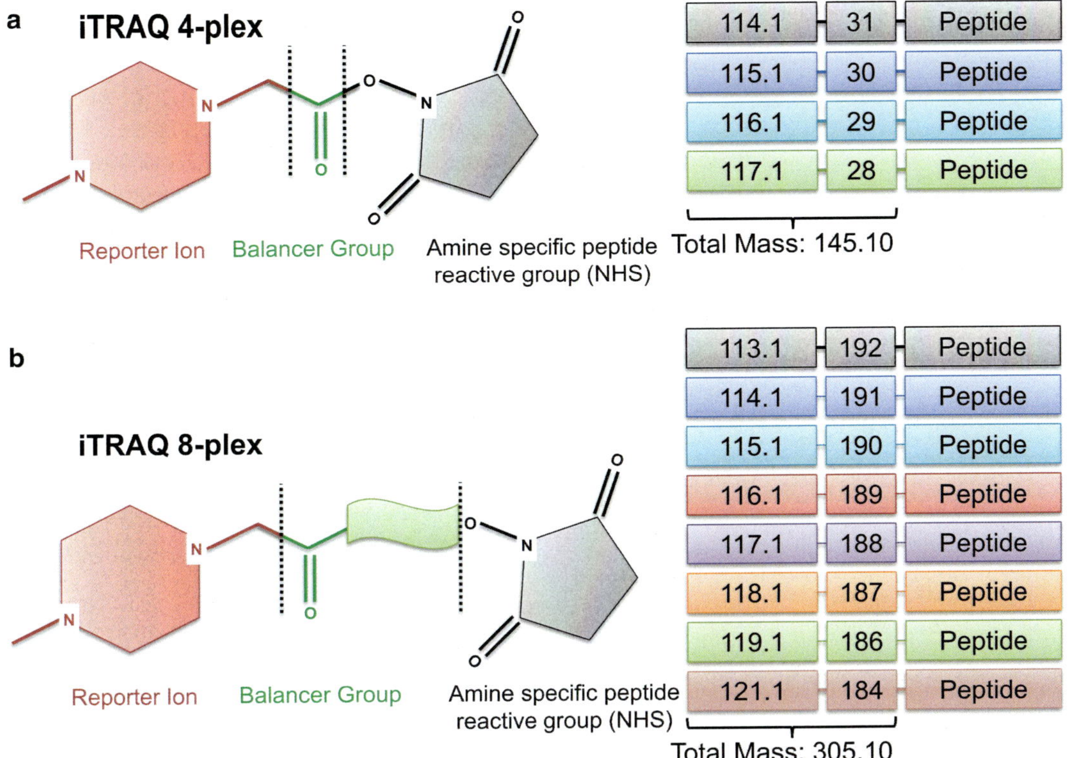

Fig. 1 Chemical structures for iTRAQ (**a**) 4-plex and (**b**) 8-plex isobaric tags. Balancer + reporter ions add up to 145 Da in 4-plex and 304 Da in 8-plex experiments. In 8-plex, reporter mass of 120 is not present as it will give erroneous quantitation since phenylalanine immonium ion is also observed at a mass of 120 Da

(NHS) ester derivative to modify the primary amine groups by linking the balancer group (carbonyl group) and a reporter group (based on *N*-methylpiperazine) to the digested peptides by forming an amide bond [29]. During MS/MS fragmentation the balancer group is lost as a neutral loss and the reporter group is released. As the labels make the peptides isobaric, at MS1 level the peptide (+isobaric tag) appears as a single peak but during MS/MS, the quantitation is achieved when the reporter ions of different masses are released and seen in low-mass region of the fragment spectrum. The advantage of iTRAQ over SILAC is that the peptide identification and quantification are concomitant at MS2 level. Due to the stable nature of this chemical addition and its stability in the mixture, the isobaric tags can be added to any sample with a high labeling efficiency which is virtually as good as that of cellular metabolic labels [1]. Isobaric labeling-based quantitation is popular due to the ease of sample multiplexing that provides high-throughput quantitation. This multiplexing allows for an easy experimental design for studying temporal dynamics and comparing several states of a wide variety of biological systems.

2 Materials and Methods

2.1 iTRAQ Workflow

This section briefly discusses the iTRAQ workflow before the data analysis methodology part in Subheading 3. Protein mixtures from different samples are extracted and digested according to lab-specific shotgun proteomics protocols. Different samples (up to 8-plex) are then labeled by separate iTRAQ reaction mixtures for differential labeling (Fig. 2). The labeled sample mixtures are subsequently combined in a 1:1 ratio for multiplexing before separation by high-performance liquid chromatography and mass spectrometry [29].

2.2 LC-MS/MS Workflow

The 8-plexed sample is now separated by any of the three strategies—

1. Separation on gel by IEF followed by MALDI mass spectrometry (MS and MS/MS of the spots).
2. Separated by gel-C-MS followed by high-resolution mass spectrometry.
3. Separation by high-performance liquid chromatography, followed by high-resolution mass spectrometry.

The data collection on the mass spectrometer is controlled through predefined workflows based on standard protocols for

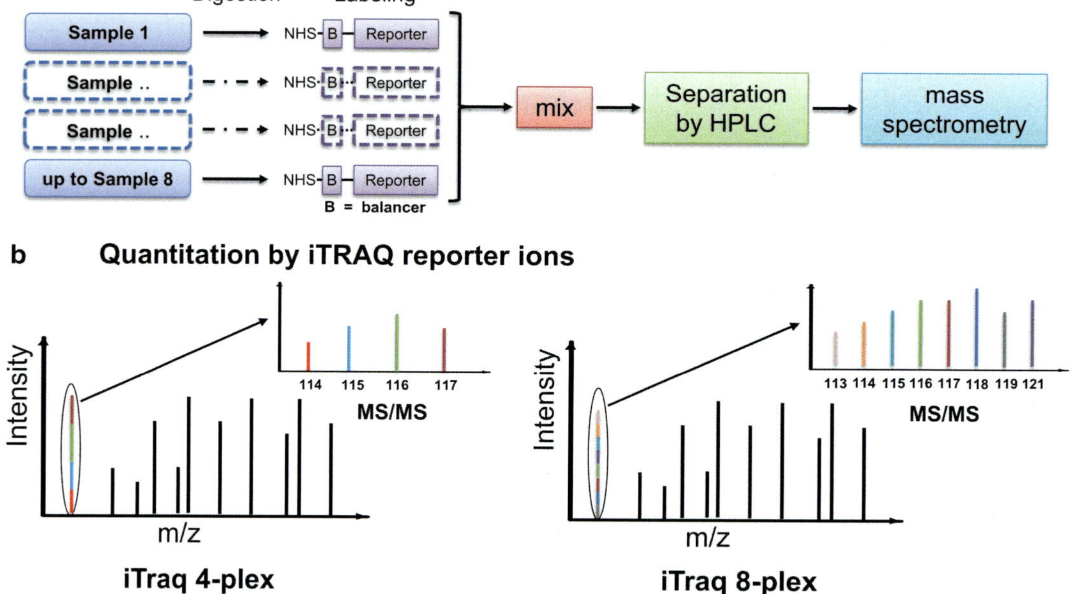

Fig. 2 (**a**) Basic iTRAQ workflow in which up to eight samples can be separately digested and labeled with iTRAQ tags, mixed together, separated by HPLC, and analyzed by mass spectrometer. (**b**) During MS/MS, the reporter ions of differential masses are released from peptide to give sample-specific quantitation of a particular peptide

iTRAQ data acquisition and generally searched through the *de facto* search engine and quantitation tool (usually proprietary) integrated with the mass spectrometer. These tools are generally opaque about their assumptions and data analysis process and thus interpretation of data becomes challenging. The next section explains the details of the analysis in a generic tool-independent framework.

3 Methods for Data Analysis

This section describes the general method for data analysis using the iTRAQ technology and the major challenges associated with it. There are plenty of tools available which can process iTRAQ data but vary in their ability to support various multiplexing (4-plex, 6-plex, or 8-plex), basic methods of data preprocessing, error and area calculations [30–32].

It should be noted that identification and quantitation are independent informatics processes until the last steps in a workflow. Not many freely available tools have a complete start-to-end transparent solution available to control all the steps in quantitation starting from raw mass spectrometry data. MSnbase [33], OCAP [34], i-Tracker [30], VEMS [35], MaxQuant [36], ProteoSuite, etc. are some of the freely available tools. A more comprehensive list is provided in Table 1.

The quantitation comprises of several steps—like data preprocessing, area and error calculation, isotopic correction, ratio calculation, calculating protein abundances, and statistical analyses for group comparisons. It should be noted that the quantitation occurs at spectra level, which represents the peptides instead of proteins. This is a side effect of shotgun proteomics as digesting proteins to peptides makes the proteome amenable for mass spectrometry. The increased complexity of sample due to numerous peptides is alleviated by HPLC. The spectra obtained now represent the quantitative readings of peptides. Some peptides are sampled more times by mass spectrometer due to multiple charge states eluting at different times and m/z and better ionization besides being abundant. Peptide ratios are compiled from these multiple spectra. Similarly, protein ratios are calculated from one or more identified and quantified peptides. Issues like heterogeneity of variance and bias need to be corrected.

3.1 Data Preprocessing and Database Search

The raw files obtained from the mass spectrometer are used in two parallel workflows which merge later—one for identification and the other for quantitation. The identification workflow involves a database search for peptide and protein identification. The schematic representation of the whole process is summarized in Fig. 3.

Table 1
List of software tools for iTRAQ quantitation

Software	Academic/ commercial	Weblink	OS environment
ProteinPilot [41]	Commercial	http://www.absciex.com/products/software/proteinpilot-software	Windows
ProteoSuite	Academic	http://www.proteosuite.org/	Windows
Multi-Q [44]	Academic	http://ms.iis.sinica.edu.tw/COmics/Software_Multi-Q.html	Web server
MFPaQ [53]	Academic	http://mfpaq.sourceforge.net/	Windows
Quant [4]	Academic	http://sourceforge.net/projects/protms/	Linux/Windows
Libra [54]	Academic	http://sashimi.svn.sourceforge.net/viewvc/sashimi/trunk/trans_proteomic_pipeline/src/Quantitation/Libra/docs/libra_info.html	Linux/Windows (cygwin)
i-Tracker [30]	Academic	http://sourceforge.net/projects/itracker/?abmode=1	Linux/Windows
IsobariQ [55]	Academic	http://norwegian-proteomics-society.uio.no/isobariq/index.html	Windows
MaxQuant [36]	Academic	http://www.maxquant.org/downloads.htm	Windows
VEMS [56]	Academic	http://yass.sdu.dk/	Windows
Census [57]	Academic	http://fields.scripps.edu/census/	Linux/Windows
OCAP [34]	Academic	https://code.google.com/p/ocap/	Linux/Windows
Isobar R [58]	Academic	http://www.bioconductor.org/packages/release/bioc/html/isobar.html	Linux/Windows
ProRata [59]	Academic	https://code.google.com/p/prorata/	Windows
MassChroQ [60]	Academic	http://pappso.inra.fr/bioinfo/masschroq/	Windows
MSnBase [33]	Academic	http://www.bioconductor.org/packages/release/bioc/html/MSnbase.html	Linux/Windows
ItraqAnalyzer [61]	Academic	http://ftp.mi.fu-berlin.de/pub/OpenMS/release1.10-documentation/html/TOPP_ITRAQAnalyzer.html	Windows
PQPQ [62]	Academic	http://www.forshed.se/jenny/index.php?n=Research.SoftwareAmpCode	Linux/Windows
Proteome discoverer	Commercial	http://www.thermoscientific.com/content/tfs/en/product/proteome-discoverer-software.html	Windows
MilQuant [63]	Academic	https://code.google.com/p/milquant/	Linux/Windows

(continued)

Table 1 (continued)

Software	Academic/ commercial	Weblink	OS environment
iTRAQPak [47]	Academic	http://cran.r-project.org/	Linux/Windows
LTQ-iQuant [64]	Academic	https://netfiles.umn.edu/users/onson001/www/LTQiQuant.html	Windows
iQuantitator [31]	Academic	http://www.biomedcentral.com/1471-2105/10/342	Linux/Windows

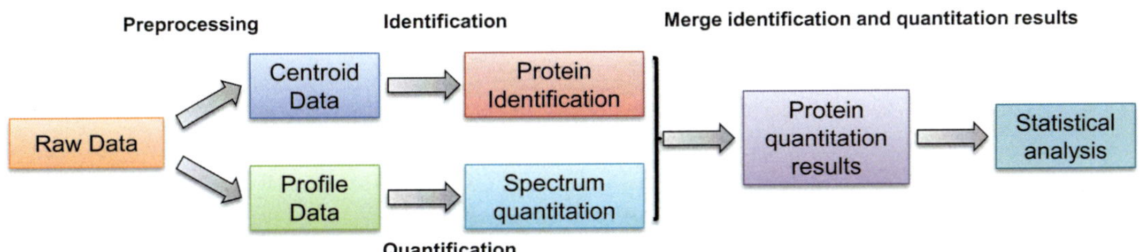

Fig. 3 Overview of iTRAQ data analysis workflow with identification and quantitation. Raw data is used for both identification (centroid data) and quantitation (profile data) in parallel workflows. The identifications are utilized to calculate peptide and protein ratios at the later steps

This involves data preprocessing to pick peaks and remove noise which is also implemented in many search algorithms [37–40]. This processing can be markedly different from quantitation, as quantitation requires minimal filtering of data for accurate area calculation. Only baseline correction and noise removal below signal-to-noise threshold are recommended at this step. While identification benefits from clean and processed spectra, quantitation may be impaired by excessive processing. Therefore, while identification should be done with centroid data, it is recommended to use profile mode data for quantitation.

There are several database search algorithms that can be used for peptide and protein identification like X!Tandem [40], Mascot [38], MassWiz [37], OMSSA [39], and Paragon [41] (within ProteinPilot). iTRAQ is set as fixed modification at N-terminal and lysine residues besides other parameters. iTRAQ is also specified as a variable modification at tyrosine residues. After search, FDR can be calculated for identified spectra (or PSM, peptide spectral matches) from the results using ProteoStats [42] or MAYU [43] tools. Once a protein list is compiled, the quantitation results can be merged for the identified and quantified proteins.

3.2 Reporter Area Calculation

This section describes the main step of isobaric quantitation where raw peak areas are integrated and error in area estimation is calculated. This step requires the identification of the reporter ion peaks

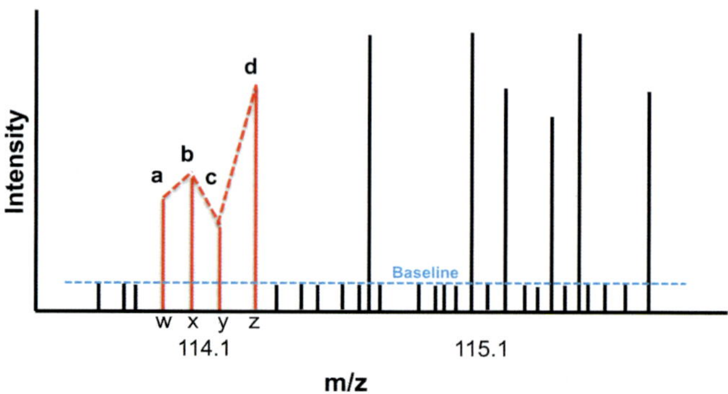

Fig. 4 Schematic of low-mass region of an MS/MS spectrum to depict peak area calculation with reporter ions and baseline (see text).

in a defined low-mass *m/z* region (~100–125 Da) within a tolerance range depending on the accuracy of the mass spectrometer. The reporter ion peak area can be calculated via three different methods (Fig. 4).

1. Trapezoid
   In trapezoid method, the area of the peaks is calculated by joining two corresponding peaks to form a trapezoid and the sum area of all such trapezoids in an *m/z* range of a reporter ion within the tolerance constitutes area of the peak (as shown in Fig. 4 for area enclosed within trapezoids bound by *abxw* + *bcyx* + *cdzy*). This method is used in i-Tracker [30] which can perform 4-plex quantitation. For 8-plex, the same group has developed an enhanced version called x-Tracker within ProteoSuite software (http://www.proteosuite.org/):

$$\text{Area} = \frac{1}{2}\times(x-w)\times(b+a)+\frac{1}{2}\times(y-x)\times(c+b)+\frac{1}{2}\times(z-y)(d+c) \qquad (1)$$

2. Sum intensity
   In this method, all the intensities for a particular reporter ion in a given *m/z* tolerance are added up to give area of that reporter ion [35]. This method works best when the data is preprocessed and filtered to remove baseline noise. Otherwise, the next method should be used. This method is generally faster as the number of peaks is lesser as are the calculations:

$$\text{Area} = a+b+c+d \qquad (2)$$

3. Baseline corrected
   The area calculated from summed intensities for a reporter ion as calculated in previous method can be improved by subtracting the base noise from the final area to obtain a *baseline-corrected area* (Fig. 4, dotted line). For this, the median of

the peaks in between the reporter masses is subtracted from the *sum intensity* calculated area [35]. This method is used in VEMS [35] and Multi-Q [44]. For noisy data, this method should be used over the previous one:

$$\text{Area} = (a + b + c + d) - (\text{median}(\text{baseline})) \tag{3}$$

3.3 Error Calculation

Once the peak area is calculated, a peak error is calculated as quantification error might be present while calculating the area. The error (E) can be calculated as (please *see* ProteinPilot help manual):

$$E = \sqrt{I} \tag{4}$$

where I represents the total intensity calculated for one reporter ion. Different error models are used in different software which may give slightly different results [4, 30, 44]. An in-depth discourse on error calculation is out of scope but can be found elsewhere [25, 45, 46]. Generally, the error model used in software comes with its documentation as no generic consensus is there. i-Tracker reference has detailed description of error calculation.

3.4 Isotopic Correction

The area calculated for each reporter ion is not the true area of that reporter ion due to overlap of the signals from different neighboring reporter ions. Due to these isotopic distortions, the peak area of one reporter ion contains an additive effect of its neighboring reporter ions. Correcting this artifact is called as *isotopic correction* to find the true area. The true area can be calculated using the parameters provided by the manufacturer for 16 purity values (for 4-plex) and 32 purity values (for 8-plex) in each batch which represent the percentages of each reporter ion that have masses differing by +2, +1, −2, and −1 Da [30]. Using the manufacturer-provided isotopic correction table (Table 2), one can calculate losses to or gains from other reporter ions (*see* **Note 1**).

After calculation of true area from the isotopic corrections, peak normalization is applied as

$$T_N = T_R / T_{\text{Total}} \tag{5}$$

where TN is the normalized peak area, TR is the true area of the reporter ion, and T_{Total} is the total of all the true area for all reporter ions.

3.5 Ratio and Error Calculation

Using the calculated areas, ratios are calculated for each reporter ion as

$$R_i = T_i / T_b \tag{6}$$

where Ti is the true area for the selected reporter ion, and Tb is the area of the base reporter ion (reporter 113.1 for iTRAQ 8-plex

Table 2
The isotopic corrections for the calculation of reporter ion area peaks as per given in protein pilot software version 4.5b

Reporter ion	−2	−1	0	2	1
113	0	0	92.87	6.89	0.24
114	0	0.94	93	5.9	0.16
115	0	1.88	93.12	4.9	0.1
116	0	2.82	93.21	3.9	0.07
117	0.06	3.77	93.29	2.88	0
118	0.09	4.71	93.29	1.91	0
119	0.14	5.66	93.33	0.87	0
121	0.27	7.44	92.11	0.18	0

experiment). As the first reporter is generally a control sample, we usually get three ratios for a 4-plex (115/114, 116/114, and 117/114) and seven ratios (114/113, 115/113, 116/113, 117/113, 118/113, 119/113, and 121/113) for an 8-plex experiment.

3.6 Combining Peptide and Protein Quantitation

This step merges the identification and quantitation results which were separately processed until now. The ratios and errors calculated at the spectra level can be combined in three ways to give the peptide ratio and error [47]:

1. Average based: All the spectra corresponding to the same peptide after a database search will be averaged to give the peptide ratio and error.
2. Median based: Peptide ratio will be the median of all the spectral ratios corresponding to that peptide. Here the median will be geometric mean, if the spectra counts are even.
3. Score based: In this particular method, ratio and error of spectra best defining a given peptide will be used as the peptide ratio and error.

Once we get the peptide ratios and errors, those can be combined to get the respective protein ratios and errors as per the following three methods:

1. Average: The average of all peptide ratios and errors is calculated as the protein ratio and error.
2. Median: For all peptide ratios and errors, median is calculated as the protein ratio and error.

3. Weighted average: Here the protein ratio (PR) [24] is calculated as

$$\mathrm{PR} = \frac{\sum_{j=1}^{n} \mathrm{PER}_j \times \mathrm{PW}_j}{\sum_{j=1}^{n} \mathrm{PW}_j} \quad (7)$$

where, PER is the peptide ratio, PW is the peptide weight, and j is the number of peptide from $1\ldots n$.

3.7 Ratio Compression and Statistical Analyses

There is a known bias of ratios being compressed towards 1 when there is simultaneous fragmentation of coeluting peptides. This effect brings about the ratios of most reporter ions closer to 1 and significant quantitative changes (fold changes) may be underestimated and excluded [48]. Proper and in-depth separation and mass spectrometry can alleviate these problems [19, 49, 50].

Another problem with isobaric tags is bias and variability on account of different intensities. There is higher error in estimating areas for low-intensity peaks and vice versa. A variance-stabilizing normalization (VSN) model has been proposed to correct for this problem [51].

Group comparisons are not covered here, but multi-run experiments can be merged using their p-values and other strategies suggested in the references [26, 52].

4 Tools

Most steps described here provide a transparent processing of iTRAQ data generally followed by many tools. These steps can also be performed using R statistical environment and several tools have been developed in R. The analysis can be performed in a software suite or a stand-alone tool. Most tools have different strengths and weaknesses. Table 1 lists various tools for iTRAQ quantitation with their download links. Users are advised to try these tools for customizing their analysis pipelines.

5 Notes

1. For isotopic corrections, to calculate the true area of the reporter ions peaks (e.g., 4-plex), we take all the 16 purity values given in the kit (as percentage) and solve a system of linear equations to apply Cramer's rule. Through the Cramer's rule determinant (D) can be calculated which will ultimately help in finding the true area for each reporter ion. If we assume 16

purity correction values as (c_1, c_2, c_3, c_4, c_5, c_6, c_7, c_8, c_9, c_{10}, c_{11}, c_{12}, c_{13}, c_{14}, c_{15}, c_{16}) for (114.1 − 2 Da, 114.1 − 1 Da, 114.1 + 1 Da, 114.1 + 2 Da, 115.1 − 2 Da, 115.1 − 1 Da, 115.1 + 1 Da, 115.1 + 2 Da, 116.1 − 2 Da, 116.1 − 1 Da, 116.1 + 1 Da, 116.1 + 2 Da, 117.1 − 2 Da, 117.1 − 1 Da, 117.1 + 1 Da, 117.1 + 2 Da). For percentage of purity of that particular reporter ion (r) we assume c_{114}, c_{115}, c_{116}, c_{117} as representative variables.

With this information, we can calculate determinant (D) using the following matrix:

$$D = \begin{vmatrix} c_{114} & c_6 & c_3 & 0 \\ c_9 & c_{115} & c_7 & c_4 \\ c_{13} & c_{10} & c_{116} & c_8 \\ 0 & c_{14} & c_{11} & c_{117} \end{vmatrix} \tag{8}$$

If D is zero or negative, then the areas calculated will become infinite or zero. In such cases, we skip the purity correction step. When it is positive, Cramer's determinant (Dr) is calculated for each reporter ion (r) as described in i-Tracker [30]. The true area (Tr) then can be calculated as:

$$T_r = D_r / D \tag{9}$$

More details can be found in reference on i-Tracker [30].

Acknowledgement

S.A. is supported by SRF grant and A.K.Y. is supported by Innovative Young Biotechnologist Award (IYBA) grant and DDRC-SFC grant from Department of Biotechnology (DBT), India. Authors acknowledge Nazmuddin Saquib for critically reviewing the manuscript, and Manu Kandpal and Vivek Arora for proofreading the manuscript.

References

1. Altelaar AF, Munoz J, Heck AJ (2012) Next-generation proteomics: towards an integrative view of proteome dynamics. Nat Rev Genet 14:35–48
2. Cox J, Mann M (2011) Quantitative, high-resolution proteomics for data-driven systems biology. Annu Rev Biochem 80:273–299
3. Bantscheff M, Hopf C, Savitski MM et al (2011) Chemoproteomics profiling of HDAC inhibitors reveals selective targeting of HDAC complexes. Nat Biotechnol 29:255–265
4. Boehm AM, Putz S, Altenhofer D et al (2007) Precise protein quantification based on peptide quantification using iTRAQ. BMC Bioinformatics 8:214
5. Ong SE, Blagoev B, Kratchmarova I et al (2002) Stable isotope labeling by amino acids in cell culture, SILAC, as a simple and accurate

approach to expression proteomics. Mol Cell Proteomics 1:376–386

6. Hsu JL, Huang SY, Chow NH et al (2003) Stable-isotope dimethyl labeling for quantitative proteomics. Anal Chem 75:6843–6852
7. Yao X, Freas A, Ramirez J et al (2001) Proteolytic 18O labeling for comparative proteomics: model studies with two serotypes of adenovirus. Anal Chem 73:2836–2842
8. Ross PL, Huang YN, Marchese JN et al (2004) Multiplexed protein quantitation in Saccharomyces cerevisiae using amine-reactive isobaric tagging reagents. Mol Cell Proteomics 3:1154–1169
9. Thompson A, Schafer J, Kuhn K et al (2003) Tandem mass tags: a novel quantification strategy for comparative analysis of complex protein mixtures by MS/MS. Anal Chem 75: 1895–1904
10. Ong SE, Mann M (2005) Mass spectrometry-based proteomics turns quantitative. Nat Chem Biol 1:252–262
11. Bantscheff M, Lemeer S, Savitski MM et al (2012) Quantitative mass spectrometry in proteomics: critical review update from 2007 to the present. Anal Bioanal Chem 404:939–965
12. Bantscheff M, Schirle M, Sweetman G et al (2007) Quantitative mass spectrometry in proteomics: a critical review. Anal Bioanal Chem 389:1017–1031
13. Vaudel M, Sickmann A, Martens L (2010) Peptide and protein quantification: a map of the minefield. Proteomics 10:650–670
14. Glibert P, Van SK, Dhaenens M et al (2014) iTRAQ as a method for optimization: enhancing peptide recovery after gel fractionation. Proteomics 14:680–684
15. Burkhart JM, Vaudel M, Zahedi RP et al (2011) iTRAQ protein quantification: a quality-controlled workflow. Proteomics 11: 1125–1134
16. Pichler P, Kocher T, Holzmann J et al (2011) Improved precision of iTRAQ and TMT quantification by an axial extraction field in an Orbitrap HCD cell. Anal Chem 83:1469–1474
17. Collins BC, Lau TY, Pennington SR et al (2011) Differential proteomics incorporating iTRAQ labeling and multi-dimensional separations. Methods Mol Biol 691:369–383
18. Unwin RD (2010) Quantification of proteins by iTRAQ. Methods Mol Biol 658:205–215
19. Ow SY, Salim M, Noirel J et al (2009) iTRAQ underestimation in simple and complex mixtures: “the good, the bad and the ugly”. J Proteome Res 8:5347–5355
20. Phanstiel D, Zhang Y, Marto JA et al (2008) Peptide and protein quantification using iTRAQ with electron transfer dissociation. J Am Soc Mass Spectrom 19:1255–1262
21. Bantscheff M, Boesche M, Eberhard D et al (2008) Robust and sensitive iTRAQ quantification on an LTQ Orbitrap mass spectrometer. Mol Cell Proteomics 7:1702–1713
22. Aggarwal K, Choe LH, Lee KH (2006) Shotgun proteomics using the iTRAQ isobaric tags. Brief Funct Genomic Proteomic 5:112–120
23. Luo R, Zhao H (2012) Protein quantitation using iTRAQ: review on the sources of variations and analysis of nonrandom missingness. Stat Interface 5:99–107
24. Gan CS, Chong PK, Pham TK et al (2007) Technical, experimental, and biological variations in isobaric tags for relative and absolute quantitation (iTRAQ). J Proteome Res 6:821–827
25. Mahoney DW, Therneau TM, Heppelmann CJ et al (2011) Relative quantification: characterization of bias, variability and fold changes in mass spectrometry data from iTRAQ-labeled peptides. J Proteome Res 10:4325–4333
26. Herbrich SM, Cole RN, West KP Jr et al (2013) Statistical inference from multiple iTRAQ experiments without using common reference standards. J Proteome Res 12:594–604
27. Choe L, D’Ascenzo M, Relkin NR et al (2007) 8-plex quantitation of changes in cerebrospinal fluid protein expression in subjects undergoing intravenous immunoglobulin treatment for Alzheimer’s disease. Proteomics 7:3651–3660
28. Dayon L, Hainard A, Licker V et al (2008) Relative quantification of proteins in human cerebrospinal fluids by MS/MS using 6-plex isobaric tags. Anal Chem 80:2921–2931
29. Wiese S, Reidegeld KA, Meyer HE et al (2007) Protein labeling by iTRAQ: a new tool for quantitative mass spectrometry in proteome research. Proteomics 7:340–350
30. Shadforth IP, Dunkley TP, Lilley KS et al (2005) i-Tracker: for quantitative proteomics using iTRAQ. BMC Genomics 6:145
31. Schwacke JH, Hill EG, Krug EL et al (2009) iQuantitator: a tool for protein expression inference using iTRAQ. BMC Bioinformatics 10:342
32. Rodriguez-Suarez E, Gubb E, Alzueta IF et al (2010) Virtual expert mass spectrometrist: iTRAQ tool for database-dependent search, quantitation and result storage. Proteomics 10:1545–1556
33. Gatto L, Lilley KS (2012) MSnbase-an R/Bioconductor package for isobaric tagged mass spectrometry data visualization, processing and quantitation. Bioinformatics 28:288–289

34. Wang P, Yang P, Yang JY (2012) OCAP: an open comprehensive analysis pipeline for iTRAQ. Bioinformatics 28:1404–1405
35. Gruhler A, Matthiesen R (2007) Quantitation with virtual expert mass spectrometrist. Methods Mol Biol 367:139–152
36. Cox J, Mann M (2008) MaxQuant enables high peptide identification rates, individualized p.p.b.-range mass accuracies and proteome-wide protein quantification. Nat Biotechnol 26:1367–1372
37. Yadav AK, Kumar D, Dash D (2011) MassWiz: a novel scoring algorithm with target-decoy based analysis pipeline for tandem mass spectrometry. J Proteome Res 10:2154–2160
38. Perkins DN, Pappin DJ, Creasy DM et al (1999) Probability-based protein identification by searching sequence databases using mass spectrometry data. Electrophoresis 20: 3551–3567
39. Geer LY, Markey SP, Kowalak JA et al (2004) Open mass spectrometry search algorithm. J Proteome Res 3:958–964
40. Craig R, Beavis RC (2004) TANDEM: matching proteins with tandem mass spectra. Bioinformatics 20:1466–1467
41. Shilov IV, Seymour SL, Patel AA et al (2007) The Paragon algorithm: a next generation search engine that uses sequence temperature values and feature probabilities to identify peptides from tandem mass spectra. Mol Cell Proteomics 6:1638–1655
42. Yadav AK, Kadimi PK, Kumar D et al (2013) ProteoStats—a library for estimating false discovery rates in proteomics pipelines. Bioinformatics 29:2799–2800
43. Reiter L, Claassen M, Schrimpf SP et al (2009) Protein identification false discovery rates for very large proteomics data sets generated by tandem mass spectrometry. Mol Cell Proteomics 8:2405–2417
44. Lin WT, Hung WN, Yian YH et al (2006) Multi-Q: a fully automated tool for multiplexed protein quantitation. J Proteome Res 5:2328–2338
45. Pan C, Kora G, Tabb DL et al (2006) Robust estimation of peptide abundance ratios and rigorous scoring of their variability and bias in quantitative shotgun proteomics. Anal Chem 78:7110–7120
46. Zhang Y, Askenazi M, Jiang J et al (2010) A robust error model for iTRAQ quantification reveals divergent signaling between oncogenic FLT3 mutants in acute myeloid leukemia. Mol Cell Proteomics 9:780–790
47. D'Ascenzo M, Choe L, Lee KH (2008) iTRAQPak: an R based analysis and visualization package for 8-plex isobaric protein expression data. Brief Funct Genomic Proteomic 7:127–135
48. Savitski MM, Mathieson T, Zinn N et al (2013) Measuring and managing ratio compression for accurate iTRAQ/TMT quantification. J Proteome Res 12:3586–3598
49. Ting L, Rad R, Gygi SP et al (2011) MS3 eliminates ratio distortion in isobaric multiplexed quantitative proteomics. Nat Methods 8:937–940
50. Ow SY, Salim M, Noirel J et al (2011) Minimising iTRAQ ratio compression through understanding LC-MS elution dependence and high-resolution HILIC fractionation. Proteomics 11:2341–2346
51. Karp NA, Huber W, Sadowski PG et al (2010) Addressing accuracy and precision issues in iTRAQ quantitation. Mol Cell Proteomics 9:1885–1897
52. Pascovici D, Song X, Solomon PS et al (2014) Combining protein ratio p-values as a pragmatic approach to the analysis of multi-run iTRAQ experiments. J Proteome Res 6: 738–746
53. Bouyssie D, de Gonzalez PA, Mouton E et al (2007) Mascot file parsing and quantification (MFPaQ), a new software to parse, validate, and quantify proteomics data generated by ICAT and SILAC mass spectrometric analyses: application to the proteomics study of membrane proteins from primary human endothelial cells. Mol Cell Proteomics 6:1621–1637
54. Deutsch EW, Shteynberg D, Lam H et al (2010) Trans-proteomic pipeline supports and improves analysis of electron transfer dissociation data sets. Proteomics 10:1190–1195
55. Arntzen MO, Koehler CJ, Barsnes H et al (2011) IsobariQ: software for isobaric quantitative proteomics using IPTL, iTRAQ, and TMT. J Proteome Res 10:913–920
56. Matthiesen R, Lundsgaard M, Welinder KG et al (2003) Interpreting peptide mass spectra by VEMS. Bioinformatics 19:792–793
57. Park SK, Yates JR, III (2010) Census for proteome quantification. Curr Protoc Bioinformatics Chapter 13:Unit-11
58. Breitwieser FP, Muller A, Dayon L et al (2011) General statistical modeling of data from protein relative expression isobaric tags. J Proteome Res 10:2758–2766
59. Pan C, Kora G, McDonald WH et al (2006) ProRata: a quantitative proteomics program for

accurate protein abundance ratio estimation with confidence interval evaluation. Anal Chem 78:7121–7131

60. Valot B, Langella O, Nano E et al (2011) MassChroQ: a versatile tool for mass spectrometry quantification. Proteomics 11:3572–3577
61. Kohlbacher O, Reinert K, Gropl C et al (2007) TOPP—the OpenMS proteomics pipeline. Bioinformatics 23:e191–e197
62. Forshed J, Johansson HJ, Pernemalm M et al (2011) Enhanced information output from shotgun proteomics data by protein quantification and peptide quality control (PQPQ). Mol Cell Proteomics 10:M111
63. Zou X, Zhao M, Shen H et al (2012) MilQuant: a free, generic software tool for isobaric tagging-based quantitation. J Proteomics 75: 5516–5522
64. Onsongo G, Stone MD, Van Riper SK et al (2010) LTQ-iQuant: a freely available software pipeline for automated and accurate protein quantification of isobaric tagged peptide data from LTQ instruments. Proteomics 10: 3533–3538

Chapter 19

Statistical Aspects in Proteomic Biomarker Discovery

Klaus Jung

Abstract

In the pursuit of a personalized medicine, i.e., the individual treatment of a patient, many medical decision problems are desired to be supported by biomarkers that can help to make a diagnosis, prediction, or prognosis. Proteomic biomarkers are of special interest since they can not only be detected in tissue samples but can also often be easily detected in diverse body fluids. Statistical methods play an important role in the discovery and validation of proteomic biomarkers. They are necessary in the planning of experiments, in the processing of raw signals, and in the final data analysis. This review provides an overview on the most frequent experimental settings including sample size considerations, and focuses on exploratory data analysis and classifier development.

Key words Feature selection, Classifier, Cross-validation, Unsupervised learning, Molecular signatures

1 Introduction

The choice of an adequate therapy for an individual patient is typically based on a clinician's judgment but also on laboratory or technically measurable characteristics. One important category of measurable characteristics is biomarkers which are used to support diagnosis, prediction, and prognosis. For approaching a definition of a proteomic biomarker one can employ that of a genomic biomarker. According to the International Conference on Harmonisation (ICH, www.ich.org), a genomic biomarker is defined as "a measureable DNA and/or RNA characteristic that is an indicator of normal biological processes, pathogenic processes, and/or response to therapeutic or other interventions" (ICH guideline E15). In terms of proteomics, a biomarker may thus be a single protein or peptide whose expression is measured in a biological sample. Nowadays, high-throughput techniques such as 2-D gel electrophoresis (2-DE), mass spectrometry (MS), or protein microarrays are frequently used for measuring expression levels of hundreds or thousands of molecular features simultaneously. Therefore, it is often possible to combine single features

Klaus Jung (ed.), *Statistical Analysis in Proteomics*, Methods in Molecular Biology, vol. 1362,
DOI 10.1007/978-1-4939-3106-4_19, © Springer Science+Business Media New York 2016

to so-called multiplex biomarkers [1, 2] or molecular signatures [3, 4], where the combination can frequently act as a "better" characteristic than individual features. Better in the sense that the error of diagnosis, prediction, or prognosis is diminished.

In the past, proteomic biomarkers, or combinations of them, have been studied in diverse diseases, e.g., different types of cancer [5, 6], and cardiovascular [7] and neurological [8, 9] diseases. These studies focused either on markers that can distinguish between different categories of cases, such as different clinical stages [5], the prediction of a therapeutic effect [6], or prognostic markers for the further course of the disease [7]. An advantage of proteomic biomarkers is that they can be measured in diverse biological samples, for example in tissue [6], blood [5], urine [10], saliva [11], or the cerebrospinal fluid [9].

However, while a large number of studies report proteomic signatures, only few of them seem to have arrived in clinical practice. In fact, most proteomic biomarkers that are routinely used for cancer diagnosis, prognosis, or prediction are single features. Pavlou et al. [12] for example list 18 features that are cleared by the US Food and Drug Administration (FDA). Furthermore, the same authors mention, among other factors, the correct use of statistical and bioinformatical methods to be highly relevant for biomarker research.

In fact, statistical and bioinformatical aspects play a very important role in analyzing the diagnostic, predictive, or prognostic power of proteomic biomarkers and for the development and evaluation of classifier models. Some reviews on statistical methods in proteomic biomarker research have been published so far, focusing for example on the data processing and classifier development, specifically in LC-MS experiments [13, 14] or experimental design specifically in SELDI-TOF mass spectrometry [15].

Here, the focus is not on a specific proteomics technique but the statistical techniques are detailed more generally. In particular, the focus is on methods for unsupervised data exploration (Subheading 2), the selection of biomarker candidates (Subheading 3), as well as the training and validation of classifier models (Subheading 4). Subheadings 3 and 4 also address some sample size issues. In Subheading 5, a short outlook on further advances of statistical methods in proteomics biomarker research is given.

For illustration of the methods, two public data sets will be employed, here, both available in packages for the statistic software R (http://www.R-project.org). The first data set was taken from the R-package "clippda" and consists of protein expression data from $n = 131$ samples of patients with liver cirrhosis, measured by means of MS [16]. The second data example was provided within the R-package "prot2D," and consists of protein expression data from $n = 12$ samples from *Pecten maximus* that were exposed to two different levels of temperature change [17].

2 Unsupervised Methods for Exploration of the Separability of Samples

Before training and validating a classifier model, it is usually helpful to explore the general separability of defined sample groups by graphical methods, or more precisely by "unsupervised" cluster methods. A number of visualization tools have been proposed for high-throughput expression data during the last two decades.

One of these tools is hierarchical clustering that was firstly used to explore DNA microarray data [18], but became then also very popular for high-throughput protein expression data, e.g., for clustering ovarian tumors by their protein expression profiles from 2-D gel electrophoresis (2-DE) [19] or lung cancer samples analyzed by MALDI-TOF mass spectrometry [20]. Hierarchical clustering algorithms are normally based on a matrix that specifies the similarity, correlation, or distance between samples. As output, the clustering algorithms produce a tree where similar samples appear as leaves at the same branch and more unsimilar samples appear at more distant branches. A hierarchical cluster tree can thus be used as a first visualization of the expression data to see whether samples can separate into different classes. If samples do not separate clearly, there is only a small chance that a classifier will yield good results. There exist different clustering algorithms to build the cluster trees which produce each very different results. Among the clustering algorithms, the "complete linkage" approach is well suited to identify groups of samples, while for example the "single linkage" approach is rather appropriate to identify outliers. To illustrate this, one artificial outlier, i.e., an extreme observation, was put to the liver data by first taking the overall mean expression for each protein and then increasing the expression for a small proportion of proteins. This outlier is hidden, when using the "complete linkage approach," but becomes uncovered when using the "single linkage" approach (Fig. 1). Besides clustering samples, cluster analysis can also be used to cluster the molecular features of a proteomics experiment, i.e., proteins or peptides. This can be helpful to get a first glance of molecular signatures that are responsible for separating the experimental groups. A cluster heatmap can be used to display clustered samples and clustered features in the same plot.

Another "unsupervised" clustering method that is very frequently used to identify groups of samples or to explore the separability is principal component analysis (PCA). Since protein expression data is usually high dimensional, i.e., a large number of proteins is studied on a rather small number of samples, the data is hard to visualize. PCA is a tool for reducing the dimensions, so that the expression data can be visualised in a two- or three-dimensional plot. Mathematically, PCA can be regarded as a transformation into another coordinate system, where the first two or three axes represent a large proportion of information from the data. Thus, in spite of the dimension reduction there is only a small

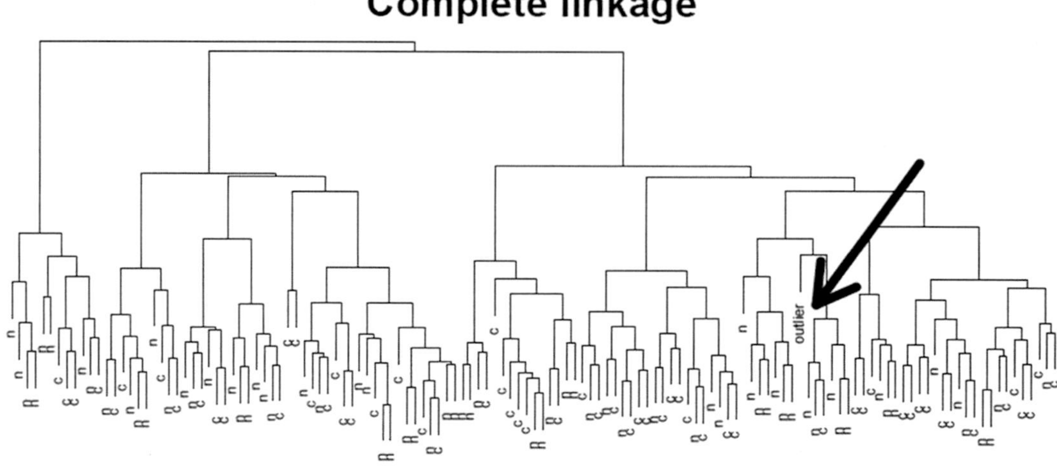

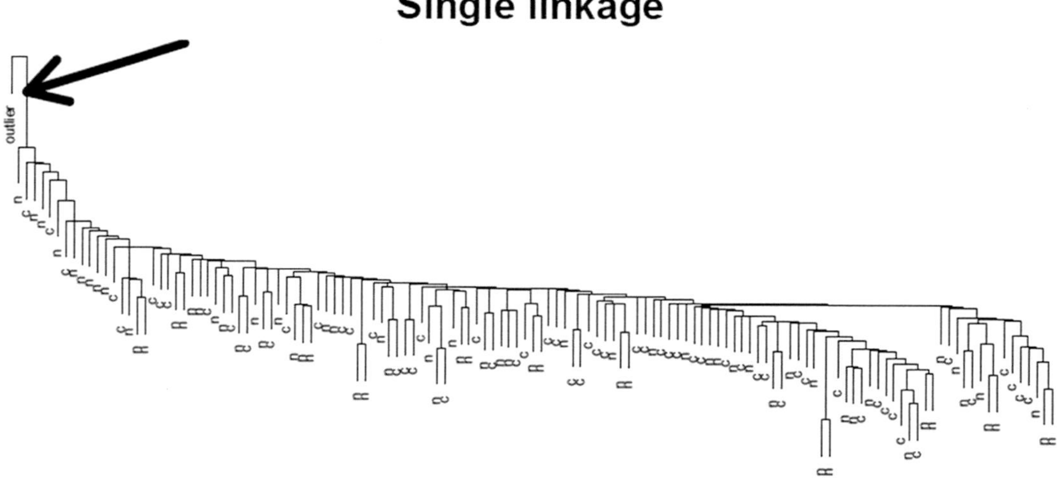

Fig. 1 Examples of hierarchical cluster trees representing the same data by two different cluster techniques. Each leaf represents one sample of the study, where samples with similar expression profiles appear close to each other. While "complete linkage" is adequate to detect a separation of groups, "single linkage" is helpful to detect outlier samples

loss of information. PCA has also very early been used for clustering protein expression data from 2-DE or MS, e.g., in experiments with bacterial samples [21, 22]. Like hierarchical clustering, PCA can be used for clustering samples as well as proteins. Figure 2 shows the PCA plot for the *P. maximus* data. It can be seen that there is no obvious separation of the samples with regard to their group annotation, i.e., the levels of temperature change.

Other supervised methods that are less frequently used in proteomics include *k*-means clustering that was used for example to identify protein expression profiles in colorectal cancers [23], or

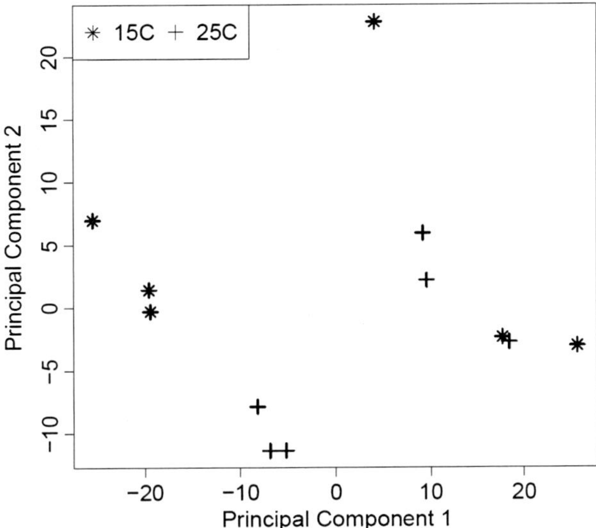

Fig. 2 Representation of high-dimensional expression data in a two-dimensional plot by means of principal components analysis (PCA). PCA is also helpful to explore whether study samples are separable into groups. In the example shown, the two experimental groups do not cluster according to the treatment they were exposed to (two different treatments of temperature change)

self-organizing maps (SOMs) that were used for clustering time profiles of protein expression data [24]. While hierarchical clustering and PCA do not assign samples to definite groups, *k*-means clustering and SOMS assign samples to a certain class.

It should be added that most of the cluster methods require a complete data matrix of expression values. However, expression matrices in proteomics are regularly incomplete; that is, for some proteins no expression level is provided. One way to handle this problem is to impute the "missing values" by utilizing the available expression data. A standard method is for example imputation by the *k*-nearest neighbor method, which was first proposed to be used for incomplete data in DNA microarray experiments [25] and then adapted for protein expression data [26]. Depending on the proportion of missing values imputation can of course influence the results more or less. A specific method for missing value imputation in the case of label-free LC-MS data was also provided by Karpievitch et al. [27].

3 Selection of Molecular Signatures and Their Biological Interpretation

Although unsupervised cluster methods can identify subgroups of proteomic features, most of the abovementioned algorithms are more or less unstable and produce different results. Therefore, in order to identify and select subsets of features that can be used as a molecular signature for certain diseases, statistical models are typically applied. The type of statistical model depends on the clinical variable of interest. Diagnostic biomarkers can be desired to distinguish between different subtypes of a disease while prognostic biomarkers are desired for forecasting the course of a disease. In the first case the aim is to distinguish different classes of samples, while in the latter case one wants to forecast a time-to-event variable (e.g., disease-free or overall survival) of an individual. In the following some standard procedures for biomarker selection are described. Further details about specific study design can also be found in Frantzi et al. [28], Pesch et al. [29], or Gosho et al. [30]. An interesting approach to guidelines for diverse study designs was proposed by a consortium of the National Cancer Institute (NCI) [31]. It should be mentioned that the selection of molecular signatures is rather a screening process for potential biomarkers and does not yield an ultimate result. Indeed, not each single feature that is selected in this process does necessarily have the power to distinguish different cases. For assessing the value of a molecular signature to distinguish between cases classifier models and their evaluation are necessary (*see* Subheading 4).

3.1 Studies with Two Groups of Patients or Samples

The most simple experimental setting is that there are two different groups of samples, and one wants to identify those protein or peptides that are differentially regulated between the two groups. Typically, research on diagnostic biomarkers starts with a discovery phase, where potential biomarkers are selected by comparing the two groups with regard to differentially expressed proteins. This discovery process can start on the tissue level; that is, the study groups are for example tumor and mucosa samples from the same patients. An easy approach to select the most relevant features is to calculate per protein the so-called fold change, a measure for the relative expression change between two groups of samples. A disadvantage of the fold change is that it does not mirror the variance of the data. Proteins with a large fold change can have a large variance making this a very uncertain finding. Therefore, statistical models which provide a p-value per protein are more preferable since they incorporate not only averages of expression levels but also variance estimates. Ad hoc, a simple two-sample test, such as the t-test or the nonparametric Mann–Whitney U test, appears to be the method of choice. In practice, more sophisticated models have been developed and are now widely used. These methods take into account that multiple tests are performed simultaneously instead of performing only one single test. For example, the linear models of Smyth [32], originally developed for DNA microarray data but also applicable for protein expression data, use a moderated version of the t-statistic. Other models for detection of differential expression specifically designed for protein expression data from MS [33], 2-DE [17], or LC-MS [34, 35] have been proposed.

In order to illustrate statistical significance and biological relevance in a "volcano" plot, –log10-transformed *p*-values are often plotted versus the log2 fold change. In the *P. maximus* data for example, several proteins show an absolute log2 fold change >1 between the two experimental groups; however not all of them are significantly differentially expressed (Fig. 3). On the other hand, some proteins show a significant *p*-value, while their strength of differential expression is rather small.

Very important to mention is the fact that the selection procedures comprise large number of simultaneously performed statistical tests. While a single statistical test can produce a false-positive conclusion, multiple tests can produce multiple false-positive conclusions. In the single statistical test, the significance level is specified to control the probability of a false decision. In the case of multiple testing *p*-values must be adjusted to control the false discovery rate (FDR), which describes the proportion of false positives among all positive finding. Standard methods for *p*-value adjustment with respect to the FDR were proposed by Benjamini and Hochberg [36] and Benjamini and Yekutieli [37].

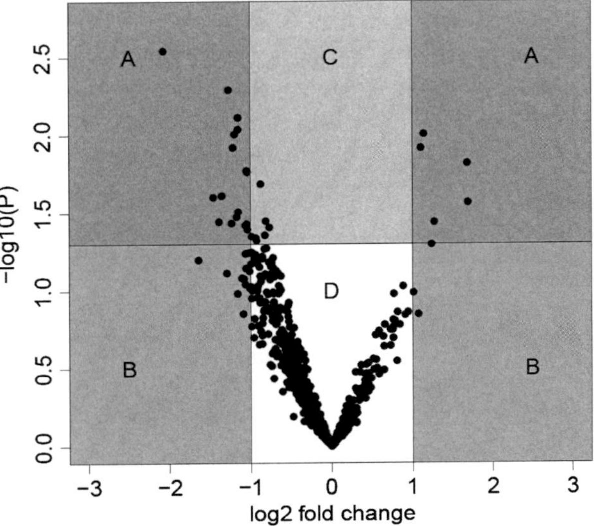

Fig. 3 Volcano plot, representing the *p*-values versus fold changes from the differential analysis of the *Pecten maximus* data. The log10 transformation of the *p*-values is chosen to obtained a zoom into the significantly regulated proteins. The plot can be divided into four regions, according to prespecified thresholds for the *p*-values and fold changes. A: proteins that are significantly different between the experimental groups and have a fold change larger than the defined threshold. B: proteins that are not significantly different but have a fold change larger than the defined threshold. C: proteins that are significantly different but have a low fold change. D: proteins that are neither significantly different nor have a large fold change

3.2 Studies with Several Groups and Multiple Experimental Factors

Frequently, the setting of an experiment or study can be more complicated; that is, the researcher might be interested in more than just two experimental groups or in multiple experimental factors. In the case that more than two groups are considered in an experiment or a study, one possible way of analysis is to do feature selection between each pair of groups. The lists of selected biomarkers can then be compared by Venn diagrams to illustrate the overlaps [38]. In some experiments, there might be multiple experimental factors, and the researchers also want to see whether there is an interaction between these factors. If the aim of the study is for example to select proteomic markers to distinguish diseased patients from healthy controls and a gender effect is of additional interest. Adequate methods for detecting interactions between experimental factors are also covered by the linear model tools of Smyth [32].

Methods for multifactorial experiments find also application in the analysis of time course data. In the related experiments the expression is observed over several points in time and compared between different groups of samples. The question is then whether the expression time profiles are different in the experimental groups, i.e., whether there is an interaction between time and group. Specific models for factorial designs and time course data have for example been proposed for label-free shotgun proteomics [39].

3.3 Selection of Prognostic Markers

In many clinical settings, it is desired to have a biomarker that can be used for the individual prognosis of a diseased person. Examples of study endpoints are the disease-free or overall survival after a cancer therapy. A statistical standard tool that can be used to correlate protein expression levels with such survival endpoints is Cox proportional hazard regression. This type of regression takes into account that some of the patients data are censored, meaning that they have not reached the event of interest at the end of the study. For patients with a censored data point it was for example observed that the patient survived 3 years from diagnosis to the end of the study, but his true death date is not known. A possibility to select prognostic markers from a survival study is to apply Cox regression to each protein. This yields not only the result of a significance test for each protein but also a measure, the hazard ratio, that relates the survival times between patients with high and low expression of the particular protein. For illustration of single proteins with regard to their effect on a survival endpoint one can use Kaplan-Meier plots that show the survival probability versus time. In the case of protein expression data it can make sense to dichotomize the groups into patients with low and high expression levels and then doing a Kaplan-Meier curve for each group. Cox regression models can also integrate more than one independent study parameter so that a group effect can be adjusted for further confounding variables.

3.4 Sample Size Issues in Studies for Biomarker Selection

Methods for sample size calculation of proteomic biomarker studies have been very rarely considered. Most proposed methods focus on the control of the FDR, i.e., the proportion of false-positive findings, under the condition of a prespecified probability for false-negative findings. When searching for example for differentially expressed proteins between two groups, one can estimate the expected FDR given the acceptable probability α for a false-positive detection of an individual protein [17]. More detailed, Cairns et al. [40] provide sample size formulae for detecting fold changes of prespecified size given the probabilities α and β for false-positive and false-negative detections, respectively. Since multiple proteins are tested simultaneously in a proteomics experiment the authors recommend a small value of α, where an appropriate level can be found by estimating the FDR in dependence of α. Further methods were proposed to calculate the sample size by additionally considering the presence of confounding effects in the two experimental groups [41]. This latter approach is based on sample size formulas for logistic regression models and considers also the effect of technical replicates. A disadvantage of this approach is, however, that equal variances in the experimental groups are assumed.

3.5 Set-Based Tests for Biological Interpretation of Molecular Signatures

When selecting proteomic biomarkers for the purpose of diagnosis, prediction, or prognosis the above-detailed procedures typically yield lists of differentially expressed proteins, or other proteomic features, or list of features that are correlated to a survival end-point. Besides the pure selection of such feature lists it is often of additional interest to obtain a biological interpretation of the list. In the genomics area, basically two types of set-based analyses are used which find so far very little attention in proteomics.

One type of method is so-called enrichment tests. Since all features of a list can have various biological, molecular, or cellular function, enrichment tests evaluate whether features of a particular function are significantly more frequent among the differentially expressed features than among the non-differential expression ones. These class of tests are therefore also called competitive tests. Enrichment tests for proteomics data are available in the software suite "Babelomics" [42]. Protein set enrichment analysis was for example applied in breast cancer studies [43]. Specifically, Cha et al. [43] could gain further functional insight into breast cancer tumorigenesis by identification of specific gene regulatory pathways that were enriched among the differentially expressed proteins.

In contrast to competitive tests are self-contained tests, or also called global test procedures, which concentrate only on the expression level of a particular set of features, e.g., all features that have the same biological function or that belong to the same biological pathway. Global test procedure were proposed for 2-DE expression data [44] and for mass spectrometric data [45]. By application of their test procedure to expression data from mouse heart tissue samples, Jung et al. [44] identified ontology terms related to biological processes in the heart muscle. Chen et al. [45] applied their method to protein expression data from a hormone therapy study and identified several biological pathways that play a role in this therapy.

4 Model Building for Diagnosis, Prediction, and Prognosis

For diagnostic purposes, the features of a molecular signature are usually combined in a classification rule, also called classifier. The classifier is a statistical model whose parameters must be trained by existing data consisting of known class levels and expression data for each sample. Since the class levels in the training data are known the related methodology is also called supervised classification in contrast to the unsupervised cluster methods where class levels are unknown and the algorithm clusters samples which have similar expression profiles. A trained supervised classifier model can be evaluated by different techniques such as cross validation or by application to an independent test set. In the following, the most popular classifier methods for high-throughput expression data, especially those used in proteomics, and the evaluation techniques will be detailed. A scheme of a typical procedure of evaluation steps in biomarker research is given in Fig. 4.

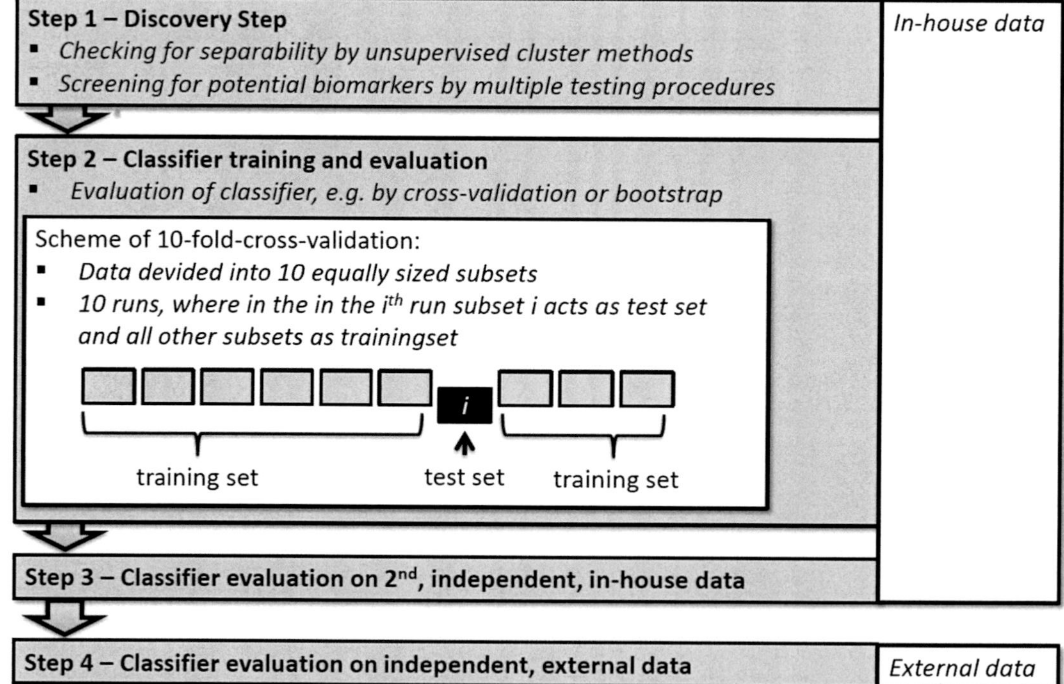

Fig. 4 Schematic presentation of a typical sequence of data analyses in proteomics biomarker research. While steps 1–3 can be performed with internal data, step 4 provides higher evidence by using external data

It should be mentioned here, that some current and future biomarker studies in the context of diagnosis, prediction, and prognosis will also include high-throughput data from different "omics" platforms such as proteomics and proteomics in parallel (*see* **Note 1**). In general, researchers who are planning biomarker studies can also orient at present reporting guideline (*see* **Note 2**).

4.1 Classifier Training

A widely used classifier method is linear discriminant analysis (LDA). Geometrically, LDA projects the data onto a hyperplane such that the ratio of the variance between the classes to the variance within the classes is maximized. Several extensions of LDA have been proposed for example diagonal quadratic and diagonal linear DA, where different covariance matrices are assumed for the different classes. LDA was for example used to classify lung cancer serum samples by their mass spectra [46].

While LDA tries to separate the samples linearly, which is frequently not possible in high-throughput data, support vector machines (SVM) are also able to separate sample points by nonlinear hyperplanes. SVMs were successfully applied to classify protein expression patterns from tuberculosis and control serum samples [47] or breast cancers of different malignity analysed by protein antibody microarrays [48].

Other classifier methods, specifically designed for proteomics data, are the technique of peak probability contrasts [49], which is especially useful for the case of multiple classes, or methods based on decision trees and random forests [50].

Several studies provide a comparison of different classifier techniques yielding, however, inconsistent recommendations. Wu et al. [51] for example apply different classifier methods to MS-bases expression data from ovarian cancer samples and conclude that random forest methods outperformed the other methods. In contrast, Dudoit et al. [52], who apply different classifier methods to high-dimensional gene expression data, found that diagonal LDA yielded lowest error rates. Wu et al. and Dudoit et al. also point out that LDA does not allow to model interactions between features (e.g., proteins) whereas random forest does so.

Most classifier techniques were originally developed for low-dimensional data, i.e. data with more samples than features. However, because protein expression data is typically high-dimensional dimension reduction may be necessary before training a classifier model. Dimension reduction can either be performed by selecting molecular signatures by multiple testing procedures as detailed above or by transformations of the data. In the first case, one would first determine a subset of proteins or peptides that is differentially expressed between the defined classes, and use the reduced data set to train the classifier. In the second case, all features remain in the data, but dimension is reduced for example by principal component analysis [53] or by the partial least squares method [54].

In the case of a survival endpoint, standard Cox regression is usually not feasible because of the high dimensionality of the data. Instead, boosting procedures have been widely used for building a prognostic model with high-throughput expression data [55, 56].

4.2 Classifier Evaluation

An important issue in the context of building diagnostic classifiers is the evaluation of their performance. In general, there are three possible steps of evaluation with an increasing grade of certainty: (1) validation within the training data, (2) validation on an independent test set generated from the same laboratory as the training data, and (3) validation on independent test sets originating from other clinics or research groups. For the first step, i.e., the within-training data validation, cross-validation and bootstrap methods are the most common approaches. The most easy form of cross-validation is leave-one-out cross-validation (LOOCV), which consists of n runs, where n is the number of samples in the training data. In each run, the classifier is trained on n-1 samples, while the remaining sample is used as test set. The measure of performance is determined separately in each run and averaged over all runs at the end. In the case that the purpose of the classifier is the assignment of the samples to different classes, the correct classification rate is the common measure of performance. An alternative to LOOCV is k-fold cross-validation which consists of $k \leq n$ runs. For k-fold CV the samples are split into k subsets and in each run k-1 of these subsets build the training data while 1 subset acts as test set. It was shown in some simulation studies that tenfold CV is preferable to LOOCV, especially in the case of high-dimensional data, since the variance of the estimated performance can be smaller [57, 58]. Besides cross-validation, bootstrap methods are frequently used for classifier performance evaluation. Bootstrap consists of B runs, where in each run a training set is drawn with replacement; that is, some samples can appear multiple times in the training set, and the remaining samples act as test set like in cross-validation. Different recommendation for the choice of the number B exists in literature [59]. A comparison of cross-validation and bootstrap methods was provided by Borra and Ciaccio [58].

Apart from the overall classification accuracy, further measures of performance are commonly used in the special case of a two class model (e.g., healthy versus diseased or therapy responders versus non-responders). These measures are sensitivity, specificity as well as positive and negative predictive values. If the samples in the one class are called the positive cases and the samples of the other class are the negative cases, sensitivity is defined as the probability of classifying a sample as positive given that it is truly positive. Vice versa, the specificity is defined as the probability of assigning a sample to the negative class, given that is it truly negative. Whether a false-negative or a false-positive classification is clinically more critical depends on the concrete situation of the studied disease and potential therapeutic alternatives [60]. In practice the clinician who might want to use the classifier result as support for a therapy decision will be more interested in the predictive values. The positive predictive value is the probability that a sample to be truly positive given that it was classified positive, and likewise the negative predictive value is defined as the probability that a sample belongs truly to the negative cases given that it was classified as negative. It should be remarked that sensitivity and specificity are independent of the prevalence, i.e., the proportion of positive cases, while the predictive values depend on the prevalence. Hence, the predictive values derived in a training set might be different in other collectives with a different distribution of positive and negative cases. In order to make classifiers comparable, measures of performance should be reported with confidence intervals [61].

Different classifiers can also be compared by receiver operator characteristic (ROC) curves. These curves plot sensitivity versus (1 – specificity). The area under the curve (AUC) is also a measures for the classification accuracy of a classifier. For the AUC, it is important to provide a confidence interval, too. Figure 5 shows two ROC curves, produced on the basis of the liver data. One plot was derived using the class probabilities of a cross-validation process, while the other one was derived using the class probabilities when the whole data was used as training and test set at the same time (the latter process is sometimes referred to as "reclassification"). The two plots show that reclassification

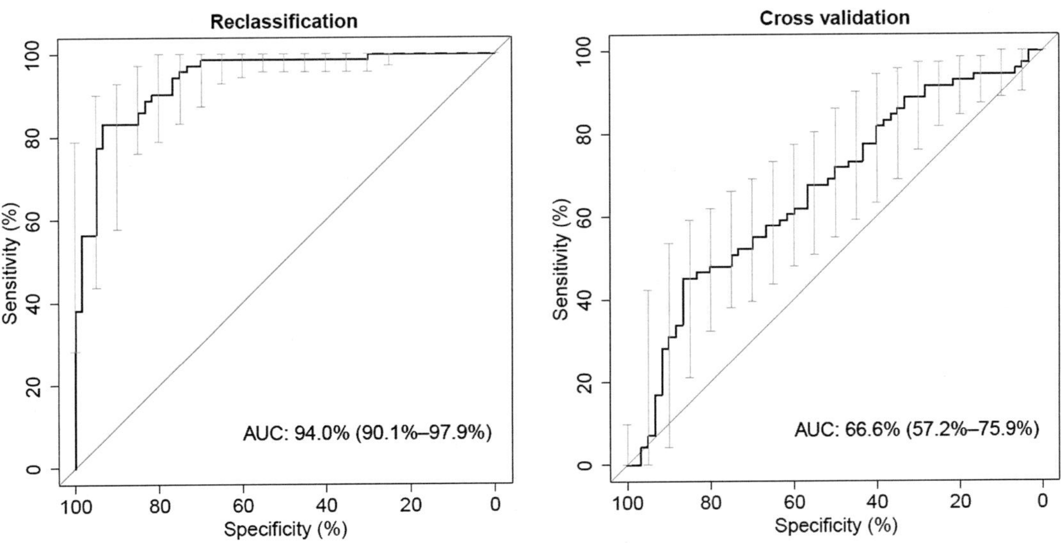

Fig. 5 ROC curves, representing sensitivity versus specificity for each possible cutoff for the class probabilities produced by a classifier. "Reclassification," where all samples are used to train and test a classifier simultaneously, usually overestimates the classifier performance in contrast to cross-validation or bootstrap procedure which provide more realistic results. Confidence intervals for the area under the curve (AUC) and the curve itself are extremely necessary to assess a classifier's performance

often yields very good but also unrealistic results compared to a proper cross-validation.

Evaluation of prognostic models for survival endpoints follows of course different measures for assessing the prediction error. A standard measure for the prediction error is the Brier score which gives the expected squared difference between true and predicted event status at a particular time [62]. For assessing this measure one can also use cross-validation of bootstrap procedures. Another measure that is regularly used for the evaluation of prognostic models is Harrell's index of concordance c [63, 64]. This index checks in a nonparametric manner whether the order of observed survival times is in concordance with the order of the predicted survival times.

While sample size methods for the selection of biomarkers concentrate on the FDR, samples size methods for building and evaluating classifiers focus on the estimation of a classifier's accuracy. The approach proposed by Fu et al. [65] consists of a classification procedure where patients are recruited sequentially and the classification rule is updated for each new patient. The procedure stops when the confidence level for the misclassification rate has reached a predefined confidence level. Figueroa et al. [66] propose a method that fits a classifier model to a set of training data in order to build a learning curve. From the classifier model and the learning curve they predict the classifiers performance for larger data sets. While the approaches of Fu et al. and Figueroa et al. focus on classifiers in general, Dobbin and Simon [67] specifically include the feature selection step that is necessary for high-dimensional expression data into their formulas for estimating the correct classification probability with regard to the sample size. Their approach covers also the case that one class in underrepresented in the population.

5 Notes

1. While most research groups mainly focused on one "omics" domain in the past, more groups study different domains simultaneously on the same samples now. Integration of genomic and proteomic biomarkers can help get a better understanding of biological mechanisms and might also help to improve classifiers. Classifier combinations have already been proposed for integrating other omics domains such as the mRNA and miRNA levels [68], and they might be helpful for integrating genomics and proteomics data.
2. Researchers for proteomic biomarkers can also orient themselves on existing guidelines which can help to make different studies comparable. The STARD (STAndards for the Reporting of Diagnostic accuracy studies) initiative for example has

proposed different recommendations of how to report the results of diagnosis studies [69]. These recommendations involve for example 25 items specifying how methods and results of a study or experiment should be described in a publication. Reporting guidelines have also been proposed in the REMARK initiative [70]. Similar like the STARD initiative the REMARK guidelines recommend 20 items specifying for what should be reported when publishing a biomarker study. Both sets of item include for example requirements for describing the aim of a study, the study collective, the methods and how to present and discuss results.

Reporting guidelines are important to make results reproducible or to qualify them for meta-analyses. Meta-analyses of high-throughput expression data were for examples proposed in the field of genomics [71], and should also be considered proteomics. However, for making meta-analyses feasible in proteomics biomarker research, public data repositories—as available for DNA microarray or next-generation-sequencing data—would be necessary. The databases "ArrayExpress" [72] and "Gene Expression Omnibus" [73] provide a platform for submitting gene expression data in standard formats. Such databases for proteomics expression data would not only enable scientists to perform meta-analyses but would also help them to reproduce the findings of other authors.

Acknowledgements

The author would like to thank Prof Olga Vitek (Northeastern University, Boston) for very helpful comments on a previous version of the manuscript.

References

1. Soares H, Chen Y, Sabbagh M et al (2009) Identifying early markers of Alzheimer's disease using quantitative multiplex proteomic immunoassay panels. Ann N Y Acad Sci 1180:56–67
2. Pan S, Chen R, Brand RE et al (2012) Multiplex targeted proteomic assay for biomarker detection in plasma: a pancreatic cancer biomarker case study. J Proteome Res 11:1937–1948
3. Baas T, Baskin CR, Diamond DL et al (2006) Integrated molecular signature of disease: analysis of influenza virus-infected macaques through functional genomics and proteomics. J Virol 80:10813–10828
4. Paweletz CP, Trock B, Pennanen M (2001) Proteomic patterns of nipple aspirate fluids obtained by SELDI-TOF: potential for new biomarkers to aid in the diagnosis of breast cancer. Dis Markers 17:301–307
5. Li J, Zhang Z, Rosenzweig J et al (2002) Proteomics and bioinformatics approaches for identification of serum biomarkers to detect breast cancer. Clin Chem 48:1296–1304
6. Brown JM, Krutzsch H, Shu H et al (2002) Proteomic analysis and identification of new biomarkers and therapeutic targets for invasive ovarian cancer. Proteomics 2:76–84
7. Wang TJ, Gona P, Larson MG et al (2006) Multiple biomarkers for the prediction of first major cardiovascular events and death. N Engl J Med 355:2631–2639

8. Hye A, Lynham S, Thambisetty M et al (2006) Proteome-based plasma biomarkers for Alzheimer's disease. Brain 129:3042–3050
9. Abdi F, Quinn JF, Jankovic J et al (2006) Detection of biomarkers with multiplex quantitative proteomic platform in cerebrospinal fluid of patients with neurodegenerative disorders. J Alzheimers Dis 9:293–348
10. Pisitkun T, Shen R-F, Knepper MA (2004) Identification and proteomic profiling of exosomes in human urine. Proc Natl Acad Sci U S A 101:13368–13373
11. Hu S, Arellano M, Boontheung P et al (2008) Salivary proteomics for oral cancer biomarker discovery. Clin Cancer Res 14:6246–6252
12. Pavlou MP, Diamandis EP, Blasutig IM (2012) The long journey of cancer biomarkers from bench to clinic. Clin Chem 59:147–157
13. Christin C, Bischoff R, Horvatovich P (2011) Data processing pipelines for comprehensive profiling of proteomics samples by label-free LC-MS for biomarker discovery. Talanta 83:1209–1224
14. Listgarten J, Emili A (2005) Statistical and computational methods for comparative proteomic profiling using liquid chromatography-tandem mass spectrometry. Mol Cell Proteomics 4:419–434
15. Caffrey RE (2010) A review of experimental design best practices for proteomics based biomarker discovery: focus on SELDI-TOF. Methods Mol Biol 641:167–183
16. Ward DG, Cheng Y, N'Kontchou G et al (2006) Changes in the serum proteome associated with the development of hepatocellular carcinoma in hepatitis C-related cirrhosis. Br J Cancer 94:287–292
17. Artigaud S, Gauthier O, Pichereau V (2013) Identifying differentially expressed proteins in 2-DE experiments: inputs from transcriptomics statistical tools. Bioinformatics 29:2729–2734
18. Eisen MB, Spellman PT, Brown PO (1998) Cluster analysis and display of genome-wide expression patterns. Proc Natl Acad Sci U S A 95:14863–14868
19. Alaiya AA, Franzén B, Hagman A et al (2002) Molecular classification of borderline ovarian tumours using hierarchical cluster analysis of protein expression profiles. Int J Cancer 98:895–899
20. Yanagisawa K, Shyr Y, Xu BJ et al (2003) Proteomic patterns of tumour subsets in non-small-cell lung cancer. Lancet 362:433–439
21. Vasseur C, Labadie J, Hébraud M (1999) Differential protein expression by Pseudomonas fragi submitted to various stresses. Electrophoresis 20:2204–2213
22. Goodacre R, Heald JK, Kell DB (1999) Characterisation of intact microorganisms using electrospray ionisation mass spectrometry. FEMS Microbiol Lett 176:17–24
23. Duncan R, Carpenter B, Main LC et al (2008) Characterisation and protein expression profiling of annexins in colorectal cancer. Br J Cancer 98:426–433
24. Zhang Y, Wolf-Yadlin A, Ross RL et al (2005) Time-resolved mass spectrometry of tyrosine phosphorylation sites in the epidermal growth factor receptor signaling network reveals dynamic modules. Mol Cell Proteomics 4:1240–1250
25. Troyanskaya O, Cantor M, Sherlock G et al (2001) Missing value estimation methods for DNA microarrays. Bioinformatics 17:520–525
26. Jung K, Gannoun A, Sitek B et al (2006) Statistical evaluation of methods for the analysis of dynamic protein expression data from a tumour study. RevStat-Stat J 4:67–80
27. Karpievitch YV, Dabney AR, Smith RD (2012) Normalization and missing value imputation for label-free LC-MS analysis. BMC Bioinformatics 13:S5
28. Frantzi M, Bhat A, Latosinska A (2014) Clinical proteomic biomarkers: relevant issues on study design & technical considerations in biomarker development. Clin Transl Med 3:7
29. Pesch B, Brüning T, Johnen G et al (2014) Biomarker research with prospective study designs for the early detection of cancer. Biochim Biophys Acta 1844:874–883
30. Gosho M, Nagashima K, Sato Y (2012) Study designs and statistical analyses for biomarker research. Sensors 12:8966–8986
31. Dancey JE, Dobbin KK, Groshen S et al (2010) Guidelines of the development and incorporation of biomarker studies in early clinical trials of novel agents. Clin Cancer Res 16:1745–1755
32. Smyth GK (2004) Linear models and empirical Bayes methods for assessing differential expression in microarray experiments. Stat Appl Genet Mol Biol 2004:Article 3
33. Ryu SY, Qian W-J, Camp DG et al (2014) Detecting differential protein expression in large-scale population proteomics. Bioinformatics 30:2741–2746
34. Clough T, Thaminy S, Ragg S et al (2012) Statistical protein quantification and significance analysis in label-free LC-MS experiments with complex designs. BMC Bioinformatics 13:S6
35. Listgarten J, Neal RM, Roweis ST et al (2007) Difference detection in LC-MC data for protein biomarker discovery. Bioinformatics 23: e198–e204

36. Benjamini Y, Hochberg Y (1995) Controlling the false discovery rate: a practical and powerful approach to multiple testing. J R Stat Soc Series B 57:289–300
37. Benjamini Y, Yekutieli D (2001) The control of the false discovery rate in multiple testing under dependency. Ann Stat 29:1165–1188
38. Hulsen T, de Vlieg J, Alkema W (2008) BioVenn—a web application for the comparison and visualization of biological lists using area-proportional Venn diagrams. BMC Genomics 9:488
39. Choi H, Fermin D, Nesvizhskii AI (2008) Significance analysis of spectral count data in label-free shotgun proteomics. Mol Cell Proteomics 7:2373–2385
40. Cairns DA, Barrett JH, Billingham LJ et al (2009) Sample size determination in clinical proteomic profiling experiments using mass spectrometry for class comparison. Proteomics 9:74–86
41. Nyangoma SO, Collins SI, Altman D et al (2012) Sample size calculations for designing clinical proteomic profiling studies using mass spectrometry. Stat Appl Genet Mol Biol 11(3)
42. A-Shahrour F, Carbonell J, Minguez P et al (2008) Babelomics: advanced functional profiling of transcriptomics, proteomics and genomics experiments. Nucleic Acids Res 36:W341–W346
43. Cha S, Imielinski MB, Rejtar T et al (2010) In situ proteomic analysis of human breast cancer epithelial cells using laser capture microdissection: annotation by protein set enrichment analysis and gene ontology. Mol Cell Proteomics 9:2529–2544
44. Jung K, Dihazi H, Bibi A et al (2014) Adaption of the global test idea to proteomics data with missing values. Bioinformatics 30:1424–1430
45. Chen LS, Paul D, Prentice RL et al (2011) A regularized Hotelling's T^2 test for pathway analysis in proteomics studies. J Am Stat Assoc 106:1345–1360
46. Baggerly KA, Morris JS, Wang J et al (2003) A comprehensive approach to the analysis of matrix-assisted laser desorption/ionization-time of flight proteomics spectra from serum samples. Proteomics 3:1667–1672
47. Agranoff D, Fernandez-Reyes D, Papdopoulos MC et al (2006) Identification of diagnostic markers for tuberculosis by proteomic fingerprinting of serum. Lancet 368:1012–1021
48. Carlsson A, Wingren C, Ingvarsson J et al (2008) Serum proteome profiling of metastatic breast cancer using recombinant antibody microarrays. Eur J Cancer 44:472–480
49. Tibshirani R, Hastie T, Narshimhan B et al (2004) Sample classification from protein mass spectrometry, by 'peak probability contrasts'. Bioinformatics 20:3034–3044
50. Geurts P, Fillet M, de Seny D et al (2005) Proteomic mass spectra classification using decision tree based ensemble methods. Bioinformatics 21:3138–3145
51. Wu B, Abbott T, Fishman D et al (2003) Comparison of statistical methods for classification of ovarian cancer using mass spectrometry data. Bioinformatics 19:1636–1643
52. Dudoit S, Fridlyand J, Speed TP (2002) Comparison of discrimination methods for the classification of tumors using gene expression data. J Am Stat Assoc 97:77–87
53. Lilien RH, Farid H, Donald BR (2010) Probabilistic disease classification of expression-dependent proteomic data from mass spectrometry of human serum. J Comput Biol 10:925–946
54. Karp NA, Griffin JL, Lilley KS (2005) Application of partial least squares discriminant analysis to two-dimensional difference gel studies in expression proteomics. Proteomics 5:81–90
55. Binder H, Allignol A, Schumacher M (2009) Boosting for high-dimensional time-to-event data with competing risks. Bioinformatics 25:890–896
56. Wang Z, Wang CY (2010) Buckly-James boosting for survival analysis with high-dimensional biomarker data. Stat Appl Genet Mol Biol 9:Article 24
57. Brage-Neto U, Dougherty ER (2004) Is cross-validation valid for small sample microarray classification? Bioinformatics 20:374–380
58. Borra S, Di Ciaccio A (2010) Measuring the prediction error. A comparison of cross validation, bootstrap and covariance penalty methods. Comput Stat Data Anal 54: 2976–2989
59. Pattengalem ND, Alipour M, Binida-Emonds ORP (2010) How many bootstrap replicates are necessary? J Comput Biol 17:337–354
60. Jung K, Grade M, Gaedcke J et al (2010) A new sensitivity-preferred strategy to build prediction rules for therapy response of cancer patients using gene expression data. Comput Methods Programs Biomed 100:132–139
61. Foody GM (2009) Classification accuracy comparison: hypothesis tests and the use of confidence intervals in evaluation of difference, equivalence and non-inferiority. Remote Sens Environ 113:1658–1663
62. Porzelius C, Schumacher M, Binder H (2010) A general, prediction error-based criterion for selecting model complexity for high-dimensional survival models. Stat Med 29: 830–838

63. Harrel FE, Lee KL (1984) Regression modelling strategies for improved prognostic prediction. Stat Med 3:143–152
64. Newson RB (2010) Comparing the predictive power of survival models using Harrell's C or Somers' D. Stata J 10:339–358
65. Fu WJ, Dougherty ER, Mallick B et al (2005) How many samples are needed to build a classifier: a general sequential approach. Bioinformatics 21:63–70
66. Figuera RL, Zeng-Treidler Q, Kandula S et al (2012) Predicting sample size required for classification performance. BMC Med Inform Decis Mak 12:8
67. Dobbin KK, Simon RM (2006) Sample size planning for developing classifiers using high-dimensional DNA microarray data. Biostatistics 8:101–117
68. Fuchs M, Beißbarth T, Wingender E et al (2013) Connecting high-dimensional mRNA and miRNA expression data for binary medical classification problems. Comput Methods Programs Biomed 111:592–601
69. Bruns DE (2003) The STARD initiative and the reporting of studies of diagnostic accuracy. Clin Chem 49:19–20
70. McShane LM, Altman DG, Sauerbrei W et al (2005) REporting recommendations for tumour MARKer prognostic studies (REMARK). Nat Clin Pract Oncol 2:416–422
71. Marot G, Mayer CD (2009) Sequential analysis for microarray data based on sensitivity and meta-analysis. Stat Appl Genet Mol Biol 8:Article 3
72. Kolesnikov N, Hastings E, Keays M et al (2015) ArrayExpress update—simplifying data submissions. Nucleic Acids Res 43: D1113–D1116
73. Edgar R, Domrachev M, Lash AE (2002) Gene Expression Omnibus: NCBI gene expression and hybridization array data repository. Nucleic Acids Res 30:207–210

Index

Klaus Jung (ed.), *Statistical Analysis in Proteomics*, Methods in Molecular Biology, vol. 1362,
DOI 10.1007/978-1-4939-3106-4, © Springer Science+Business Media New York 2016

I

K

L

M

N

O

P

Q

R

Zeitfracht Medien GmbH
Ferdinand-Jühlke-Straße 7
99095 Erfurt, Deutschland
produktsicherheit@kolibri360.de